Fuzzy Randomness

Bernd Möller · Michael Beer

Springer

Berlin
Heidelberg
New York
Hong Kong
London
Milan
Paris
Tokyo

Bernd Möller
Michael Beer

Fuzzy Randomness

Uncertainty in Civil Engineering and Computational Mechanics

With 234 Figures

Springer

Univ.-Prof. Dr. -Ing. habil. Bernd Möller
Dr. -Ing. Michael Beer

Institute of Structural Analysis
(Lehrstuhl für Statik)
Dresden University of Technology
Mommsenstraße 13
01062 Dresden
Germany

E-mail: moeller@rcs.urz.tu-dresden.de
E-mail: mb18@rcs.urz.tu-dresden.de

ISBN 3-540-40294-2 Springer-Verlag Berlin Heidelberg New York

Cataloging-in-Publication Data applied for
Bibliographic information published by Die Deutsche Bibliothek.
Die Deutsche Bibliothek lists this publication in the Deutsche Nationalbibliografie; detailed bibliographic data is available in the Internet at <http://dnb.ddb.de>.

Springer-Verlag Berlin Heidelberg New York
Springer-Verlag is a part of Springer Science+Business Media

springeronline.com

Typesetting: Data conversion by author
Cover-design: Struve & Partner, Heidelberg
Printed on acid-free paper 02 / 3020

Preface

Everything is vague to a degree
you do not realize till you have tried
to make it precise.

Bertrand Russell

Nature has never attended a course in probability theory. This poignant philosophical statement underlines the fact that the phenomenon of uncertainty cannot be described by means of probability theory alone. Data and models encountered in the natural sciences and engineering are more or less characterized by uncertainty. With the inclusion of interval variables and random variables classical mathematical models are available for the treatment of uncertainty.

In recent years work on the formulation of new mathematical models for describing uncertainty has intensified. This work is based on chaos theory, convex modeling, fuzzy set theory, and fuzzy randomness. The aim of these uncertainty models is to handle data exhibiting, in particular, nonstochastic properties and informal uncertainty.

Fuzzy randomness is a generalized uncertainty model that permits the simultaneous consideration of stochastic, lexical, and informal uncertainty. Initial drafts for the uncertainty model "fuzzy randomness" were developed about 25 years ago. This book considers these developments and enhances fuzzy randomness in such a way that it may be applied in the fields of civil engineering and computational mechanics. These further scientific developments are based on the results of research work carried out under the supervision of the leading author at the Institute of Structural Analysis of Dresden University of Technology with the much-appreciated financial support of the "Deutsche Forschungsgemeinschaft (DFG)" (*German Research Foundation*).

Uncertainty models are the topic of controversial discussion in scientific circles. For this reason the phenomenon of uncertainty is introduced in the first chapter. By way of examples an attempt is made to promote an understanding of uncertainty and to guide readers from a deterministic way of thinking towards the acceptance of uncertainty. The mathematical fundamentals of fuzziness and fuzzy randomness are then introduced in Chap. 2. Particularly with regard to applications, the sections dealing with fuzzy functions and fuzzy random functions are certain to be of special interest. The reader is expected to be in command of the knowledge gained in a basic university mathematics course, with the inclusion of stochastic elements.

A specification of uncertainty in any particular case is often difficult. For this reason Chaps. 3 and 4 are devoted solely to this problem. The derivation of fuzzy variables for representing informal and lexical uncertainty reflects the subjective assessment of objective conditions in the form of a membership function. Techniques for modeling fuzzy random variables are presented for data that simultaneously exhibit stochastic and nonstochastic properties.

The application of fuzzy randomness is demonstrated in three fields of civil engineering and computational mechanics: structural analysis, safety assessment, and design.

The methods of fuzzy structural analysis and fuzzy probabilistic structural analysis developed in Chap. 5 are applicable without restriction to arbitrary geometrically and physically nonlinear problems. The most important forms of the latter are the Fuzzy Finite Element Method (FFEM) and the Fuzzy Stochastic Finite Element Method (FSFEM).

Fuzzy randomness may also be combined with a variety of methods based on reliability theory. This is demonstrated in Chap. 6 by considering the further development of the First Order Reliability Method (FORM) to obtain the Fuzzy First Order Reliability Method (FFORM). As a result of the safety assessment the existing uncertainty is evident in the fuzzy failure probability, which now becomes an uncertain parameter. A general design concept for uncertain structural parameters is finally developed in Chap. 7.

Writing a book always requires the assistance of colleagues. The authors wish to thank Professor Wolfgang Graf for many fruitful discussions, and the research coworkers at the Institute of Structural Analysis, Dr-Ing. Andreas Hoffmann, Dr-Ing. Nguyen Song Ha, Dipl.-Ing. Martin Liebscher, and Dipl.-Ing. Jan-Uwe Sickert for their valuable support in the preparation of the manuscript. We are also grateful to Dr Ian Westwood (PhD, Civil Engineering) for translating Chaps. 1, 3, and 4, and for his assistance in the formulation of additional parts of the manuscript in English.

Finally, we wish to thank the publishers "Springer-Verlag" for their much-appreciated editorial work and printing of the book.

Dresden, September 2003 Bernd Möller Michael Beer

Contents

Abbreviations

arithmetics

a, ..., z	variables, a, ..., z $\in \mathbb{R}$
$\underline{a}, ..., \underline{z}$	vectors
$\underline{A}, ..., \underline{Z}$	matrices
$\sum$	sum
$\prod$	product
ln	natural logarithm
ld	logarithm to base 2

analysis

(... , ...)	ordered n-tuple, elements of a column matrix
(... , ...)	open interval
[... , ...]	closed interval
$\mid ... \mid, \| ... \|$	absolute value, norm
$\times$	Cartesian product
sup	supremum (least upper bound, maximum)
min	minimum
sup min	max-min operator
max	maximum
lim	limes, limit
∞	infinity
Δ	difference
d	differentiation
∂	partial differentiation
$\int$	integration
$\rightarrow$	mapping

set theory

$\underline{\mathbf{A}}, ..., \underline{\mathbf{Z}}$	fundamental sets
$\underline{A}, ..., \underline{Z}$	crisp sets
$\underline{A}^C, ..., \underline{Z}^C$	complement of $\underline{A}$, ..., $\underline{Z}$ with regard to the assigned fundamental set $\underline{\mathbf{A}}$; ...; $\underline{\mathbf{Z}}$
$\mathbb{N}$	natural numbers
$\mathbb{R}$	real numbers
$\mathbb{R}^n$	n-dimensional Euclidean space
{... , ...}	set of ..., elements of a set
$\in$	element of
$\notin$	not element of
$\varnothing$	empty set
$\cap$	intersection
$\cup$	union
$\setminus$	set difference, ... without ...
$\subseteq$	subset of
$\mathfrak{P}(...)$	power set
$\mathfrak{M}(...)$	family of sets
$\mathfrak{S}(...)$	σ-algebra
$M_{\mathfrak{M}}(...)$	set function

measure theory

$M^s_{\mathfrak{M}}(...)$	(crisp) measure
$M^u_{\mathfrak{M}}(...)$	uncertain measure
V(...)	distribution function
[... , ...]	measure space
E	event
H	Shannon's entropy
H_u	modified Shannon's entropy, Shannon's uncertainty measure

logic

$\wedge$	conjunction, logical *and*
$\vee$	alternative, logical *or*
$\vert$	for which the following holds
$\Rightarrow$	implication; if ... , then
$\Leftrightarrow$	equivalence; if and only if ...

$\forall$	universal quantifier, for all
$\exists$	existential quantifier, there exists

fuzzy set theory

$\sim$	fuzziness
$\underline{\tilde{a}}, \ldots, \underline{\tilde{z}}$	fuzzy vectors
$<a, b, c>$	fuzzy triangular number with the interval bounds of the support a and c and the mean value b
$<a, b, c, d>$	fuzzy trapezoidal interval with the interval bounds of the support a and d and the interval bounds of the mean interval b and c
$\mu(\underline{x})$	membership function of the fuzzy vector $\underline{\tilde{x}}$
$\underline{\tilde{A}}, \ldots, \underline{\tilde{Z}}$	fuzzy sets
$\underline{\tilde{A}}^{C}, \ldots, \underline{\tilde{Z}}^{C}$	complement of $\underline{\tilde{A}}; \ldots; \underline{\tilde{Z}}$
$\tilde{E}$	fuzzy event
α_i	α-level for $\alpha_i = \mu$
$\underline{A}_{\alpha_i}, \ldots, \underline{Z}_{\alpha_i}$	α-level sets for $\alpha = \alpha_i$
$\tilde{x}(\underline{\tilde{t}}), \tilde{x}(\underline{t})$	fuzzy function
$\tilde{x}(\underline{\theta})$	fuzzy field
$\tilde{x}(\tau)$	fuzzy process
$\underline{\theta}$	spatial coordinates
τ	time coordinate
$\tilde{x}(\underline{t}) = x(\underline{\tilde{s}}, \underline{t})$	bunch parameter representation of a fuzzy function
$\underline{\tilde{s}}$	fuzzy bunch parameter vector
$x(\underline{t}) \in \tilde{x}(\underline{t})$	trajectory of $\tilde{x}(\underline{t})$
$X_{\alpha}(\underline{t})$	α-function set

fuzzy structural analysis

$\tilde{x}_i$	fuzzy input variables including fuzzy model parameters
$\tilde{z}_j$	fuzzy result variables
$\underline{\mathbf{X}}$, x-space	space of the fuzzy input variables and fuzzy model parameters
$\underline{\mathbf{Z}}$, z-space	space of the fuzzy result variables
$\underline{\tilde{X}}$	fuzzy input set
$\underline{\tilde{Z}}$	fuzzy result set
$\underline{X}_{\alpha_i}$	crisp input subset for $\alpha = \alpha_i$
$\underline{Z}_{\alpha_i}$	crisp result subset for $\alpha = \alpha_i$

fuzzy probabilistics

$\underline{A}, ..., \underline{Z}$	real-valued random vectors
$\underline{\tilde{A}}, ..., \underline{\tilde{Z}}$	fuzzy random vectors
$\underline{X}_j$	original of the fuzzy random vector $\underline{\tilde{X}}$
$E\tilde{X}_i$, $\tilde{m}_{x_i}$	fuzzy expected value of the i-th component of $\underline{\tilde{X}}$
$D^2\tilde{X}_i$, $\tilde{\sigma}^2_{x_i}$	fuzzy variance of the i-th component of $\underline{\tilde{X}}$
$\sqrt{D^2\tilde{X}_i}$, $\tilde{\sigma}_{x_i}$	fuzzy standard deviation of the i-th component of $\underline{\tilde{X}}$
$\tilde{P}(E)$	fuzzy probability of the event E
$\tilde{F}(\underline{x})$	fuzzy probability distribution function of $\underline{\tilde{X}}$
$\tilde{f}(\underline{x})$	fuzzy probability density function of $\underline{\tilde{X}}$
$\underline{\tilde{X}}(\underline{\tilde{t}}), \underline{\tilde{X}}(\underline{t})$	fuzzy random function
$\underline{\tilde{X}}(\underline{\theta})$	fuzzy random field
$\underline{\tilde{X}}(\tau)$	fuzzy random process
$\underline{\theta}$	spatial coordinates
τ	time coordinate
$\underline{\tilde{X}}(\underline{t}) = \underline{X}(\underline{\tilde{s}}, \underline{t})$	bunch parameter representation of a fuzzy random function
$\underline{\tilde{s}}$	fuzzy bunch parameter vector
$\underline{X}_j(\underline{t})$	original function
$\underline{X}_\alpha(\underline{t})$	random α-function set
$\tilde{\bar{x}}_i$	fuzzy mean of the i-th component of a sample belonging to $\underline{\tilde{X}}$
$\tilde{s}^2_{x_i}$	fuzzy variance of the i-th component of a sample belonging to $\underline{\tilde{X}}$
$\tilde{s}_{x_i}$	fuzzy standard deviation of the i-th component of a sample belonging to $\underline{\tilde{X}}$

fuzzy probabilistic safety assessment

$\underline{\mathbf{X}}$, x-space	original space of the basic variables
$\underline{\mathbf{Y}}$, y-space	standard normal space
$g(\underline{x})$	limit state surface in $\underline{x}$
$h(\underline{y})$	limit state surface in $\underline{y}$
$l(\underline{y})$	linearized limit state surface in $\underline{y}$
*perm_*	permissible
*req_*	required

$\Phi^{NN}(\underline{y})$	probability distribution function of the standard normal distribution
$\varphi^{NN}(\underline{y})$	probability density function of the standard normal distribution
$\underline{\tilde{x}}_B$	fuzzy design point in x-space
$\underline{\tilde{y}}_B$	fuzzy design point in y-space
$\tilde{\beta}$	fuzzy reliability index
$\tilde{P}_f$	fuzzy failure probability

structural design based on clustering

$d_1, ..., d_n$	permissible design parameters
$\overline{d}_1, ..., \overline{d}_n$	nonpermissible design parameters
$\underline{D}^n$	space of the design parameters
$r_1, ..., r_m$	restricted parameters
$\overline{r}_1, ..., \overline{r}_m$	nonrestricted parameters
$\underline{R}^m$	space of the restricted parameters
CT_h	design constraint
C_v	cluster
$\tilde{x}_1^{[v]}, ..., \tilde{x}_n^{[v]}$	modified fuzzy input variables, alternative design variants
$\tilde{z}_1^{[v]}, ..., \tilde{z}_m^{[v]}$	modified fuzzy result variables

1 Introduction

1.1 The Phenomenon of Uncertainty

For quantifying physical parameters such as geometry, material, or loading parameters, real numbers or integers are mainly applied, i.e., a *deterministic data model* is applied. In order to describe randomness resulting from imprecise readings or fluctuating ambient conditions, measurements are repeated and lumped together in a concrete data sample. Mathematical statistics offers methods for describing data samples with the aid of random variables. A common approach for this purpose is to specify a probability distribution function in order to obtain a *stochastic data model*.

With a deterministic and stochastic data model at his disposal, does the engineer have an adequate instrument for quantifying data?

If the term *uncertainty* is introduced and defined according to [22] as the gradual assessment of the truth content of a proposition, doubt arises as to whether the truth content may be stated with sufficient accuracy using each of the data models in all cases. For example, each measurement is more or less uncertain. This uncertainty can neither be accounted for by the deterministic nor the stochastic data model [194] owing to the fact that each sample element is considered as a crisp (real or integer) number in stochastic mathematics. This uncertainty is unavoidable and may always be made evident by the correct choice of scale. Viertl [194] thus introduces nonprecise numbers and nonprecise vectors.

Only vague information is often available concerning the magnitude of a physical parameter, e.g., in the form of lexical descriptions such as "very large", "average", or "weak". Also, the truth content of such descriptions cannot be accounted for by either the deterministic or the stochastic data model. It is thus clear that an extension of the data models is necessary.

Just how difficult it is to correctly state the truth content of a set of measured values using the methods of statistics is demonstrated by the following numerical experiment.

The universe **X** is given with an expected value E(X) = 3 and a variance VAR(X) = 1 for a theoretically exact normal distribution. 40 000 data samples, each containing n = 100 elements, are drawn; this represents a relatively large sample size for engineering problems. Although the distribution type for the universe is known, a normal distribution (ND), a logarithmic normal distribution

(LND), and a Gumbel distribution (GD) are assumed as alternatives for the data samples drawn.

For each data sample the parameters of the three assumed distributions are estimated by means of the maximum likelihood method and the expected value, variance, and quantile values are finally computed.

The 40 000 data samples yield 40 000 realizations for the parameters of the distribution function. These are classified and yield an approximation for the unknown density functions of the expected value, variance, and quantile values according to the Glivenko theorem. As may be seen in Figs. 1.1 and 1.2, the assumption of different types of probability distribution functions alone leads to different moments and quantile values.

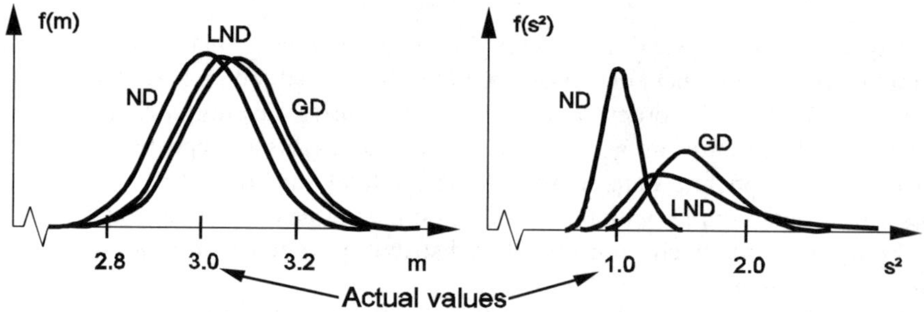

Fig. 1.1. Distributions of the expected values and variances

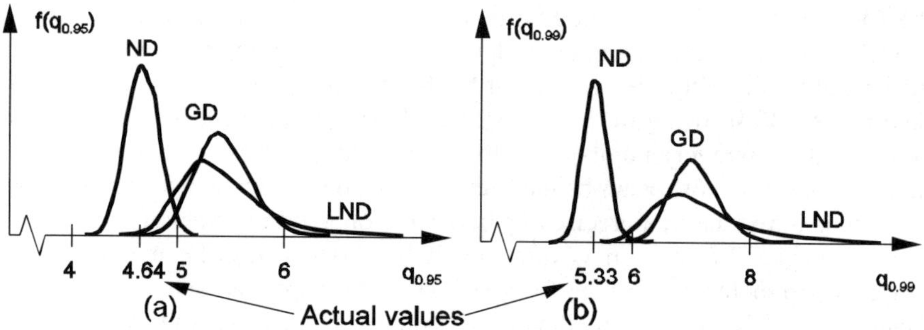

Fig. 1.2. Distributions of the 95%- (a) and 99%-quantile values (b)

This result is naturally of no surprise to mathematicians. The results clearly demonstrate the preconditions that must be satisfied in order to correctly account for uncertainty using the stochastic data model.

An attempt is now made to improve the distribution parameters by means of a Bayesian estimation. The parameter distributions obtained from the 40 000 data samples are chosen as prior distributions for a Bayesian estimation. For the assumed normal and logarithmic normal distributions 10 000 data samples, each containing n = 100 elements, are drawn again. As shown in Fig. 1.3 the variance of the distribution of quantile values reduces. The frequently applied modal values of these distributions only change very slightly.

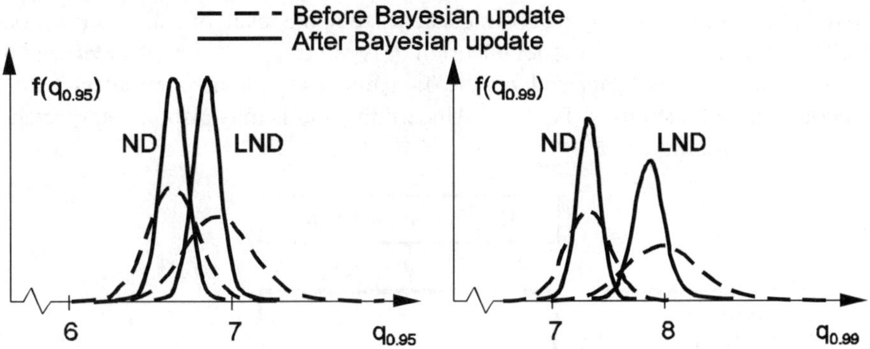

Fig. 1.3. Distributions of quantile values resulting from a Bayesian estimation

Although the important precondition of constant reproduction conditions is complied with in this numerical experiment, it is evident that considerable informal uncertainty exists if the distribution type of the universe cannot be correctly stated.

In civil-engineering applications, however, only small data samples are available in general. Reproduction conditions frequently change when sample elements are generated. This is the case, for example, when specimens for determining concrete strength collected from different construction sites are lumped together to form a single sample.

On the basis of these considerations it follows that under nonconstant reproduction conditions, small sample sizes, uncertain and non-numerical data, or noncompliance with the i.i.d. (identically independently distributed) paradigm, considerable uncertainty exists, which may only be described using extended uncertainty models.

The development of extended uncertainty models and their practical application in civil engineering form the subject matter of this book.

Besides the phenomenon of data uncertainty, the phenomenon of model uncertainty also exists in civil engineering. In algorithmic terms, model uncertainty is far more difficult to account for, and is interpreted as a separate concept.

As a criterion for distinguishing between data uncertainty and model uncertainty the assignment of uncertainty to the input parameters of a computational model or to the model itself is adopted. A distinction is realized by the introduction of the *model concept*.

A *model* may be interpreted as a set of elements and rules that maps input variables onto result variables. A model is thus a self-contained entity that processes information.

Model uncertainty is uncertainty in the mapping. This is induced by uncertain parameters (structural parameters) that act exclusively *within the model* and are thus referred to as *uncertain model parameters*. Uncertain model parameters are not explicitly mapped onto result variables; they only influence the mapping. Model uncertainty arises in the abstraction process, the result of which is the model itself. A model possessing model uncertainty is referred to as an *uncertain model*. Such a model is characterized by the fact that crisp input parameters lead to uncertain model responses (Fig. 1.4). Uncertain models may also be interpreted as uncertain algorithms.

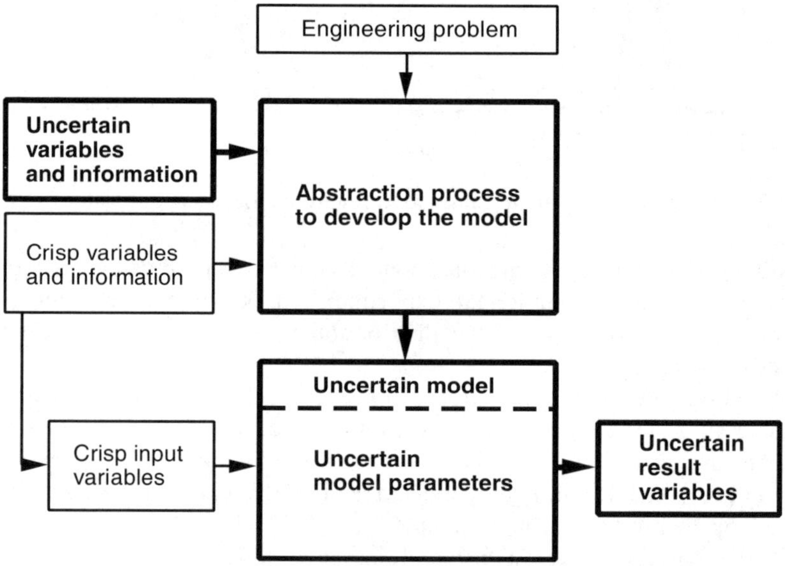

Fig. 1.4. Model uncertainty

Data uncertainty is uncertainty in the input variables. All uncertain variables (structural parameters) that are explicitly introduced into a model as input variables are referred to as *uncertain input variables*. As input variables *external to the model*, they have no influence on the model itself, but are mapped onto result variables with the aid of the model. Data uncertainty is not accounted for in the abstraction process of model construction (Fig. 1.5).

The problem lies in the choice of the model for a specific case. The specification of *model limits* is a subjective decision based on objective information. A model may be comprised of several submodels and may also be a submodel of a superordinate model system. The models of a model system are ranked hierarchically, whereby interactive relationships may exist (Fig. 1.6).

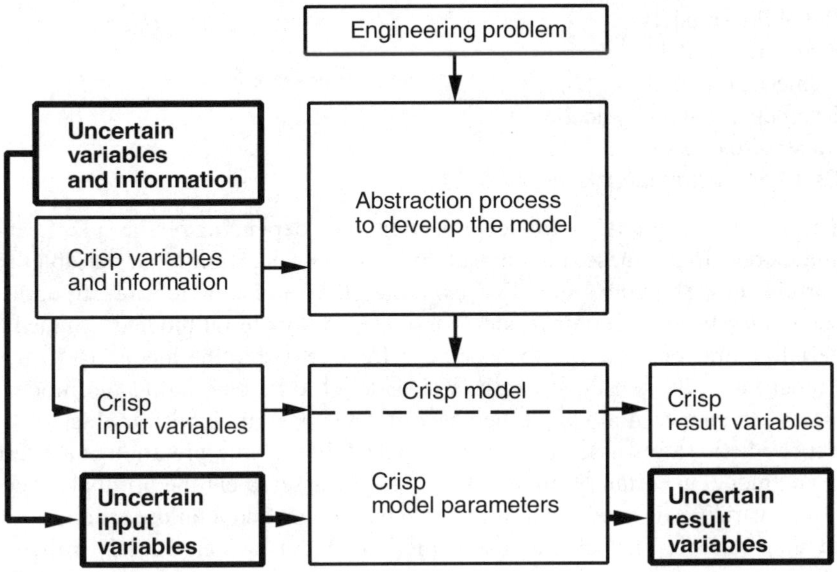

Fig. 1.5. Data uncertainty

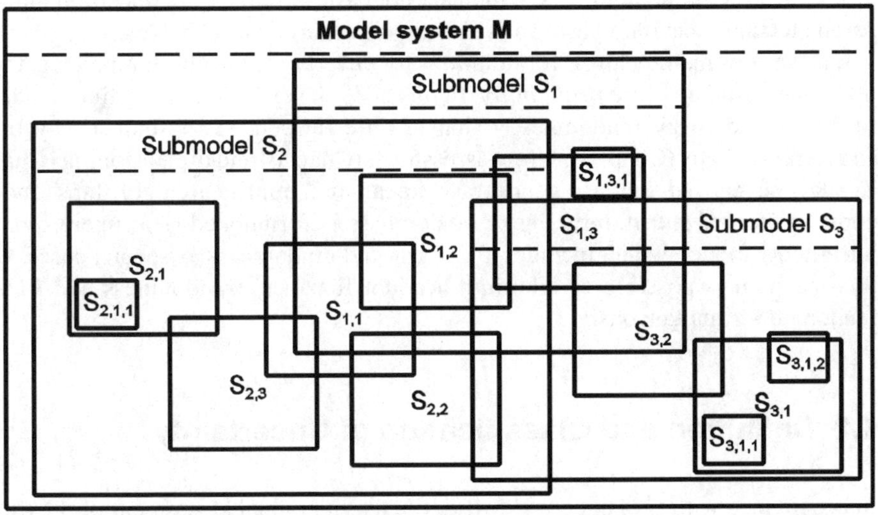

Fig. 1.6. Model system with hierarchically ranked and interactively coupled submodels

In civil engineering, mathematical descriptions of real objects and relationships of varying degrees of mathematical complexity may be described in the form of a model. Examples of models in civil engineering are:

- Reliability models
- Structural models
- Numerical models
- Geometrical and physical models
- Material models
- Error-estimation models

Uncertain parameters are model dependent, i.e., depending on the specification of the model limits, these must be subdivided into uncertain input variables and uncertain model parameters. For example, the uncertain tensile strength of concrete may be an uncertain model parameter of a material model, provided the material parameter is interpreted as being effective within the model itself. If, on the other hand, the tensile strength is considered to be external to the model in respect of the material model, it then becomes an uncertain input variable.

Probabilistic safety assessment leads to an additional interpretation of the terms data and model uncertainty. In this type of safety assessment the original space of the basic variables is subdivided into a survival region and a failure region by the limit state surface. The selected mechanical model in each case solely influences the limit state surface.

Uncertainty in the limit state is thus model uncertainty, whereas uncertainty in the structural parameters selected as basic variables is data uncertainty. All uncertain structural parameters, which are not basic variables, are uncertain model parameters; all uncertain basic variables are uncertain input variables.

The various mathematical formulations for uncertainty are also lumped together under the term *imprecise probability*. In this book the extended uncertainty models fuzziness and fuzzy randomness (Chap. 2) are introduced in such a way that structural analysis (Chap. 5), safety assessment (Chap. 6) and dimensioning (Chap. 7) may be carried out using arbitrary linear and nonlinear algorithms under consideration of uncertainty. Fuzzy randomness is formulated as a superordinate uncertainty model, which includes fuzziness and randomness as special cases. By this means it is possible to take into account fuzziness, randomness and fuzzy randomness simultaneously.

1.2 Definition and Classification of Uncertainty

In accordance with [22] uncertainty is defined as a gradual assessment of the truth content of a proposition, e.g., in relation to the occurrence of an event. Uncertainty may be classified according to its type (Fig. 1.7).

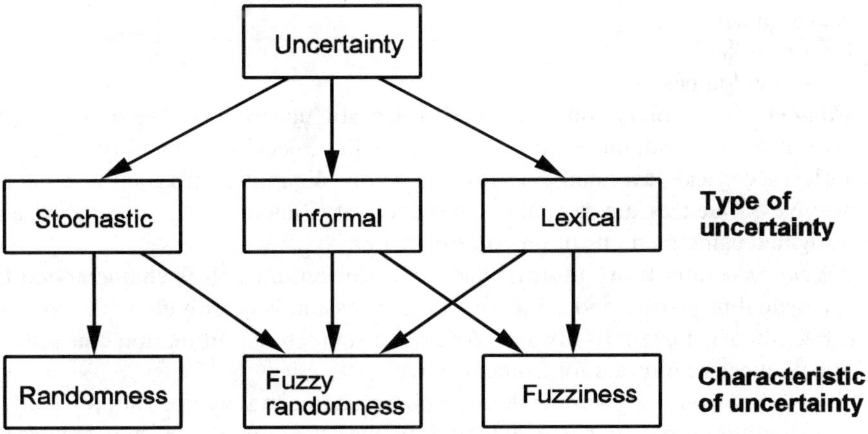

Fig. 1.7. Classification of uncertainty

The *type of uncertainty* may be

- Stochastic
- Informal, or
- Lexical

in nature.

Stochastic uncertainty is present when the random results of an experiment under identical boundary conditions may be observed almost indefinitely and may be described deterministically with regard to its occurrence.

Informal uncertainty results from an information deficit. This occurs, e.g., when only a small number of observations are available, the boundary conditions (apparently) are subject to arbitrary fluctuations or the system overview is incomplete.

Lexical uncertainty is uncertainty that is quantified by linguistic variables. In order to take account of this in models the linguistic variables must be transformed onto a numerical scale.

Different methods are available for the mathematical description and quantification of uncertainty. Possible methods include, e.g., probability theory [15, 107], interval algebra [2, 129], convex modeling [14, 49], the method of subjective probability [41, 202], chaos theory [84], fuzzy set theory [10, 212] and the theory of fuzzy random variables [90, 96]. In a mathematical formulation a characteristic is assigned to the uncertainty.

The *uncertainty characteristic* is described by

- Randomness
- Fuzziness, and
- Fuzzy randomness

Randomness is solely concerned with stochastic uncertainty. Uncertainty with the characteristic randomness (not concerned with subjective probability approach) satisfies statistical laws and possesses a purely objective information content; subjective influences are not taken into account. Randomness is described and investigated using methods of probability theory.

Fuzziness results from informal and lexical uncertainty. It is characterized by nonstatistical properties and subjective influences and is dealt with on the basis of fuzzy set theory. Fuzziness always possesses a subjective information content and often an objective information content as well.

Fuzzy randomness is present when the observed uncertainty does not fulfil all of the preconditions of stochastic uncertainty, but does exhibit partial stochastic properties. For example, the observed event, the boundary conditions or the observations may possess informal or lexical uncertainty, or the number of observations is insufficiently large for a statistical evaluation free of doubt. The treatment of fuzzy randomness is based on the theory of fuzzy random variables. The information content of uncertainty with the characteristic fuzzy randomness is mainly objective, although subjective influences may be simultaneously taken into account.

Each characteristic of uncertainty may occur in a discrete as well as a continuous form. The numerical treatment of the mathematical-mechanical model for structural assessment depends on the characteristic of the uncertainty.

1.3 Examples of Data and Model Uncertainty

A consideration of data and model uncertainty and the application of realistic simulation models make it possible, e.g., to fully exploit the load-bearing reserves of constructions, to take account of additional aspects in safety assessment, or to obtain more comprehensive information regarding load-bearing behavior.

The following examples from different fields of civil engineering illustrate where and how data and model uncertainty may occur in existing and planned constructions.

Geotechnical Engineering. The nonhomogeneity of the subsoil and the removal of soil samples only at particular points may lead to a considerable uncertainty in the definition of soil parameters, which thus exhibit informal uncertainty. Knowledge based on experience must often be drawn upon in design considerations, e.g., in the form of linguistic assessments such as "high load-bearing capacity" or "prone to subsidence".

Although uncertainty may be reduced by more extensive investigations, it cannot be eliminated altogether. For example, the expert valuation of the foundation soil for a planned construction site with an area of 47 230 m^2 was based on the results of 19 boreholes, five trial borings, and four light and eight heavy exploratory piledrives. Although this is indicative of a very meticulous soil investigation, intervals rather than crisp numbers were stated for foundation soil parameters. The stiffness modulus, for example, was stated to lie in the range 80–200 MN/m^2, and the layer thicknesses were also provided in the form of intervals. The engineering consultants were obviously aware of the existing informal uncertainty and adopted the interval as the data model.

Considerable uncertainty was also apparent in the size and distribution of earth pressure, the position of slip surfaces as well as in the load-bearing capacity and load-bearing behavior of anchors, piles, sheet piling, and other preconstruction measures. The prognoses for subsidence, settlement, and load transfer in the soil are uncertain.

Assessment of Existing Structures. Structures are subject to time-dependent effects such as material damage, system modification, or a change in loading, which result in time-dependent changes in the load-bearing capacity and structural reliability. These time-dependent processes exhibit pronounced data and model uncertainty. The changes in loading that occur during the service life of a structure are difficult to predict and are thus uncertain.

The assessment of time-dependent damage or the time-dependent level of safety is characterized by considerable uncertainty. In addition to the choice of a suitable model to account for uncertainty, the structural-mechanical effects must be described with sufficient accuracy in the numerical simulation.

The geometry of existing structures is essentially only measurable on the basis of the exposed structure; parts of the structure covered remain unaccounted for. The arrangement and cross sections of reinforcement are not fully verifiable. Support conditions can only be approximately estimated, while only vague assumptions may be made in many cases concerning foundations. Boundary and coupling conditions for structural components cannot be determined with sufficient accuracy, while degrees of constraint or local subsidence behavior may often only be determined on the basis of poorly founded assumptions.

Only a small number of material specimens are often available. Material damage resulting from the removal of samples remains unaccounted for, and the historical design documents only provide a limited insight into actual material properties. The load history and the change in material properties over the service life of the structures are generally unknown.

Old arched bridges made of natural stone are good examples of the latter. For the safety assessment of a natural stone arched bridge for road-traffic loading based on the First Order Reliability Method (FORM), it was required, e.g., to determine the masonry strength from only three mortar and four stone specimens and to model the latter as a random variable. Due to the fact that the bridge is designed as a rubble stone masonry construction, its strength is mainly determined by the compressive strength of the mortar. Only the cohesive strength was measured, however. By means of an approximate conversion three values were determined for the cube compressive strength of the mortar. The uncertainty in this case is based on the extremely sparse information content of only three measurements as well as the approximation values for the compressive strength. The uncertainty resulting from the information deficit becomes apparent when interval estimates are carried out under the assumption of a normal distribution for the expected value and the standard deviation of the mortar compressive strength. For a confidence level of 90% the actual expected value is found to lie in the interval [12.0, 19.8], and the standard deviation in the interval [1.3, 10.3]. On the basis of six additional mortar specimens and a direct measurement of their compressive strength the informal uncertainty of the expected value reduces to the interval [12.9, 16.8] and to [1.6, 4.9] for the standard deviation. Considering the high sensitivity of the results of probabilistic safety assessments in relation to the distribution assumptions for the basic variables, the residual uncertainty is still considerable, and, if neglected, may lead to erroneous estimates.

Numerical Simulation of Blasting Processes. The numerical simulation of blasting processes relates to multibody systems and contact problems. The prediction of a blasting process by a priori simulation involves data and model uncertainty. Uncertain observations, measurements, information deficits, opinions, and expert valuations determine blasting scenarios. Uncertain input data include, e.g., the position of the demolition charge, the charge quantity, the detonation sequence, the arrangement and quantity of reinforcement, or structural weakening.

The collapse process possesses model uncertainty. The collapse mechanisms, involving cascaded impact, contact, deformation and rupture processes, may only be realistically assessed when material, geometrical, and external load parameters are accounted for in the multibody system model. The uncertain results, e.g., drop radius, distance from neighboring buildings, degree of demolition or spread areas, also possess data uncertainty.

Textile Strengthening of Structures. The introduction of new materials and technologies is also accompanied by data and model uncertainty. In the development of textile-reinforced concrete and wood an awareness of the governing uncertainty, e.g., regarding the bond between filaments in textile yarn (model uncertainty) or the determination of sensitive material parameters (data uncertainty), requires enhanced uncertainty-dependent safety concepts.

In the case of textile-reinforced plane structures multidimensional uncertainty dependent on the position vector also arises, which must be accounted for using uncertain functions. The uncertain material behavior of new materials requires the formulation of uncertain processes.

Urban and Pavement Engineering. The construction and maintenance of almost all infrastructure constructions are characterized by inherent uncertainty. Uncertainty is a characteristic of, e.g., the state of supply and sewer lines, demand forecasts for drinking water, gas and district heating, forecasts of rainwater runoff as well as the interaction between cost trends for media and consumption. This is of importance for, e.g., pipeline dimensioning and service-life predictions of pipeline systems. The mechanical behavior of road constructions also exhibits uncertainty, e.g., with regard to rubble layers and their anticipated loading.

Hydraulic Engineering. The uncertainty of water levels, precipitation quantities and distributions, water-catchment areas, flow velocities, flow conditions, as well as the behavior of transported material and bank defenses or ice pressure on hydraulic structures has an effect on, e.g., the dimensioning of dams or runoff channels as well as floodwater forecasts.

Further Examples. Examples of uncertainty may be quoted for all disciplines of civil engineering. Without comment, the following key words are given: damage forecasts, earthquake effects, wind effects, risk assessment, pollution transport in water and air, and terrorist attacks.

1.4 On the State of Development of Uncertainty Models

Uncertainty models are gaining increasing importance in engineering sciences, and the intensive research in this field worldwide is a sure sign of wider application in the future. Developments so far are outlined under the three aspects (1) modeling of uncertainty, (2) quantification of uncertainty, and (3) structural analysis and safety assessment.

Modeling of Uncertainty. The usefulness of computational results not only depends on the quality of the computational model for a particular purpose but also on the appropriate, mathematical modeling of uncertainty [7, 8, 137, 169, 170].

The remarkably advanced state-of-the-art of uncertainty modeling is clearly illustrated, e.g., by publications [6, 52, 177]. The most highly developed model to date is the probabilistic uncertainty model. In combination with methods of mathematical statistics [21, 128] the established techniques of probability theory have found wide application in this field [171]. With the introduction of subjective probability [202], developments have extended beyond classical statistics and the objectivity of the probability measure. Using Bayesian methods [16] in particular, an attempt is made to include subjective information and expert valuations in probability calculations [61, 156, 195]. Extensions have also been investigated in order to account for nonstochastic uncertainty [34, 83]. Interval mathematics [2, 43] and convex modeling [14, 50] offer additional modeling techniques, although these only permit a binary assessment of the membership of elements to a set. More effective is the theory of rough sets [143], which takes into account the gray zone between membership and nonmembership. Fuzzy set theory [10, 22, 155, 205, 212] additionally permits the assessment of gray-zone elements of a set based on a normalized scale ranging between zero and unity. The comprehensive modeling of uncertainty was first made possible by the development of the theory of fuzzy random variables [90, 96, 97], see also [87, 88, 91, 204]. Randomness and fuzziness are thereby taken into consideration simultaneously; real-valued random variables and fuzzy parameters are included as special cases. Chaos theory [3, 84] follows a different model concept of uncertainty, in which nonpredictable system behavior is investigated (for stable systems within specific margins). Generally speaking, the starting point for uncertainty modeling is a classification of the various types of uncertainty [51, 139, 191].

Fuzzy set theory and the theory of fuzzy random variables are presently the subject of controversial discussion in the scientific literature [52] and are often judged very conservatively. The literature sources [14, 33, 43, 54, 86, 89, 105, 110, 130, 131, 135, 181, 192] bear testimony, however, to increasing acceptance.

Quantification of Uncertainty. Depending on the characteristic of uncertainty, different types of information are required for quantification purposes.

For the purpose of defining the membership functions of *fuzzy variables*, i.e., for fuzzification, generally valid algorithms and methods do not exist; a subjective assessment is made under consideration of objective information [17, 22, 24, 25, 37, 39]. Fuzzification is a problem-specific quantification method, which is essentially based on estimations by experts. A number of remarks on fuzzification are given, e.g., in [194].

A precondition for the reliable statistical description of a *random variable* is the availability of extensive data in the form of samples [40, 128]. The parameters and the type of the probability distribution of a random variable are determined and evaluated using the methods of mathematical statistics. Estimation theory provides a well-developed basis for the specification of probability distributions and parameters [21, 128]. Test methods for hypothetically assumed distributions may be found in the wide range of literature on test theory [21, 128].

Statistical estimates are based on the properties of an observed sample and extrapolate on the statistical properties of the universe from which the sample was taken. A wide range of point and interval estimators are available for this purpose, which operate parametrically or nonparametrically. A distinction must be made as to whether or not a priori information is included in the estimation.

If *no a priori information* is available on the random variable to be described, an estimation may be made, e.g., on the basis of the method of moments, the maximum likelihood method, or the empirical distribution. The choice of a suitable method is problem specific and is based on the essential properties of the estimator (unbiasedness, consistency, and efficiency). Point estimations only yield a crisp number for the value sought (e.g., parameter of a distribution, probability, or quantile value). A disadvantage of the latter is that no information is obtained concerning the accuracy of the estimation. Especially for small sample sizes, considerable deviations from the actual value (e.g., of a parameter) may result [40, 128], see also [111]. An interval estimation on the other hand provides additional information concerning the accuracy of an estimation.

In the case of a prescribed confidence level (probability) an interval is computed for the variable to be estimated, which includes the actual value of this variable with the desired probability. It should be mentioned, however, that a confidence interval is not uniquely defined by the specification of the confidence level. In addition, it is also necessary to define the interval type, e.g., as a one-sided or special two-sided interval. Numerical-estimation methods are preferably applied for interval estimations as these avoid the complicated and frequently nonrealizable analytical determination of the distribution of an estimator. Reference is made here to the resampling methods bootstrap [48, 179] and jack-knife [64, 179]. Further developments and solutions for selected problems of estimation and test theory are given in [30, 44, 74, 78, 213].

Estimation methods based on *a priori information* concerning a random variable (e.g., with regard to the distribution type and the range of values of parameters), and that improve the latter on the basis of an up-to-date sample, may also be applied for point and interval estimation. The estimation is based on Bayes theorem, i.e., conditional probabilities are used [195]. A comprehensive account of Bayesian methods is given in [16]. Bayesian methods are essentially applied for parameter estimations. A priori information and subjective evaluations are hereby included as prior distributions of the parameters to be determined or improved. On the basis of an up-to-date sample the prior distributions – analogous to the maximum likelihood estimation approach – are improved, i.e., converted into posterior distributions. On the basis of the latter the improved parameters are computed as crisp numbers (point estimation, e.g., as an expected value) or as a confidence interval (interval estimation). The foregoing comments on point and interval estimations also apply in this case. The confidence intervals may be determined directly from the posterior density. In addition, it is also possible to compute predictive distributions for the random variable from the posterior distribution. The analytical determination of the posterior distributions is problematic, however, as these cannot be assigned to a particular type of distribution in many cases. In order to solve this problem analytically, conjugate prior distributions are adopted [16, 58]. A prior density corresponding to the density structure of the attribute (of the random variable) is hereby adopted. This means that the posterior density also exhibits this structure and may thus be described analytically. In many applications the numerical determination of the posterior density as an empirical distribution is considered to be advantageous.

The further development and application of Bayesian methods underlines their importance for probabilistic-based uncertainty modeling [138, 140]. For example, resampling methods in combination with Bayesian estimations [28, 29] are suggested for this purpose. In [214] a Bayesian estimation method based on [213] is presented, which operates without information on the shape of the sought distribution and may be applied to small as well as large sample sizes. This method is extended to two-dimensional distributions in [215]. Applications of Bayesian methods may be found, e.g., for service-life estimations in [193], especially under consideration of fluctuating boundary conditions (partial repair) [148]. A Bayesian method for the probabilistic assessment of the state of pipeline systems (in town infrastructure) based on limited and incomplete data is given in [31]. A Bayesian method for quantifying uncertainty in the choice of mechanical and statistical models is outlined in [209] and applied for reliability analysis in fatigue problems. An application of Bayesian inference for estimating system reliability under consideration of expert valuations is published in [156]. A method for the Bayesian estimation of the elasticity modulus of concrete is given in [61]. Special mention is made of the development of robust Bayesian analysis [81]. A large number of prior distributions are thereby included in the Bayesian estimation, thus permitting the consideration of a realistic spread of results. The set of prior distributions is not limited independently of data. The subjective nonprobabilistic quantifiable uncertainty in the specification of the prior distributions is accounted for in the analysis according to the principle of a worst-case study; the reliability of the results improves.

Statistical estimation methods with and without consideration of a priori information are also characterized by a range of *problems*. A precondition in many cases is that the distribution type of the universe or individual parameters of the sought distribution are already known. These preconditions are included in the estimation as assumptions, the validity of which cannot be generally verified or only verified with uncertainty. A disadvantage of Bayesian estimation methods is that (often very uncertain) assumptions must be made for the prior distributions of the parameters, and hence highly subjective uncertainty based on valuations by experts are irreversibly lumped together with objective information obtained from samples. An essential problem of statistics, which cannot be accounted for by randomness, but which is clearly nonprobabilistic in character, is highlighted in [196]: "Model choice is a fundamental problem in data analysis."

All statistical estimation methods are restricted to uncertainty with the characteristic randomness; randomness is also exclusively presumed for estimated and assumed parameters. This precondition for statistical estimation methods is formulated as an assumption for the mathematical model; its validity for specific applications cannot be verified. Systematic errors and nonstochastic uncertainty in the data and boundary conditions cannot be taken into account. Examples of the latter are dealt with, e.g., in [30, 112, 141, 142, 211]. A probabilistic safety assessment with limited information on the adopted random variables is described, e.g., in [42].

The results of a statistical parameter estimation are generally only approximation values for the sought parameters; all prognoses derived from the latter, e.g., regarding structural behavior or structural reliability, contain nonquantifiable inaccuracies. These criticisms are highlighted in [56] as an incentive for future research.

Major advancements are anticipated through the extension of statistical methods in order to take account of *nonstochastic uncertainty*. The rapidly expanding research in this field bears testimony to this conjecture [7, 11, 93, 194, 195]. The increasing application of fuzzy methods combined with statistical approaches based on a selection of publications is discussed in [177].

Initial methods already exist for the analysis of uncertain data [11, 194]. The application of resampling methods is also being followed; in [199] a bootstrap method is applied for the evaluation of fuzzy data. A method for evaluating fuzzy data with the aid of a generalized histogram is presented in [18]. In [66] and [67] fuzzy parameters are applied for hypothesis testing in order to fuzzify the transition between the rejection and acceptance of a hypothesis for a given significance level. Investigations on the testing of fuzzy hypotheses are also being carried out [4].

Fuzzy probabilistics still comprises the essential subject matter of fundamental research [88, 90, 91]. Characteristic values and parameters of *fuzzy random variables* are defined in [87, 204]; statistical laws and time-dependent problems are investigated in [98, 160] and [151], see also [197]. A fuzzy random variable may be described by its fuzzy probability distribution function. A method for determining uncertain distribution functions is given in [194] and [195].

Bayesian methods have also been extended by the inclusion of fuzzy variables. A contribution to Bayesian statistics including uncertainty is given, e.g., in [177]. Basic considerations relating to Bayesian methods for uncertain data are dealt with in [194] and [195]. Methods are proposed in the latter that lead to uncertain posterior distributions and uncertain predictive distributions. Uncertain prior distributions are also discussed. A Bayesian estimation yielding uncertain confidence intervals is also presented for the evaluation of uncertain distributions.

Only a limited number of contributions may be found in the literature concerning applications that make use of a coupling of fuzzy and Bayesian methods. In [34] a fuzzy Bayesian method is proposed for the reliability analysis of structures, in which the functional values of the inputted conditional probability distributions are interpreted as probabilities of occurrence of fuzzy events. Although this enables nonstochastic uncertainty to be accounted for, it is not possible to map the latter onto safety prognoses; the failure probability and reliability index are crisp numbers. In [83] the structural reliability of existing constructions is assessed with the aid of a Bayesian method, in which observations (e.g., cracks in the concrete) are quantified using fuzzy numbers and introduced as a fuzzy data set into the Bayesian theorem. A combination with kriging, in which a priori information is accounted for with the aid of the Bayesian theorem and uncertainty is described using fuzzy variables, is described in [9] (kriging = an interpolation technique based on the theory of regionalized variables, which, besides the domain of values to be estimated, also yields information on its local quality). A Bayesian test of fuzzy hypotheses is discussed in [188], while in [159] the application of a fuzzy Bayesian method for decision making in control problems is presented, in which the results are defuzzified. A consideration of the uncertain loading of a mechanical system for the purpose of system control is dealt with in [94].

Although the fundamental principles and applications of statistical methods and associated problems are dealt with in the very wide range of literature on mathematical statistics, the effects of uncertain results are not discussed. Due to the fact that the effects of input parameters cannot be estimated a priori in nonlinear problems, the a priori choice of a suitable (e.g., conservative) quantification method is nonrealizable, i.e., all possibilities must be examined.

It is the *task of the engineer* to interpret statistical results and to apply his findings to the formulation of reliable input and model parameters for structural analysis and safety assessment. This also applies to fuzzification methods; the specification of the membership function is the task of the engineer. Structural analysis and safety assessment are only capable of yielding realistic results, provided uncertainty is comprehensively, appropriately, and reliably quantified [170].

Structural Analysis and Safety Assessment Considering Uncertainty. In structural analysis and safety assessment the existing data and model uncertainty may be taken into account using models based on probability theory, fuzzy set theory or the theory of fuzzy random variables [1, 73, 99, 104, 135, 170]. Technical codes base on the semiprobabilistic safety concept with partial safety factors.

Numerous high-performance algorithms are available for *probabilistic safety assessment and structural analysis* [57, 108, 144, 145, 152, 167, 168, 169, 170]; a wide range of basic literature dealing with this topic is available for engineers, e.g., [107, 166, 182]. Uncertain variables are described as random variables, random fields, or random processes. Their probability distributions are either presumed to be known or are estimated using statistical methods. In the case of multidimensional random vectors, stochastic fields, and processes, additional relationships must be analyzed and described. Generally speaking, care must be taken to determine which structural parameters possess uncertainty [170]. Uncertainty is generally not only present in parameters external to the model such as, e.g., applied loads, but also in parameters included in the computational model, i.e., in resistance values (e.g., material and geometrical parameters). For this purpose the method of stochastic finite elements [23, 185] offers an appropriate modeling tool. Basic principles of this method are given, e.g., in [62, 63, 65, 79, 80, 101, 172, 183]. With regard to the solution of probabilistic problems, a basic distinction must be made between analytical methods and simulation methods [168]. According to current developments, simulation methods (Monte Carlo simulation and further developments thereof) [170, 173, 174] are clearly favored, especially due to the considerable amount of computational effort required by analytical methods.

Basic considerations regarding the Monte Carlo simulation are given in [55, 71, 157]. Further developments comprise the subject matter of entire conference proceedings [173]; the contributions [53] and [146] are worthy of special mention. Further important work in this field may be found in the contributions presented in [174]. An overview of new developments is given in [171] while a comprehensive collection of publications may be found in [167]. Problems dealt with in the wide-ranging literature include, e.g., dynamic problems [100, 163, 170, 184], eigenvalue problems [187], service-life analyses [147], and stability investigations [165].

The further development of simulation methods is essentially aimed at reducing the numerical effort, as measured by the number of necessary deterministic calculations. An increase in computational efficiency is a precondition for the application of simulation methods in the linear and nonlinear analysis of real structures with many degrees of freedom; this promotes the acceptance of simulation methods in engineering practice [108, 171]. For the safety assessment of systems exhibiting nonlinear behavior a wide range of publications exists, in which, e.g., the response surface method is applied [26, 167]. Moreover, several variance-reducing methods have been developed for efficiently estimating failure probability. These include, amongst others, importance sampling, adaptive sampling, directional sampling, stratified sampling, Latin hypercube sampling, and combinations of the latter [5, 55, 170, 175].

Intensive research has been carried out in recent years to take account of nonstochastic uncertainty in civil engineering. Various algorithms have been developed for fuzzy structural analysis based on *fuzzy set theory*. Most of the work in this field is based on the vertex method [45, 130, 131, 135]. Due to the fact that only the interval bounds or corner points of the inputted α-level sets are taken into consideration in the analysis, these applications are restricted to monotonic problems. If the solution is determinable by a boundary search of the α-level sets, the gradient method may be applied [19, 20]. Further applications are described in [1, 73, 136], combinations with finite element approaches are discussed in [132, 153]. A knowledge of special properties of the mapping is always a precondition in such applications. In [72] a transformation method is presented as a further development of the vertex method. Although this method is no longer restricted to monotonic problems, it may lead to an underestimation of the uncertainty of the fuzzy results due to the a priori specification of the points to be analyzed in the space of the input values. In this method the numerical effort increases exponentially with the number of fuzzy input variables.

In safety assessment different techniques are used to take account of fuzzy values; a general overview is given in [82]. On the basis of possibility theory [47] fuzzy variables may be treated as basic variables. The possibility measure or the necessity measure are thereby applied for the quantitative description of the safety level [32, 39]. Fuzzy limit states resulting from fuzzy variables are dealt with, e.g., in [36, 104] and [161]; these may be computed with the aid of fuzzy models and may be evaluated on either a possibilistic or a probabilistic basis [47, 212]. In [99] fuzzy methods are considered in combination with the finite element method. Fuzziness and randomness are modeled separately and are coupled through a common integral evaluation to yield an uncertain safety prognosis. An application of fuzzy set theory for safety assessment in stability problems is presented in [162], a fuzzy model for life prediction considering fatigue is dealt with in [12]. Fuzzy methods have been successfully applied in many fields for the purpose of optimization [105, 181] as well as system and process control. In [164] decision-making models are applied in structural design. These applications are based on the principle of *approximate reasoning,* which is applied to construct inference mechanisms for decision-making purposes. Approximate reasoning is dealt with comprehensively in the basic literature on fuzzy set theory [10, 22, 52, 68, 155, 207, 208].

Only a few examples of the application of *fuzzy probabilistics* are available in the literature. A concept for fuzzy probabilistic safety assessment is presented in [104]. An optimization method using random sets, which may be considered to be α-level sets of fuzzy random variables, is outlined in [192]. A proposal for determining fuzzy-stochastic system responses in linear dynamic problems is presented in [210]. A further application is discussed in [70].

2 Mathematical Basics for the Formal Description of Uncertainty

In this chapter some selected theoretical basics and enhancements concerning the definition and the treatment of uncertainty are stated and explained. These explanations do not represent an overall and detailed mathematical overview but only summarize fundamentals that may be helpful for understanding the developed methods and engineering applications considered subsequently.

2.1 Fuzziness - Definitions and Arithmetic

The set theoretical description of fuzziness stated in the following is based on a variety of fundamentals [10, 22, 40, 60, 139, 155, 205, 212]. Regarding the application to engineering problems special properties of fuzzy sets are considered. For improved clarity only the one-dimensional case is treated, an extension to multidimensional problems can be realized in an elementary way.

2.1.1 Crisp Set and Fuzzy Set

In [60] the term *set* is defined using the formulation stated by Cantor. After this a set is *a combination of particular, well-distinguishable objects (which are called "elements" of the set) to an ensemble*.

In classical set theory the membership of elements in relation to a set is assessed in binary terms according to a crisp condition. An element either belongs or does not belong to the set, the boundary of the set is crisp. As a further development of classical set theory, fuzzy set theory permits the gradual assessment of the membership of elements in relation to a set; this is described with the aid of a membership function. A fuzzy set is defined as follows (Fig. 2.1):

If $\mathbf{X}$ represents a fundamental set and x are the elements of this fundamental set, to be assessed according to an (lexical or informal) uncertain proposition and assigned to a subset A of $\mathbf{X}$, the set

$$\tilde{A} = \{(x, \mu_A(x)) \mid x \in \mathbf{X}\} \tag{2.1}$$

is referred to as the uncertain set or fuzzy set on **X**. $\mu_A(x)$ is the membership function (characteristic function) of the fuzzy set $\tilde{A}$. The fuzzy set $\tilde{A}$ is also referred to as a *fuzzy variable* $\tilde{x}$, or in the multidimensional case as fuzzy vector $\underline{\tilde{x}}$.

The membership function may be continuous (the elements of the set are not countable), or the set contains only discrete elements assessed by membership values (the elements of the set are countable). The following holds for the functional values of the membership function $\mu_A(x)$

$$\mu_A(x) \geq 0 \quad \forall x \in \mathbf{X} . \tag{2.2}$$

The functional values $\mu_A(x)$ are the greater the more the assessment criterion is fulfilled. If

$$\sup_{x \in \mathbf{X}} \left[\mu_A(x) \right] = 1 \tag{2.3}$$

holds, the membership function (and the fuzzy set $\tilde{A}$) is referred to as *normalized.* In this case the membership function describes the mapping of the fundamental set **X** onto the interval [0, 1]. Subsequently, exclusively normalized membership functions are considered.

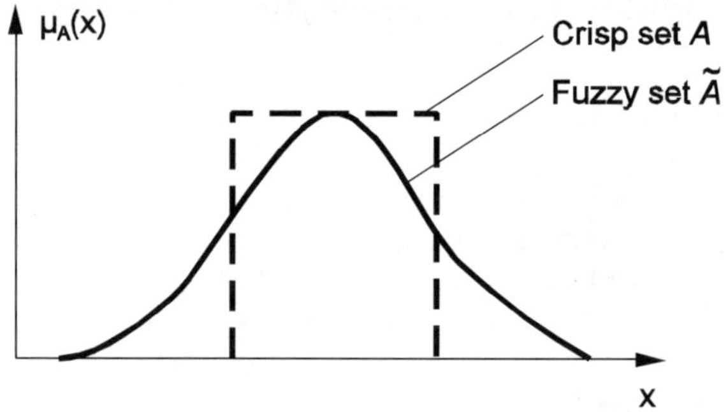

Fig. 2.1. Comparison of a crisp set and a fuzzy set

The *support* $S(\tilde{A})$ of a fuzzy set $\tilde{A}$ is a crisp set. This contains the elements

$$S(\tilde{A}) = \left\{ x \in \mathbf{X} \mid \mu_A(x) > 0 \right\} . \tag{2.4}$$

In Fig. 2.2 the support of a normalized fuzzy set is indicated.

A fuzzy set $\tilde{A}$ is referred to as convex if its membership function $\mu_A(x)$ monotonically decreases on each side of the maximum value, i.e., if

$$\mu_A(x_2) \geq \min[\mu_A(x_1), \mu_A(x_3)] \ \forall x_1, x_2, x_3 \in \mathbf{X} \text{ with } x_1 \leq x_2 \leq x_3 \tag{2.5}$$

applies.

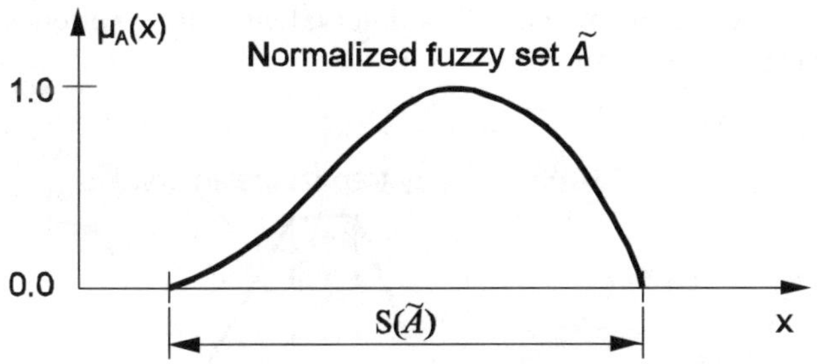

Fig. 2.2. Normalized fuzzy set $\tilde{A}$ with its support S($\tilde{A}$)

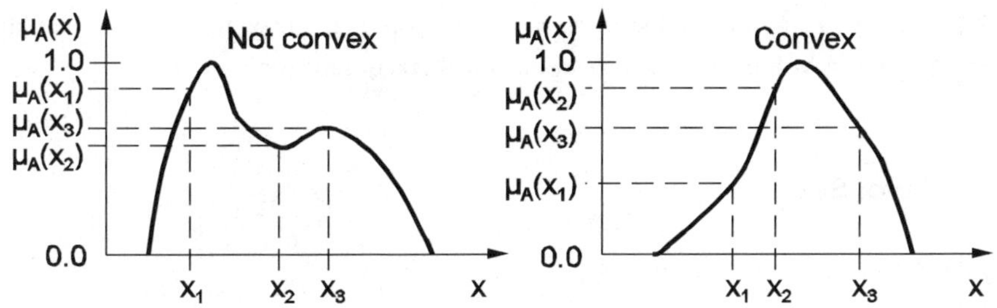

Fig. 2.3. Fuzzy set $\tilde{A}$, not convex and convex

2.1.2 Fuzzy Number

A *fuzzy number* $\tilde{a}_Z$ is a convex, normalized fuzzy set $\tilde{A} \subseteq \mathbb{R}$ whose membership function is at least segmentally continuous and has the functional value $\mu_A(x) = 1$ at precisely one of the x values. This point x is referred to as the mean value of the fuzzy number.

A *fuzzy interval* $\tilde{a}_I$ is a normalized fuzzy set $\tilde{A} \subseteq \mathbb{R}$ with a mean interval, i.e., all elements in this interval $x \in [x_1, x_2]$ possess the membership $\mu_A(x) = 1$. Likewise, the membership function $\mu_A(x)$ must be convex and at least segmentally continuous.

According to the definition of fuzzy sets it is distinguished between continuous and discrete fuzzy numbers and fuzzy intervals.

Fuzzy numbers and fuzzy intervals with linear membership functions between $\mu = 0$ and $\mu = 1$ (Fig. 2.4) can be represented by the following simple notation:

- Fuzzy triangular number $\tilde{a}_Z = \langle x_1, x_2, x_3 \rangle$
- Fuzzy trapezoidal interval $\tilde{a}_I = \langle x_1, x_2, x_3, x_4 \rangle$

Fuzzy triangular numbers are determined by specifying the smallest and the largest values x_1 and x_3 (interval bounds of the support) as well as the value x_2 (mean value) belonging to the functional value $\mu_A(x_2) = 1$. When specifying a fuzzy trapezoidal interval x_2 and x_3 define the bounds of the interval with the functional value $\mu_A(x) = 1$ (mean interval).

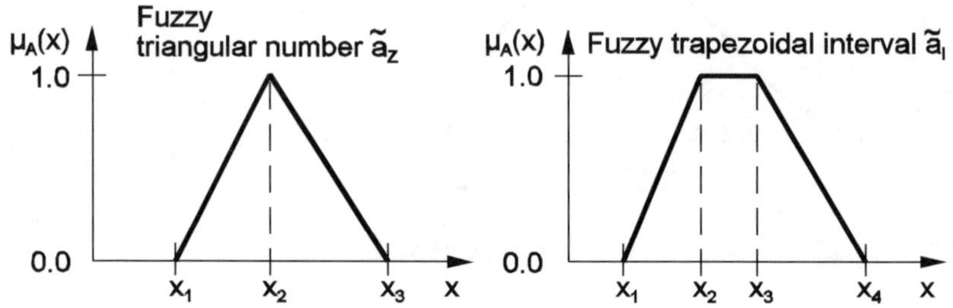

Fig. 2.4. Fuzzy triangular number $\tilde{a}_Z = <x_1, x_2, x_3>$ and fuzzy trapezoidal interval $\tilde{a}_I = <x_1, x_2, x_3, x_4>$ with linear branches of the membership function

2.1.3 α-level Set

From the fuzzy set $\tilde{A}$ the crisp sets

$$A_{\alpha_k} = \{x \in \mathbf{X} \mid \mu_A(x) \geq \alpha_k\} \tag{2.6}$$

may be extracted for real numbers $\alpha_k \in (0, 1]$. These crisp sets are called *α-level sets*. All α-level sets A_{α_k} are crisp subsets of the support $S(\tilde{A})$. For several α-level sets of the same fuzzy set $\tilde{A}$ the following holds (Fig. 2.5):

$$A_{\alpha_k} \subseteq A_{\alpha_i} \quad \forall \alpha_i, \alpha_k \in (0,1] \text{ with } \alpha_i \leq \alpha_k. \tag{2.7}$$

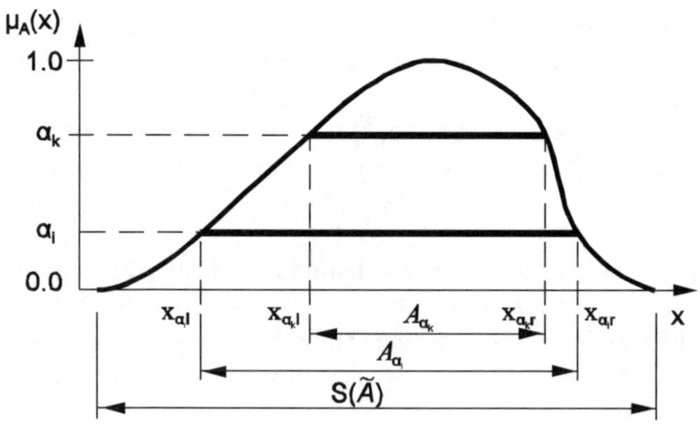

Fig. 2.5. α-level sets

If the fuzzy set $\tilde{A}$ is convex, each α-level set A_{α_k} is a connected interval $[x_{\alpha_k l}, x_{\alpha_k r}]$ in which

$$x_{\alpha_k l} = \min[x \in \mathbf{X} \mid \mu_A(x) \geq \alpha_k], \tag{2.8}$$

$$x_{\alpha_k r} = \max[x \in \mathbf{X} \mid \mu_A(x) \geq \alpha_k]. \tag{2.9}$$

2.1.4 Linguistic Variables

A *linguistic variable* v_l is defined by the quadruple

$$v_l = \{T, \mathbf{X}, G, S\} \tag{2.10}$$

with the elements

- T: set of the terms t of the linguistic variable referred to as the term set
- $\mathbf{X}$: fundamental set including at least all physically relevant numerical elements $x \in \mathbf{X}$
- G: set of syntactic rules on whose basis further terms may be deduced
- S: set of semantic rules that assign fuzzy sets on the fundamental set $\mathbf{X}$ each to one element $t \in T$ (mapping rule).

For example, the set T may comprise the terms *very low*, *low*, *medium*, *high*, and *very high*, which could be used to assess the stiffness of a material like, e.g., foundation soil or the degree of connections between structural members like, e.g., of end-plate shear connections between steel girders at frame corners (Fig. 2.6).

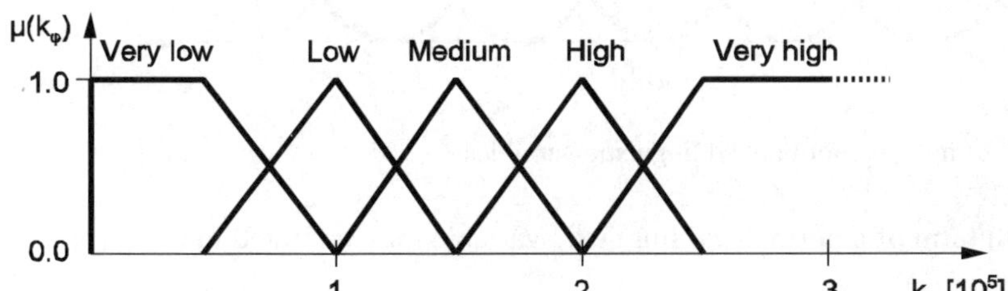

Fig. 2.6. Terms of the linguistic variable stiffness of the rotational spring at a steel frame corner

For a better manageability of linguistic variables and their application in various cases the concept from Eq. (2.10) is generalized. A *normalized linguistic variable* v_{ln} is defined by

$$v_{ln} = \{T_G, I_E, S\}, \tag{2.11}$$

with

- T_G: set of all terms t_G of the normalized linguistic variable including the syntactic rules

- I_E: unit interval [0, 1] representing the normalized fundamental set
- S: set of semantic rules that assign fuzzy sets on I_E each to one element $t_G \in T_G$.

A normalized linguistic variable is to be understood as being the set of n mappings from the unit interval I_E with $e \in [0, 1]$ into the value range of the membership functions $\mu_i \in [0, 1]$ of the terms t_{Gi} with i = 1, ..., n. Each of the n mappings represents one term t_{Gi}. The membership function $\mu_i(e)$ indicates the subjectively assessed membership degrees for the assignment of values e to the i-th term. Two examples of normalized linguistic variables are shown in Fig. 2.7.

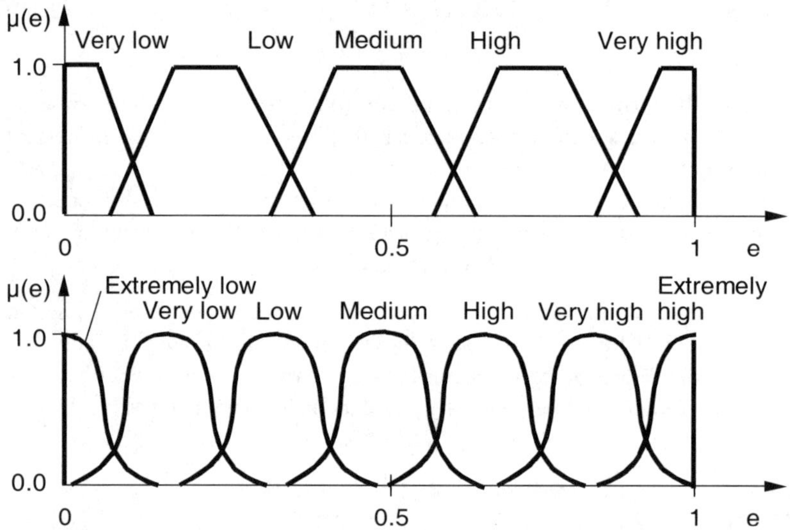

Fig. 2.7. Examples of normalized linguistic variables

The i-th term of a normalized linguistic variable may be stated in the form

$$t_{Gi} = \{(e, \mu_i(e)) \mid e \in [0, 1]; \mu_i(e) \in [0, 1]\}; \quad i = 1, ..., n, \tag{2.12}$$

where n denotes the number of terms.

Each term t_{Gi} is to be transformed into the value range of the crisp set

$$\underline{X}_k = \{x \mid x \in [x_{kl}, x_{kr}]\} \tag{2.13}$$

containing all reasonable values for the fuzzy structural parameter $\tilde{x}_k$. This is realized by transforming each element $e \in [0, 1]$ onto one element $x_k \in [x_{kl}, x_{kr}]$ and leads to the membership function $\mu(x_k)$ for the fuzzy structural parameter $\tilde{x}_k$.

The transformation between e and x_k may be given by the linear relationship

$$e = \frac{x_k - x_{kl}}{x_{kr} - x_{kl}}; \quad \mu_i(e) = \mu(x_k), \tag{2.14}$$

or may be described by a nonlinear dependency like, e.g.,

$$e = \left(\frac{x_k - x_{kl}}{x_{kr} - x_{kl}}\right)^v \quad ; \quad \mu_i(e) = \mu(x_k) \ ; \quad v > 0 \ ; \quad v \neq 1, \tag{2.15}$$

as shown in Fig. 2.8.

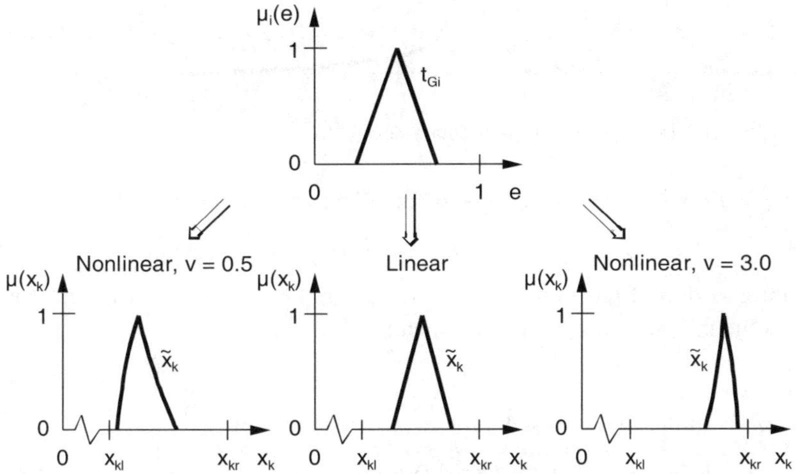

Fig. 2.8. Transformation of a normalized linguistic variable

2.1.5 Set Theoretical Operations

Set theoretical operations on fuzzy sets are defined with the aid of their membership functions $\mu_A(x)$. This represents an extension of classical set theory.

The *inclusion* of fuzzy sets is defined as follows. If two fuzzy sets $\tilde{A}$ and $\tilde{B}$ specified by their membership functions $\mu_A(x)$ and $\mu_B(x)$ are given on $\mathbf{X}$ and $\mu_A(x) \leq \mu_B(x)$ holds for all $x \in \mathbf{X}$, then $\tilde{A}$ is contained in $\tilde{B}$

$$\tilde{A} \subseteq \tilde{B}. \tag{2.16}$$

The *complement* $\tilde{A}^C$ of the fuzzy set $\tilde{A}$ (Fig. 2.9) is defined by

$$\tilde{A}^C = \{(x, \mu_{A^C}(x)) \mid x \in \mathbf{X}; \mu_{A^C}(x) = 1 - \mu_A(x)\}. \tag{2.17}$$

Complementation of the fuzzy set $\tilde{A}$ means the negation of the assessment that had been used for assigning the elements x to $\tilde{A}$. In the case that linguistic variables are used as valuation scale, the complement must not be developed by using the antonym but has to be formed by a preceding "it is wrong that ...". If $\tilde{T}_S$ denotes the fuzzy set assessing a structure as being *safe*, then the complementary set $\tilde{T}_S^C$ expresses: "It is wrong that the structure is safe". The complementary set $\tilde{T}_S^C$ is not the set assessing a structure as being *unsafe*.

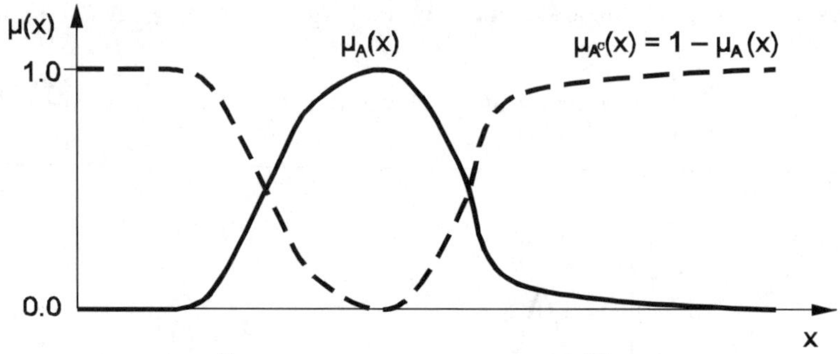

Fig. 2.9. Fuzzy set $\tilde{A}$ and fuzzy complementary set $\tilde{A}^C$

The *intersection* $\tilde{D}$ of the fuzzy sets $\tilde{A}$ and $\tilde{B}$ on **X** is obtained from

$$\tilde{D} = \tilde{A} \cap \tilde{B} = \{(x, \mu_D(x)) \mid x \in \mathbf{X}; \mu_D(x) = \min\,[\mu_A(x), \mu_B(x)]\}. \tag{2.18}$$

Thus the intersection of the fuzzy set $\tilde{A}$ and its complement $\tilde{A}^C$ – different from classical set theory – is not obtained as an empty set (Fig. 2.10)

$$\tilde{A} \cap \tilde{A}^C \neq \varnothing. \tag{2.19}$$

The *union* $\tilde{V}$ of the fuzzy sets $\tilde{A}$ and $\tilde{B}$ on **X** results from

$$\tilde{V} = \tilde{A} \cup \tilde{B} = \{(x, \mu_V(x)) \mid x \in \mathbf{X}; \mu_V(x) = \max\,[\mu_A(x), \mu_B(x)]\}. \tag{2.20}$$

If the union of the fuzzy set $\tilde{A}$ and its complement $\tilde{A}^C$ is computed on the fundamental set **X**, then – again different from classical set theory – this does not lead to the fundamental set again (Fig. 2.10)

$$\tilde{A} \cup \tilde{A}^C \neq \mathbf{X}. \tag{2.21}$$

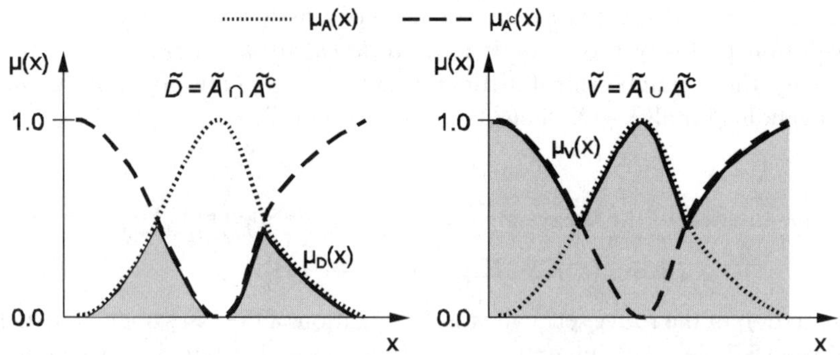

Fig. 2.10. Intersection $\tilde{D}$ and union $\tilde{V}$ of the fuzzy set $\tilde{A}$ and its complement $\tilde{A}^C$

On the basis of the definitions concerning intersection, union, and complementation the following laws of classical set theory remain valid also for fuzzy sets; whereby $\tilde{A}$, $\tilde{B}$, and $\tilde{C}$ denote fuzzy sets on **X**.

- Commutativity:

$$\tilde{A} \cap \tilde{B} = \tilde{B} \cap \tilde{A}; \quad \tilde{A} \cup \tilde{B} = \tilde{B} \cup \tilde{A} \tag{2.22}$$

- Associativity:

$$(\tilde{A} \cap \tilde{B}) \cap \tilde{C} = \tilde{A} \cap (\tilde{B} \cap \tilde{C}); \quad (\tilde{A} \cap \tilde{B}) \cap \tilde{C} = \tilde{A} \cap (\tilde{B} \cap \tilde{C}) \tag{2.23}$$

- Distributivity:

$$\tilde{A} \cap (\tilde{B} \cup \tilde{C}) = (\tilde{A} \cap \tilde{B}) \cup (\tilde{A} \cap \tilde{C}); \quad \tilde{A} \cup (\tilde{B} \cap \tilde{C}) = (\tilde{A} \cup \tilde{B}) \cap (\tilde{A} \cup \tilde{C}) \tag{2.24}$$

- Adjunctivity:

$$\tilde{A} \cap (\tilde{B} \cup \tilde{A}) = \tilde{A}; \quad \tilde{A} \cup (\tilde{B} \cap \tilde{A}) = \tilde{A} \tag{2.25}$$

- De Morgan laws:

$$(\tilde{A} \cap \tilde{B})^{C} = \tilde{A}^{C} \cup \tilde{B}^{C}; \quad (\tilde{A} \cup \tilde{B})^{C} = \tilde{A}^{C} \cap \tilde{B}^{C} \tag{2.26}$$

- Idempotence:

$$\tilde{A} \cap \tilde{A} = \tilde{A}; \quad \tilde{A} \cup \tilde{A} = \tilde{A} \tag{2.27}$$

- Monotonicity:

$$\tilde{A} \subseteq \tilde{B} \Rightarrow \tilde{A} \cap \tilde{C} \subseteq \tilde{B} \cap \tilde{C}; \quad \tilde{A} \subseteq \tilde{B} \Rightarrow \tilde{A} \cup \tilde{C} \subseteq \tilde{B} \cup \tilde{C} \tag{2.28}$$

- Operations with the empty set $\varnothing$ and the fundamental set $\mathbf{X}$:

$$\tilde{A} \cap \varnothing = \varnothing; \quad \tilde{A} \cap \mathbf{X} = \tilde{A}; \quad \tilde{A} \cup \varnothing = \tilde{A}; \quad \tilde{A} \cup \mathbf{X} = \mathbf{X} \tag{2.29}$$

Owing to the noncompliance with the complementary relations in Eqs. (2.19) and (2.21) the algebra of fuzzy sets is not a Boolean set algebra. Alternative definitions of the set operations complementation, intersection, and union of fuzzy sets, which do not make use of the operators *min* and *max* to generate the membership functions of the results, are not considered here. Further investigations concerning operations with fuzzy sets are discussed in [10, 22, 155, 205, 212].

2.1.6 Cartesian Product

The fuzzy sets $\tilde{A}_1, ..., \tilde{A}_n$ on the fundamental sets $\mathbf{X}_1, ..., \mathbf{X}_n$ form the *Cartesian product*

$$\tilde{K} = \tilde{A}_1 \times \tilde{A}_2 \times ... \times \tilde{A}_n. \tag{2.30}$$

From the fundamental sets $\mathbf{X}_1, ..., \mathbf{X}_n$ the product space $\underline{\mathbf{X}} = \mathbf{X}_1 \times \mathbf{X}_2 \times ... \times \mathbf{X}_n$ (x-space) is formed, whose coordinate axes are perpendicular to one another. The Cartesian product $\tilde{K}$ comprises all combinations of the elements $x_1, ..., x_n$ of the fuzzy sets $\tilde{A}_1, ..., \tilde{A}_n$. With the aid of the minimum operator membership values $\mu_K(\underline{x}) = \mu_K(x_1, ..., x_n)$ are assigned to one n-tuple $(x_1, ..., x_n)$ each. The Cartesian product $\tilde{K}$ thus represents a fuzzy set in the product space $\underline{\mathbf{X}}$ with the membership function $\mu_K(\underline{x})$

$$\tilde{K} = \left\{\left(\underline{x} = (x_1, \ldots, x_n), \mu_K(\underline{x}) = \mu_K(x_1, \ldots, x_n)\right) \mid x_i \in \mathbf{X}_i; \mu_K(\underline{x}) = \min_{i=1,\ldots,n} [\mu_{A_i}(x_i)]\right\}. \quad (2.31)$$

Figure 2.11 shows an example of the Cartesian product formed by the fuzzy set $\tilde{A}_1$ and $\tilde{A}_2$.

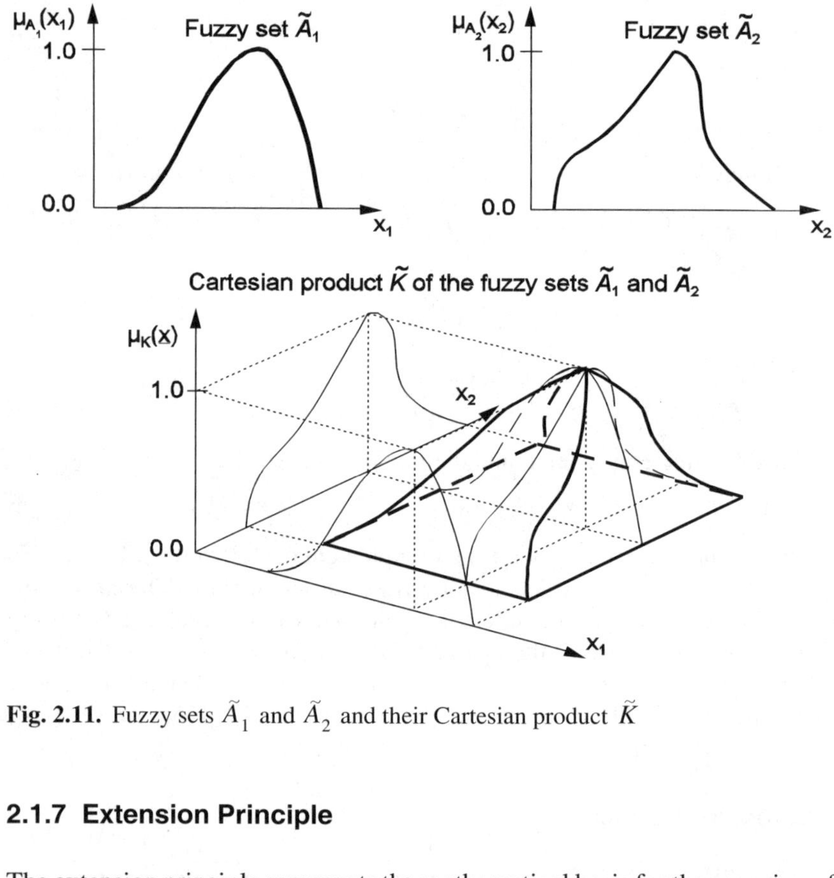

Fig. 2.11. Fuzzy sets $\tilde{A}_1$ and $\tilde{A}_2$ and their Cartesian product $\tilde{K}$

2.1.7 Extension Principle

The extension principle represents the mathematical basis for the mapping of fuzzy sets into a result space.

The fuzzy sets $\tilde{A}_1, \ldots, \tilde{A}_n$ on the fundamental sets $\mathbf{X}_1$, ..., $\mathbf{X}_n$ are linked with the aid of the Cartesian product. Thus an n-dimensional space (x-space) with the coordinate axes x_1, ..., x_n being perpendicular to one another is established. In this space the Cartesian product of the fuzzy sets $\tilde{A}_1, \ldots, \tilde{A}_n$ represents a fuzzy set with an n-dimensional membership function $\mu_K(x_1, \ldots, x_n)$, which is referred to as fuzzy input set $\underline{\tilde{X}}$ (Fig. 2.12). The extension principle specifies the mapping of the fuzzy input set $\underline{\tilde{X}}$ into a new fundamental set $\mathbf{Z}$ with the aid of the mapping $z = f(x_1, \ldots, x_n)$ (Fig. 2.13). On the fundamental set $\mathbf{Z}$ the fuzzy set $\tilde{B}$ together with its membership function $\mu_B(z)$ is gained (Eq. (2.32)). The membership values $\mu_B(z)$ are computed with the aid of the max-min operator (Eq. (2.33)).

On this basis the *extension principle* can be defined:

Given are:
- n+1 fundamental sets $\mathbf{X}_1, ..., \mathbf{X}_i, ..., \mathbf{X}_n, \mathbf{Z}$
- n fuzzy sets $\tilde{A}_i$ on the fundamental sets $\mathbf{X}_i$ with the membership functions $\mu(x_i) = \mu_{Ai}(x_i)$; i = 1, ..., n
- the mapping $\mathbf{X}_1 \times ... \times \mathbf{X}_i \times ... \times \mathbf{X}_n \rightarrow \mathbf{Z}$ with $z = f(x_1, ..., x_n)$ and $z \in \mathbf{Z}$

The mapping $f(\tilde{A}_1, ..., \tilde{A}_n)$ leads to the fuzzy set $\tilde{B}$ on $\mathbf{Z}$ (Fig. 2.12)

$$\tilde{B} = \{(z, \mu_B(z)) \mid z = f(x_1, ..., x_n); z \in \mathbf{Z}; (x_1, ..., x_n) \in \mathbf{X}_1 \times ... \times \mathbf{X}_n\}, \tag{2.32}$$

with the membership function

$$\mu(z) = \mu_B(z) = \begin{cases} \sup\limits_{z = f(x_1, ..., x_n)} \min\,[\mu(x_1), ..., \mu(x_n)], & \text{if} \exists z = f(x_1, ..., x_n) \\ 0 & \text{otherwise} \end{cases}. \tag{2.33}$$

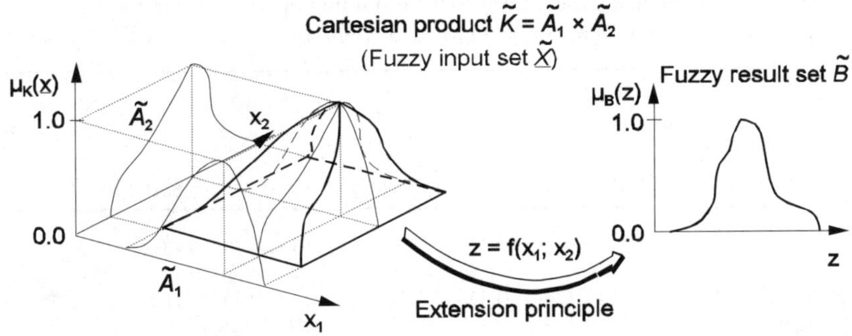

Fig. 2.12. Mapping of the fuzzy sets $\tilde{A}_1$ and $\tilde{A}_2$ onto the fuzzy result set $\tilde{B}$ with the aid of the extension principle

Example 2.1. The application of the extension principle is demonstrated by means of the mapping of the two discrete fuzzy sets $\tilde{A}_1 \subseteq \mathbf{X}_1$ and $\tilde{A}_2 \subseteq \mathbf{X}_2$ onto the fuzzy result set $\tilde{B} \subseteq \mathbf{Z}$. The discrete elements of the fuzzy sets $\tilde{A}_1$ and $\tilde{A}_2$ are recorded together with their membership values $\mu(x_1)$ and $\mu(x_2)$ in Fig. 2.13. Table 2.1 contains these value pairs in the first column and in the top line.

The mapping is determined by the function

$$z = f(x_1, x_2) = 3 \cdot x_1 - x_2 + 5. \tag{2.34}$$

The therewith described mapping of $\tilde{A}_1$ and $\tilde{A}_2$ onto $\tilde{B}$ does not represent a one-to-one mapping. The results of the mapping for each possible combination of elements from $\tilde{A}_1$ and $\tilde{A}_2$ are plotted in Table 2.1. Exactly one element z (element of the fuzzy result set $\tilde{z} = \tilde{B}$) is assigned to each combination (x_1, x_2). In the *first step*, the membership values $\mu(z)$ are computed using the minimum operator.

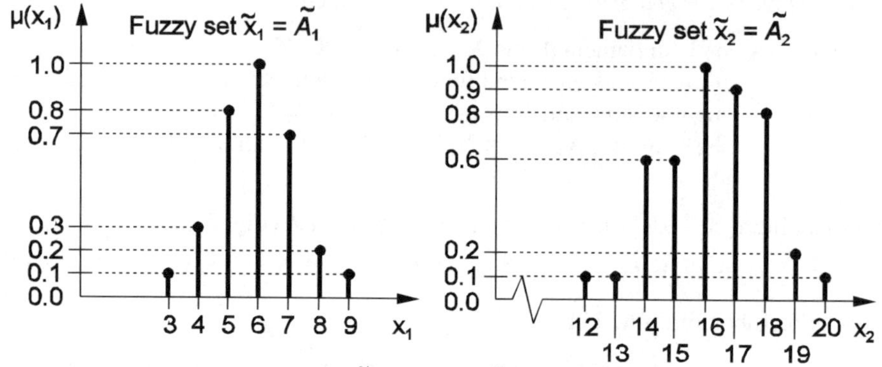

Fig. 2.13. Discrete fuzzy sets $\tilde{x}_1 = \tilde{A}_1$ and $\tilde{x}_2 = \tilde{A}_2$

Table 2.1. Discrete fuzzy sets $\tilde{A}_1$ and $\tilde{A}_2$ and discrete fuzzy result set $\tilde{B}$ as value pairs $(x_i, \mu(x_i))$ and $(z, \mu(z))$ with $\mu(z)$ according to the minimum operator (all plotted values $\mu(z)$) and according to the max-min operator (marked elements)

$\tilde{A}_2$				$\tilde{A}_1$			
	(3, 0.1)	(4, 0.3)	(5, 0.8)	(6, 1.0)	(7, 0.7)	(8, 0.2)	(9, 0.1)
(12, 0.1)	(2, 0.1)	(5, 0.1)	(8, 0.1)	(11, 0.1)	(14, 0.1)	(17, 0.1)	(20, 0.1)
(13, 0.1)	(1, 0.1)	(4, 0.1)	(7, 0.1)	(10, 0.1)	(13, 0.1)	(16, 0.1)	(19, 0.1)
(14, 0.6)	(0, 0.1)	(3, 0.3)	(6, 0.6)	(9, 0.6)	(12, 0.6)	(15, 0.2)	(18, 0.1)
(15, 0.6)	(−1, 0.1)	(2, 0.3)	(5, 0.6)	(8, 0.6)	(11, 0.6)	(14, 0.2)	(17, 0.1)
(16, 1.0)	(−2, 0.1)	(1, 0.3)	(4, 0.8)	(7, 1.0)	(10, 0.7)	(13, 0.2)	(16, 0.1)
(17, 0.9)	(−3, 0.1)	(0, 0.3)	(3, 0.8)	(6, 0.9)	(9, 0.7)	(12, 0.2)	(15, 0.1)
(18, 0.8)	(−4, 0.1)	(−1, 0.3)	(2, 0.8)	(5, 0.8)	(8, 0.7)	(11, 0.2)	(14, 0.1)
(19, 0.2)	(−5, 0.1)	(−2, 0.2)	(1, 0.2)	(4, 0.2)	(7, 0.2)	(10, 0.2)	(13, 0.1)
(20, 0.1)	(−6, 0.1)	(−3, 0.1)	(0, 0.1)	(3, 0.1)	(6, 0.1)	(9, 0.1)	(12, 0.1)

Different combinations of elements from $\tilde{A}_1$ and $\tilde{A}_2$ lead to identical results z, e.g., the value z = 3. Thus initially, each of these elements z possesses several, mostly different membership values. For z = 3, e.g., the membership values $\mu = 0.3$, $\mu = 0.8$, and $\mu = 0.1$ are obtained.

In the *second step*, with the aid of the maximum operator the largest of these membership values is assigned to the element z, i.e., the element z = 3 gets μ = 0.8. In Table 2.1 the according value pairs (z, μ(z)) are marked; these form the discrete fuzzy result set $\tilde{z} = \tilde{B}$ (Fig. 2.14).

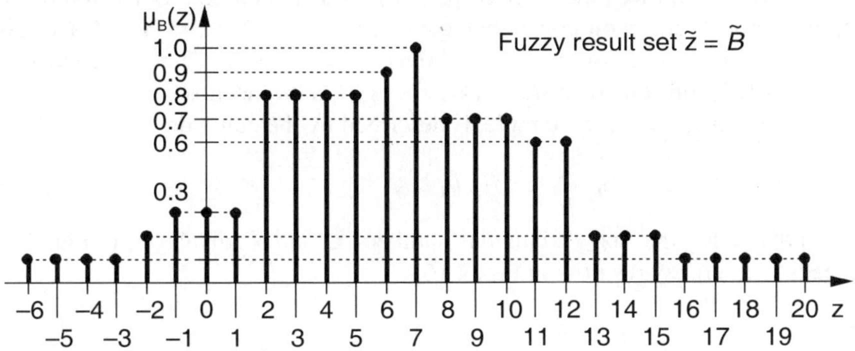

Fig. 2.14. Discrete fuzzy result set $\tilde{B}$

For some special mappings, like, e.g., addition or subtraction, the curve of the membership function μ(z) of the fuzzy result set $\tilde{z} = \tilde{B}$ can be computed in a closed form (it is presumed that the membership functions of the fuzzy sets $\tilde{A}_1, \ldots, \tilde{A}_n$ can also be stated in a closed form). The (extended) addition of the fuzzy triangular numbers $\tilde{x}_1$ = <2, 4, 5> and $\tilde{x}_2$ = <3, 5, 8>, for example, again yields the fuzzy triangular number $\tilde{z}_1$ = <5, 9, 13>, their (extended) multiplication leads to the fuzzy result $\tilde{z}_2$ with a nonlinear membership function (Fig. 2.15).

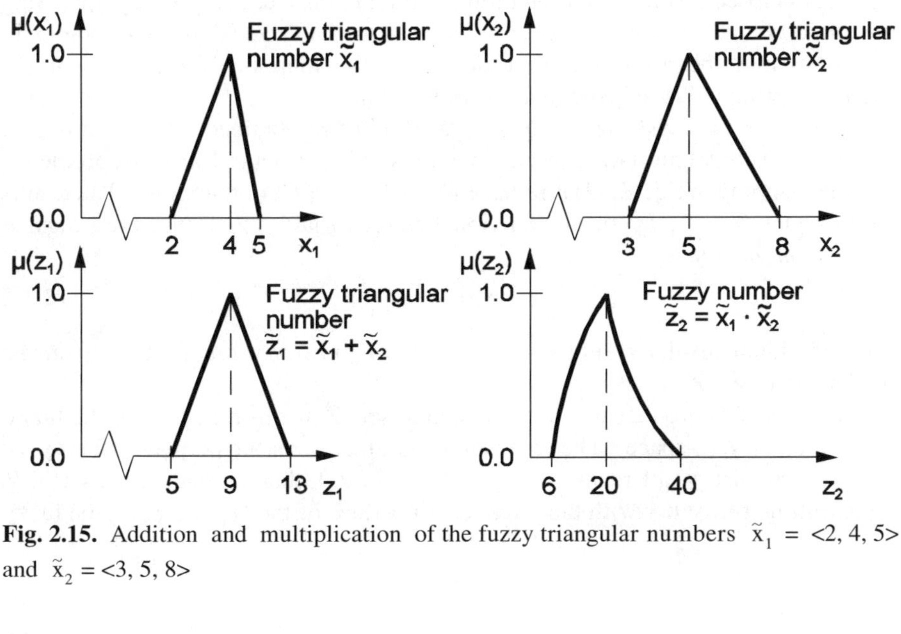

Fig. 2.15. Addition and multiplication of the fuzzy triangular numbers $\tilde{x}_1$ = <2, 4, 5> and $\tilde{x}_2$ = <3, 5, 8>

More complicated mappings generally require a numerical determination of the membership functions $\mu(z_i)$. When the fuzzy sets $\tilde{A}_1, ..., \tilde{A}_n$ are continuous fuzzy sets, the number of points that have to be evaluated in the space of the associated fundamental sets $\mathbf{X}_1$, ..., $\mathbf{X}_n$ (x-space) according to Eqs. (2.32) and (2.33) is unlimited. If, in addition, the x-space possesses more than one dimension or the mapping is not strictly monotonic, then the mapping $\mathbf{X}_1 \times \mathbf{X}_2 \times ... \times \mathbf{X}_n \rightarrow \mathbf{Z}$ does not represent a one-to-one mapping and is thus not reversible. Thus, in general, a quasi-infinite number of points $(x_1, ..., x_n)$ is numerically evaluated. A nonmonotonic mapping, for example, is described by the equation

$$\tilde{z} = (\tilde{x}_1 - 3)^2 + (\tilde{x}_2 - 6)^2 + 2. \tag{2.35}$$

When introducing the fuzzy triangular numbers $\tilde{x}_1$ and $\tilde{x}_2$ displayed in Fig. 2.15 this leads to the fuzzy result $\tilde{z}$ in Fig. 2.16.

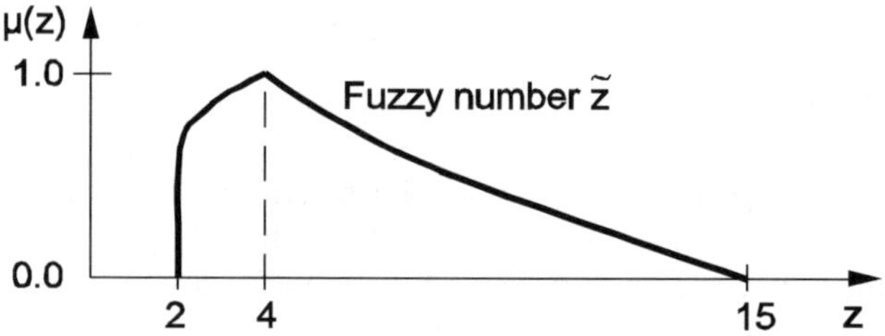

Fig. 2.16. Fuzzy result $\tilde{z}$ of the mapping according to Eq. (2.35)

The mapping of the extension principle may be given in a closed form as well as in a nonclosed form. Nonclosed or implicit mappings, e.g., as usually appearing in structural analysis, demand a numerical treatment of the extension principle. In the following, the algorithm that describes the mapping $f(\tilde{A}_1,...,\tilde{A}_n)$ of the extension principle is referred to as *mapping model*.

In structural analysis generally several results are obtained. The definition of the mapping model must thus be extended for solving structural-analysis problems.

The mapping in Eq. (2.32) is replaced by a discrete set of mappings. The results are m-tuples $(z_1, ..., z_m)$ of the individual result variables z_j. Hence the *extended mapping model* reads

$$(z_1, ..., z_m) = f(x_1, ..., x_n). \tag{2.36}$$

The individual results z_j represent elements of the fuzzy result sets $\tilde{Z}_j$ on the fundamental sets $\mathbf{Z}_j$.

The Cartesian product of the fundamental sets $\mathbf{Z}_j$ forms the space of the fuzzy result vectors $\underline{\mathbf{Z}}$ (z-space). The associated coordinate axes are perpendicular to one another. The set of all result m-tuples $(z_1, ..., z_m)$ characterizes a subset of $\underline{\mathbf{Z}}$ representing – together with the membership values of the $(z_1, ..., z_m)$ – the fuzzy result set $\underline{\tilde{Z}}$. *Therewith the extension principle is interpreted as being a general mapping rule describing the mapping of the fuzzy input set* $\underline{\tilde{X}}$ *onto the fuzzy result set* $\underline{\tilde{Z}}$

$$\underline{\tilde{X}} \rightarrow \underline{\tilde{Z}}. \tag{2.37}$$

When the extension principle is applied to the mapping of α-level sets A_{i,α_k} of the fuzzy sets $\tilde{A}_i$ (with the same α_k for each $\tilde{A}_i$) into the result space, the following holds:

The membership values $\mu(x_i)$ of each element x_i of the α-level sets A_{i,α_k} of the fuzzy variables $\tilde{x}_i = \tilde{A}_i$ comply with the condition $\mu(x_i) \geq \alpha_k$. For each point $\underline{x} = \{x_1, ..., x_n\}$ with $x_i \in A_{i,\alpha_k}$, $i = 1, ..., n$ the minimum operator yields the membership value $\mu(\underline{x}) = \min\,[\mu(x_i);\ i = 1, ..., n]$. After the computation of the elements $z_j = f_j(\underline{x})$ of the fuzzy result variables $\tilde{z}_j = \tilde{B}_j$ the membership values are determined with the aid of the maximum operator. Consequently, it is ensured that $\mu(z_j) \geq \alpha_k$ also holds. All computed elements z_j belong to the α-level set B_{j,α_k}.

For all points $\underline{x}$ that contain at least one x_i with $\mu(x_i) < \alpha_k$, i.e., $x_i \notin A_{i,\alpha_k}$, follows from the minimum operator that $\mu(\underline{x}) < \alpha_k$ holds. Each element z_j of the fuzzy result $\tilde{z}_j$ computed from these points $\underline{x}$ possesses – before the application of the maximum operator – the membership value $\mu(z_j) < \alpha_k$. Only if the mapping of points $\underline{x}$ with $x_i \in A_{i,\alpha_k}$ for all $i = 1, ..., n$ yields exactly the same element z_j, membership values $\mu(z_j) \geq \alpha_k$ are obtained by applying the maximum operator.

From this it follows that the α-level sets B_{j,α_k} of the fuzzy result variable $\tilde{z}_j = \tilde{B}_j$ are completely described by the mapping of the α-level sets A_{i,α_k} into the result space. The extension principle with the max-min operator may be applied to individual α-levels. For each α-level α_k the extension principle describes the mapping of the α-level set $\underline{X}_{\alpha_k}$ of the fuzzy input set $\underline{\tilde{X}}$ onto the α-level set $\underline{Z}_{\alpha_k}$ of the fuzzy result set $\underline{\tilde{Z}}$

$$\underline{X}_{\alpha_k} \rightarrow \underline{Z}_{\alpha_k}. \tag{2.38}$$

2.1.8 Interaction between Fuzzy Variables

Interaction is defined as being the mutual dependency of fuzzy variables. An interactive relationship may be formulated directly (explicitly) or indirectly (implicitly). Regarding its cause two types of interaction are distinguished.

Interaction between fuzzy variables may exist *a priori* (first type). In this case the existing dependencies between these variables must already be considered when generating the fuzzy input set. The shape of the membership function $\mu(\underline{x})$ of the fuzzy input set $\underline{\tilde{X}}$ in the product space $\mathbf{\underline{X}}$, which is developed from the Cartesian product $\tilde{K} = \tilde{A}_1 \times \tilde{A}_2 \times ... \times \tilde{A}_n$, is influenced by the interaction relationships. An interactive dependency between the $\tilde{A}_1, ..., \tilde{A}_n$ affects by constraining $\underline{\tilde{X}}$ in comparison to $\tilde{K}$ (without interaction), because the demands regarding the membership functions of the fuzzy variables $\tilde{x}_i = \tilde{A}_i$ must strictly be met. Hence the fuzzy input set $\underset{\sim}{\tilde{X}}$ (with interaction) is a subset of $\tilde{K}$ (without interaction) (Fig. 2.17)

$$\underset{\sim}{\tilde{X}} \subseteq \tilde{K}. \tag{2.39}$$

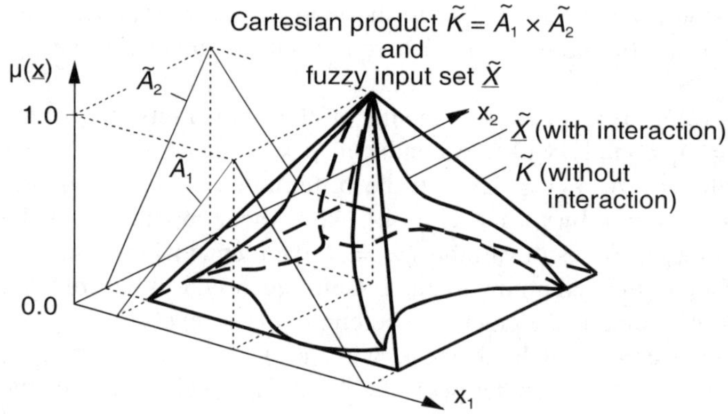

Fig. 2.17. Cartesian product of the fuzzy variables $\tilde{A}_1$ and $\tilde{A}_2$ without interaction and fuzzy input set $\underset{\sim}{\tilde{X}}$ with interaction (first type)

The second type of interaction occurs *within the mapping*. If the fuzzy result is stepwise computed (more than one step) from the $\tilde{x}_i = \tilde{A}_i$, then intermediate results are obtained in each of these steps. These intermediate results are linked to the fuzzy variables $\tilde{x}_i$ according to the associated part of the mapping model. In the case that several $\tilde{x}_i$ affect the intermediate results simultaneously, these intermediate results are no longer independent of each other, i.e., interactive dependencies exist between them.

When the mapping model is stepwise applied, these interactive dependencies must be considered. Nonobservance of this interaction leads to the defect that, starting from the intermediate results, *nonpermissible* parameter combinations are generated and processed forward. Thereby additional, artificial uncertainty is introduced into the mapping in each step.

With a rising number of mapping steps the uncertainty of the intermediate results increases exponentially, the solution "diverges" – the results are deficient [19, 20]. Algorithms of structural analysis generally contain such mapping models that must be applied stepwise, e.g., solution algorithms for (differential) equation systems or eigenvalue problems.

Example 2.2. The effect of both types of interaction is demonstrated by means of the mapping of the two fuzzy triangular numbers $\tilde{x}_1$ and $\tilde{x}_2$ onto the fuzzy result $\tilde{z}$ within two mapping steps.

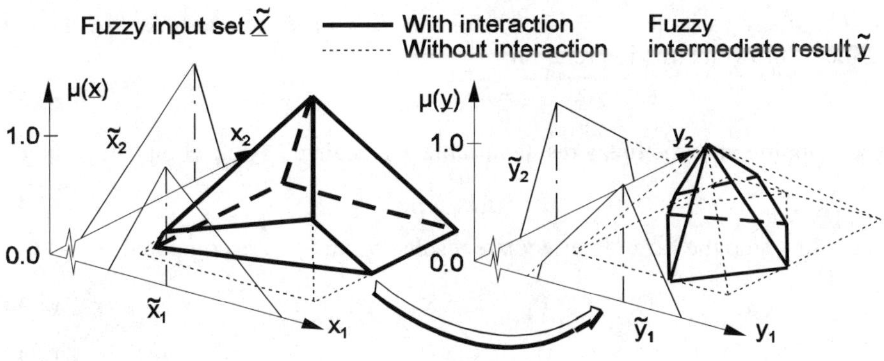

Fig. 2.18. Mapping of the fuzzy variables $\tilde{x}_1$ and $\tilde{x}_2$ onto the fuzzy intermediate result $\tilde{y}$

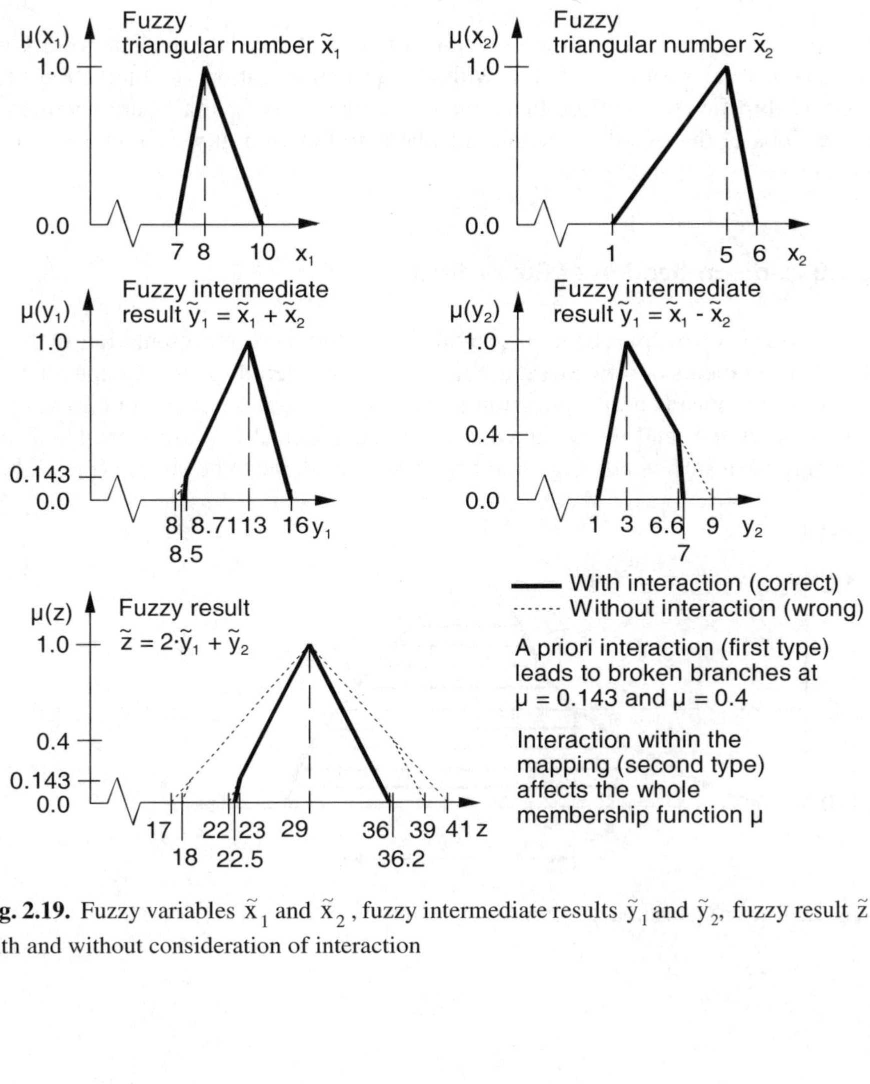

Fig. 2.19. Fuzzy variables $\tilde{x}_1$ and $\tilde{x}_2$, fuzzy intermediate results $\tilde{y}_1$ and $\tilde{y}_2$, fuzzy result $\tilde{z}$ with and without consideration of interaction

The fuzzy variables $\tilde{x}_1 = <7, 8, 10>$ and $\tilde{x}_2 = <1, 5, 6>$ are linked by the prescribed a priori interaction relation

$$\tilde{x}_1 - 2 \cdot \tilde{x}_2 \leq 4. \tag{2.40}$$

The mapping onto the fuzzy result variable $\tilde{z}$ is realized by the composite function

$$\tilde{z} = f(\tilde{x}_1, \tilde{x}_2). \tag{2.41}$$

In the first step the fuzzy intermediate results $\tilde{y}_1$ and $\tilde{y}_2$ are computed

$$\tilde{y}_1 = \tilde{x}_1 + \tilde{x}_2, \tag{2.42}$$

$$\tilde{y}_2 = \tilde{x}_1 - \tilde{x}_2. \tag{2.43}$$

The fuzzy result $\tilde{z}$ is obtained in the second step

$$\tilde{z} = 2 \cdot \tilde{y}_1 + \tilde{y}_2. \tag{2.44}$$

Figures (2.18) and (2.19) show a comparison of the fuzzy intermediate results and the fuzzy final results with and without the consideration of interaction. The membership functions of the fuzzy intermediate results $\tilde{y}_1$ and $\tilde{y}_2$ are obtained as projections of the two-dimensional membership function $\mu(y_1, y_2)$ on the coordinate planes $y_1 = 0$ and $y_2 = 0$.

2.1.9 α-discretization of Fuzzy Sets

The concept of α-discretization provides an alternative representation of fuzzy sets. The elements of a fuzzy set are no longer considered separately one after the other, which means a discretization of the support, but the membership scale is now discretized, and the assigned α-level sets (Sect. 2.1.3) are considered. For this purpose a sufficiently high number of α-levels have to be chosen (Fig. 2.20).

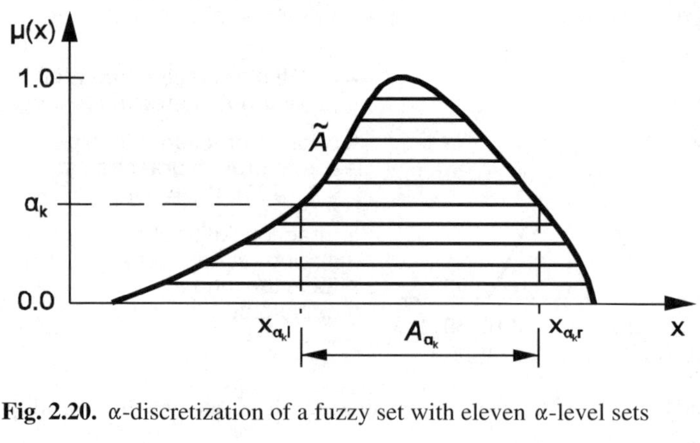

Fig. 2.20. α-discretization of a fuzzy set with eleven α-level sets

The α-level sets A_{i,α_k}, i = 1, ..., n of the fuzzy sets $\tilde{A}_1, ..., \tilde{A}_i, ..., \tilde{A}_n$ form the n-dimensional crisp subspace $\underline{X}_{\alpha_k}$ of the x-space. For $\alpha_k \to +0$ the crisp support subspace $\underline{X}_{\alpha_k \to +0}$ is obtained. The crisp subspace $\underline{X}_{\alpha_k}$ assigned to the α-level α_k is illustrated by means of three fuzzy variables $\tilde{x}_1 = \tilde{A}_1$, $\tilde{x}_2 = \tilde{A}_2$ and $\tilde{x}_3 = \tilde{A}_3$ in Fig. 2.21. If the fuzzy variables $\tilde{x}_i$ are convex fuzzy sets without any interactive dependency, they form an n-dimensional hypercuboid. Nonconvex fuzzy variables $\tilde{x}_i$ lead to a disjoint subset $\underline{X}_{\alpha_k}$. If interaction exists, the shape of the subspace $\underline{X}_{\alpha_k}$ generally departs from the shape of the n-dimensional hypercuboid; the formation of "voids" in $\underline{X}_{\alpha_k}$ is possible.

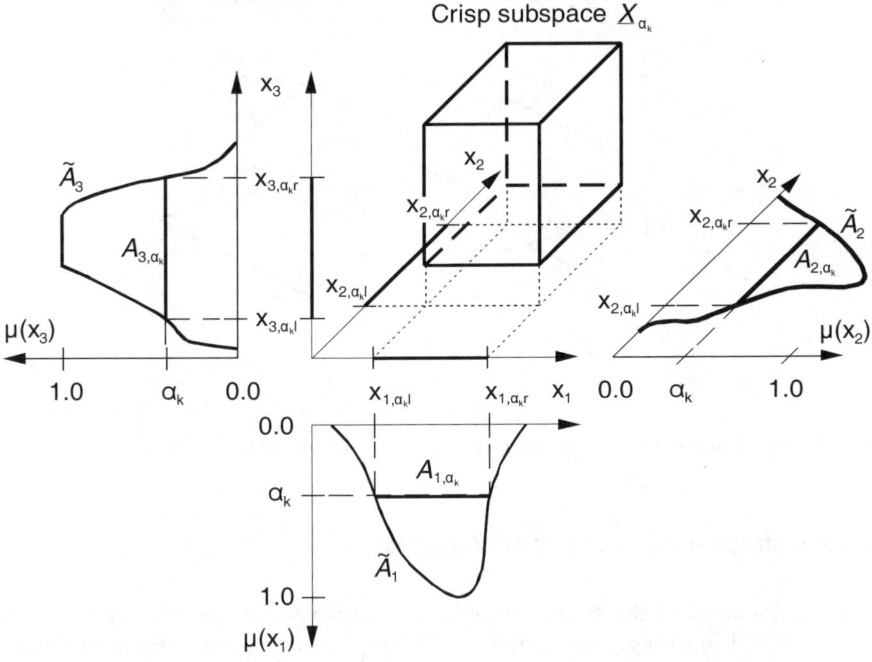

Fig. 2.21. Forming the crisp subspace $\underline{X}_{\alpha_k}$ for the α-level α_k, no interaction

Between the two subspaces $\underline{X}_{\alpha_i}$ and $\underline{X}_{\alpha_k}$ for α_i and α_k the following relationship holds

$$\underline{X}_{\alpha_i} \subseteq \underline{X}_{\alpha_k} \quad \forall\ \alpha_i, \alpha_k \mid \alpha_i, \alpha_k \in (0, 1]\,,\ \alpha_i \geq \alpha_k. \tag{2.45}$$

It follows that all subspaces $\underline{X}_{\alpha_k}$ are contained in the support subspace $\underline{X}_{\alpha_k \to +0}$. If α_k may take on all real values in the interval (0, 1], the entirety of $\underline{X}_{\alpha_k}$ then forms the fuzzy input set $\underline{\tilde{X}}$; α_k is equal to the membership values of the subspaces $\underline{X}_{\alpha_k}$. For selected values $\alpha_k \in (0, 1]$ the fuzzy input set $\underline{\tilde{X}}$ is discretized (Fig. 2.22).

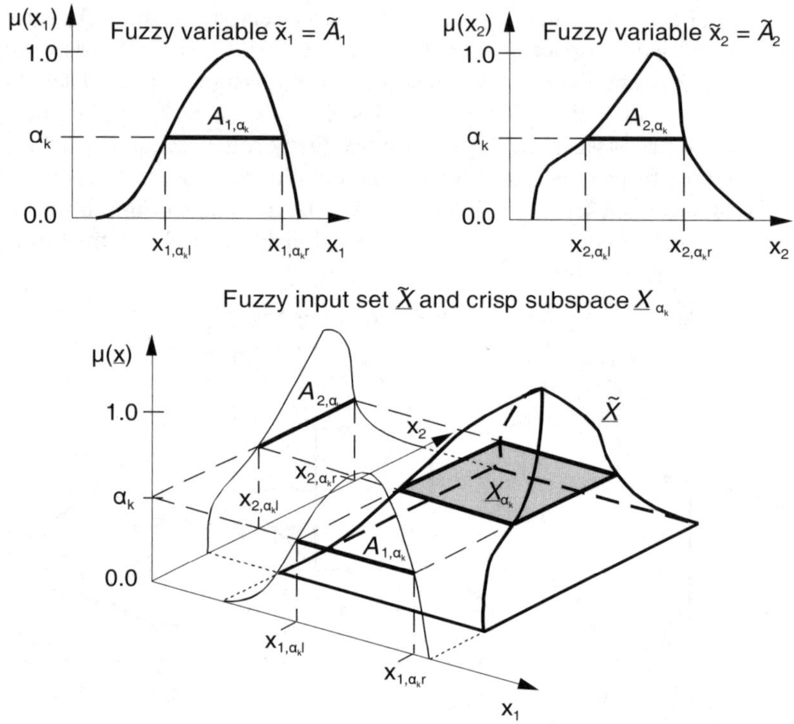

Fig. 2.22. Fuzzy input set $\underline{\tilde{X}}$ and crisp subspace $\underline{X}_{\alpha_k}$ for two fuzzy variables

2.1.10 Defuzzification of Fuzzy Variables

The *defuzzification* of the fuzzy variable $\tilde{z}_j$ represents the mapping of $\tilde{z}_j$ into a subspace which no longer possesses the "dimension fuzziness", the information content of $\tilde{z}_j$ decreases. This results in the crisp value z_{j0}.

For the conversion of the fuzzy value $\tilde{z}_j$ into the crisp value z_{j0} the uncertainty of $\tilde{z}_j$ is evaluated. For this purpose several defuzzification algorithms are available [22, 155]. Choosing a suitable procedure and necessary specifications is a subjective decision that affects the results. In the following, three defuzzification algorithms are discussed, which are based on different ideas.

Centroid Method. This defuzzification method yields the crisp result value searched for as the center of the area below the membership function of the fuzzy variable $\tilde{z}_j$

$$z_{j0} = \int_{z_j} z_j \cdot \mu(z_j)\,dz_j \cdot \left[\int_{z_j} \mu(z_j)\,dz_j\right]^{-1}. \tag{2.46}$$

The crisp result z_{j0} may be influenced by weighting several chosen α-levels and so may be matched to the requirements in the particular case. If, for example, the weighting factors are defined increasingly with rising values α_k, then the defuzzification result tends to the deterministic solution according to the α-level $\alpha_k = 1$.

Level Rank Method. The concept of this method (after Rommelfanger [155]) is based on the α-discretization (Sect. 2.1.9). The membership scale of the fuzzy variable $\tilde{z}_j$ is discretized with the aid of r chosen α-levels, and then the arithmetic mean of the interval centers of the α-level sets is computed as defuzzification result (Fig. 2.23)

$$z_{j0} = \frac{1}{r} \cdot \sum_{k=1}^{r} \frac{z_{j,\alpha_k l} + z_{j,\alpha_k r}}{2} . \tag{2.47}$$

Depending on the number and the position of the chosen α-levels the defuzzification result varies. A weighting may be introduced here by altering the density of the α-levels in certain membership ranges.

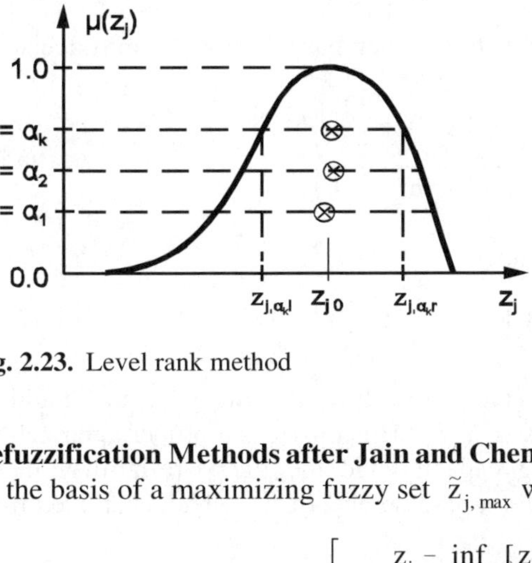

Fig. 2.23. Level rank method

Defuzzification Methods after Jain and Chen. Jain assesses the fuzzy variable $\tilde{z}_j$ on the basis of a maximizing fuzzy set $\tilde{z}_{j,\max}$ with the membership function

$$\mu(z_{j,\max}) = \left[\frac{z_j - \inf\limits_{z_j \in \tilde{z}_j} [z_j]}{\sup\limits_{z_j \in \tilde{z}_j} [z_j] - \inf\limits_{z_j \in \tilde{z}_j} [z_j]} \right]^k . \tag{2.48}$$

The positive real exponent k is to be defined depending on the particular application. For example, the specification k = 0.5 expresses an aversion to large values z_j, whereas k = 2 signals a preference for large values z_j (Fig. 2.24).

From the maximizing fuzzy set $\tilde{z}_{j,\max}$ the crisp valuation number

$$\mu_r(z_j) = \sup_{z_j \in \tilde{z}_j} \min \left[\mu(z_j), \mu(z_{j,\max}) \right] \tag{2.49}$$

is computed. It lies within the interval [0, 1] and may be interpreted as being a membership value stating the degree with which the fuzzy result should be assessed regarding the large values $z_j \in \tilde{z}_j$. This is reasonable when large values z_j are adverse or dangerous, like, e.g., when assessing fuzzy failure probability, fuzzy displacements, or fuzzy internal forces.

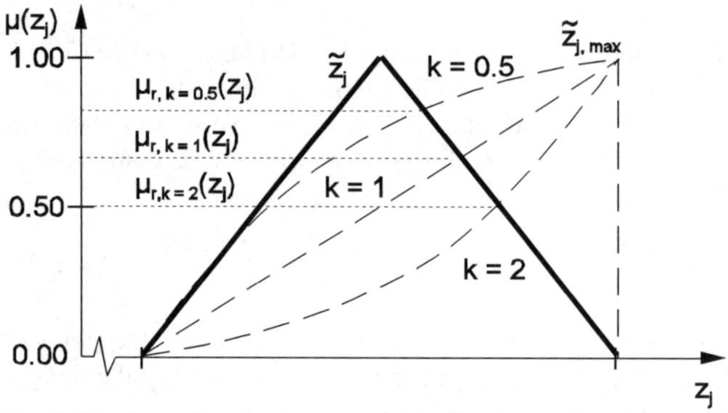

Fig. 2.24. Assessing the fuzzy variable $\tilde{z}_j$ with the maximizing fuzzy set after Jain

In order to enhance Jain's basic method, Chen has additionally introduced the minimizing fuzzy set $\tilde{z}_{min}$ with

$$\mu(z_{j,min}) = \left[\frac{\sup\limits_{z_j \in \tilde{z}_j} [z_j] - z_j}{\sup\limits_{z_j \in \tilde{z}_j} [z_j] - \inf\limits_{z_j \in \tilde{z}_j} [z_j]} \right]^k . \tag{2.50}$$

This leads to the crisp valuation number

$$\mu_l(z_j) = \sup_{z_j \in \tilde{z}_j} \min \left[\mu(z_j), \mu(z_{j,min}) \right] . \tag{2.51}$$

The value $\mu_l(z_j)$ expresses the degree with which the fuzzy result should be assessed regarding the small values $z_j \in \tilde{z}_j$. This provides a proper approach for assessing, e.g., fuzzy load-bearing capacity and the fuzzy reliability index. Furthermore, Chen suggests a "medium assessment" of $\tilde{z}_j$, which is derived from $\mu_r(z_j)$ and $\mu_l(z_j)$

$$\mu_m(z_j) = 0.5 \left[\mu_r(z_j) + (1 - \mu_l(z_j)) \right] . \tag{2.52}$$

This modified form represents a suitable basis for assessing, e.g., fuzzy structural responses and fuzzy safety measures on serviceability level.

On the basis of Jain's and Chen's approaches the fuzzy variable $\tilde{z}_j$ may be *defuzzified* as follows. The valuation interval [0, 1], which comprises all possible numbers $\mu_r(z_j)$, $\mu_l(z_j)$ and $\mu_m(z_j)$, is mapped onto those parts of the support of $\tilde{z}_j$, which lie on the left or on the right side of the mean value $z_{j\,\alpha=1}$ of $\tilde{z}_j$. This leads to the defuzzification results

$$z_{j\,0\,r} = z_{j,\alpha=0\,r} - \mu_r \left(z_{j,\alpha=0\,r} - z_{j,\alpha=1} \right) , \tag{2.53}$$

$$z_{j\,0\,l} = \mu_l\,(z_{j,\alpha=1} - z_{j,\alpha=0\,l}) + z_{j,\alpha=0\,l}\,, \tag{2.54}$$

$$z_{j\,0\,m} = \mu_m\,(z_{j,\alpha=0\,r} - z_{j,\alpha=0\,l}) + z_{j,\alpha=0\,l}\,. \tag{2.55}$$

Alternatively, the values $z_{j\,0\,r}$ and $z_{j\,0\,l}$, which are assigned to the valuation numbers $\mu_r(z_j)$ and $\mu_l(z_j)$, may be applied as defuzzification results

$$z_{j0\,r} = z_j \mid \mu(z_j) = \mu_r(z_j)\ , \tag{2.56}$$

$$z_{j0\,l} = z_j \mid \mu(z_j) = \mu_l(z_j)\ . \tag{2.57}$$

Defuzzification of fuzzy variable $\tilde{z}_j$ (e.g., intermediate results or final results of fuzzy structural analysis or safety assessment with fuzzy variables) may lead to a loss of important information. However, it may serve as a helpful tool to derive decisions, and it may also be applied to auxiliary variables and intermediate results of unimportant influence and may thus simplify computations in the sense of an approximation solution.

2.1.11 Fuzzy Functions

2.1.11.1 Definition of Fuzzy Functions, Fuzzy Processes, and Fuzzy Fields

A fuzzy function may be explained by extending the definition of a classical function. A classical function is a single-valued mapping of the elements t from the fundamental set $\mathbf{T} \subseteq \mathbb{R}$ onto the elements x of the fundamental set $\mathbf{X} \subseteq \mathbb{R}$. It may be denoted by

$$x(t)\colon\ \mathbf{T} \rightarrow \mathbf{X}\ , \tag{2.58}$$

where $t \in \mathbf{T}$ represents the arguments of the function and $x \in \mathbf{X}$ indicates the functional values or results. The set $\mathbf{T}$ is referred to as argument domain and $\mathbf{X}$ denotes the range of values of $x(t)$.

In accordance with Eq. (2.58), the classical function x(t) may also be explained as a set of functional values $x_t \in \mathbf{X}$ that belong to specified variables $t \in \mathbf{T}$

$$x(t) = \left\{x_t = x(t)\ \ \forall t \mid t \in \mathbf{T}\right\}. \tag{2.59}$$

For the enhancement of the classical function fuzziness may be considered in different ways:

- $x(\tilde{t})$ crisp mapping of fuzzy variables $\tilde{t} \subseteq \mathbf{T}$
- $\tilde{x}(t)$ uncertain mapping of crisp variables $t \in \mathbf{T}$
- $\tilde{x}(\tilde{t})$ uncertain mapping of fuzzy variables $\tilde{t} \subseteq \mathbf{T}$

The following definition of fuzzy functions comprises these three cases. Given are:

- The fundamental sets $\mathbf{T} \subseteq \mathbb{R}$ and $\mathbf{X} \subseteq \mathbb{R}$
- The set $\mathbf{F(T)}$ of all fuzzy variables $\tilde{t}$ on the fundamental set $\mathbf{T}$
- The set $\mathbf{F(X)}$ of all fuzzy variables $\tilde{x}$ on the fundamental set $\mathbf{X}$

An uncertain mapping of $\mathbf{F(T)}$ onto $\mathbf{F(X)}$ that assigns exactly one $\tilde{x} \in \mathbf{F(X)}$ to each $\tilde{t} \in \mathbf{F(T)}$, respectively, is referred to as a *fuzzy function* denoted by

$$\tilde{x}(\tilde{t}): \mathbf{F(T)} \mathrel{\tilde{\rightarrow}} \mathbf{F(X)} \, . \tag{2.60}$$

The fuzzy function $\tilde{x}(\tilde{t})$ leads for each $\tilde{t} \in \mathbf{F(T)}$ to the fuzzy result $\tilde{x}_t = \tilde{x}(\tilde{t})$ with $\tilde{x}_t \in \mathbf{F(X)}$. It represents the uncertain mapping of the fuzzy variables $\tilde{t} \in \mathbf{F(T)}$ onto the fuzzy variables $\tilde{x}_t \in \mathbf{F(X)}$.

The fuzzy function $\tilde{x}(\tilde{t})$ may thus also be interpreted as being a set of fuzzy results or fuzzy functional values $\tilde{x}_t \in \mathbf{F(X)}$ belonging to specified $\tilde{t} \in \mathbf{F(T)}$

$$\tilde{x}(\tilde{t}) = \left\{ \tilde{x}_t = \tilde{x}(\tilde{t}) \;\; \forall \tilde{t} \mid \tilde{t} \in \mathbf{F(T)} \right\} . \tag{2.61}$$

Therewith the following two special cases are comprised:

Special case 1. For the crisp mapping of fuzzy variables $\tilde{t} \in \mathbf{F(T)}$ Eq. (2.60) reads

$$x(\tilde{t}): \mathbf{F(T)} \rightarrow \mathbf{F(X)} \, . \tag{2.62}$$

The function $x(\tilde{t})$ describes the crisp assignment of precisely one uncertain $\tilde{x} \in \mathbf{F(X)}$ to each uncertain $\tilde{t} \in \mathbf{F(T)}$. Equation (2.61) changes to

$$x(\tilde{t}) = \left\{ \tilde{x}_t = x(\tilde{t}) \;\; \forall \tilde{t} \mid \tilde{t} \in \mathbf{F(T)} \right\} . \tag{2.63}$$

Special case 2. Considering the fuzzy mapping of crisp variables $t \in \mathbf{T}$, Eq. (2.60) results in

$$\tilde{x}(t): \mathbf{T} \mathrel{\tilde{\rightarrow}} \mathbf{F(X)} \, . \tag{2.64}$$

The fuzzy function $\tilde{x}(t)$ assigns exactly one uncertain $\tilde{x} \in \mathbf{F(X)}$ to each crisp $t \in \mathbf{T}$. From Eq. (2.61) follows

$$\tilde{x}(t) = \left\{ \tilde{x}_t = \tilde{x}(t) \;\; \forall t \mid t \in \mathbf{T} \right\} . \tag{2.65}$$

A fuzzy function with discrete arguments, i.e., a discrete fuzzy function, is defined on a finite or denumerable set $\mathbf{F(T)}$. It may be described by a sequence of fuzzy variables $\{\tilde{x}_1 = \tilde{x}(\tilde{t}_1), \tilde{x}_2 = \tilde{x}(\tilde{t}_2), \ldots\}$. Conversely, each sequence of fuzzy variables may be interpreted as being a fuzzy function with discrete arguments. If $\mathbf{F(T)}$ is a continuous set, $\tilde{x}(\tilde{t})$ then represents a fuzzy function with continuous arguments, i.e., a continuous fuzzy function. In the multidimensional case, i.e., for $\mathbf{F(\underline{T})}$ with $\underline{\mathbf{T}} \subseteq \mathbb{R}^m$, the properties concerning the arguments may also occur in a mixed form, several coordinates of $\underline{\mathbf{T}}$ may then be linked to properties being different from the others.

In view of a convenient style for common engineering applications and without restricting universal validity the following explanations are presented in the form of special case 2.

If the fundamental set **T** represents the time coordinate, the fuzzy function $\tilde{x}(t)$ is referred to as a *fuzzy process* denoted by $\tilde{x}(\tau)$ with $\tau \in \mathbf{T}$.

If the set **T** comprises exceptionally spatial coordinates, i.e., in the three-dimensional case $\underline{\theta} = (\theta_1, \theta_2, \theta_3) \in \underline{\mathbf{\Theta}} \subseteq \mathbb{R}^3$, the fuzzy function $\tilde{x}(t)$ is called the *fuzzy field* characterized by $\tilde{x}(\underline{\theta})$.

In structural analysis the fundamental set $\underline{\mathbf{T}}$ may contain both the time coordinate τ and the spatial coordinates $\underline{\theta}$ and may thus possess four dimensions. In this case the assigned fuzzy function is denoted by $\tilde{x}(\underline{t}) = \tilde{x}(\underline{\theta}, \tau)$ with $\underline{t} = (\underline{\theta}, \tau)$.

In general, the multidimensional fundamental set $\underline{\mathbf{T}}$ may comprise further dimensions, like, e.g., temperature or moisture degree, in addition to time and spatial coordinates. These dimensions are described by the coordinates φ_i and incorporated into the vector $\underline{t} = (\underline{\theta}, \tau, \underline{\varphi})$ of all coordinates. Thereby $\underline{t}$ represents a vector in the parameter space $\underline{\mathbf{T}} \subseteq \mathbb{R}^m$.

On this basis fuzzy processes and fuzzy fields may be treated as special cases of the fuzzy function $\tilde{x}(\underline{t})$.

Bunch Parameter Representation. A fuzzy function $\tilde{x}(\underline{t})$ may be formulated depending on *fuzzy bunch parameters* $\underline{\tilde{s}}$ and crisp arguments $\underline{t}$

$$\tilde{x}(\underline{t}) = x(\underline{\tilde{s}}, \underline{t}) . \tag{2.66}$$

In the *general case* Eq. (2.65) then reads

$$\tilde{x}(\underline{t}) = x(\underline{\tilde{s}}, \underline{t}) = \left\{ \tilde{x}_t = x(\underline{\tilde{s}}, \underline{t}) \ \forall \underline{t} \mid \underline{t} \in \underline{\mathbf{T}} \right\} . \tag{2.67}$$

For each crisp bunch parameter vector $\underline{s} \in \underline{\tilde{s}}$ with the assigned membership value $\mu(\underline{s})$ a crisp function $x(\underline{t}) = x(\underline{s}, \underline{t}) \in \tilde{x}(\underline{t})$ with $\mu(x(\underline{t})) = \mu(\underline{s})$ is obtained. In contrast to Eq. (2.61), the fuzzy function $\tilde{x}(\underline{t})$ may thus be represented by the fuzzy set of all real-valued functions $x(\underline{t}) \in \tilde{x}(\underline{t})$ with $\mu(x(\underline{t})) = \mu(x(\underline{s}, \underline{t})) = \mu(\underline{s})$

$$\tilde{x}(\underline{t}) = x(\underline{\tilde{s}}, \underline{t}) = \left\{ \left(x(\underline{t}), \mu(x(\underline{t})) \right) \mid x(\underline{t}) = x(\underline{s}, \underline{t}) ; \mu(x(\underline{t})) = \mu(\underline{s}) \ \forall \underline{s} \in \underline{\tilde{s}} \right\} , \tag{2.68}$$

which may be generated from all possible real vectors $\underline{s} \in \underline{\tilde{s}}$. For every $\underline{t} \in \underline{\mathbf{T}}$ each of the crisp functions $x(\underline{t})$ takes values that are simultaneously contained in the associated fuzzy functional values $\tilde{x}(\underline{t})$. The real functions or elements $x(\underline{t})$ of $\tilde{x}(\underline{t})$ are defined for all $\underline{t} \in \underline{\mathbf{T}}$, they are referred to as *trajectories*.

Trajectories of a fuzzy function with discrete arguments are sequences of real numbers, whereas trajectories of a fuzzy function with continuous arguments generally represent continuous functions of $\underline{t}$. Trajectories of a discrete fuzzy function with continuous arguments are described by step functions.

Example 2.3. The one-dimensional fuzzy function

$$\tilde{x}(t) = x(\underline{\tilde{s}}, t) = \tilde{a} \cdot \sin(\tilde{\omega}_0 \, t + \tilde{b}) \tag{2.69}$$

with the fuzzy bunch parameters $\tilde{a}$ = <0.9, 1.0, 1.1> (fuzzy amplitude), $\tilde{\omega}_0$ = <0.9, 1.0, 1.1> (fuzzy frequency), and $\tilde{b}$ = <–0.1, 0.0, 0.1> (fuzzy phase angle) is considered. Within the fuzziness of the fuzzy bunch parameter vector $\tilde{\underline{s}} = (\tilde{a}, \tilde{\omega}_0, \tilde{b})$ different crisp points $\underline{s} \in \tilde{\underline{s}}$ with the associated membership values $\mu(\underline{s})$ may be chosen to generate various trajectories. As $\tilde{\underline{s}}$ represents a continuous fuzzy set, an indefinite number of trajectories exist. For illustrating the fuzzy function in Fig. 2.25 four crisp parameter vectors are determined: $\underline{s}_1 = (1.0, 1.0, 0.0)$ with $\mu(\underline{s}_1) = 1.0$, $\underline{s}_2 = (0.95, 0.95, -0.05)$ with $\mu(\underline{s}_2) = 0.5$, $\underline{s}_3 = (1.05, 1.05, 0.05)$ with $\mu(\underline{s}_3) = 0.5$, and $\underline{s}_4 = (1.1, 1.1, 0.1)$ with $\mu(\underline{s}_4) = 0.0$.

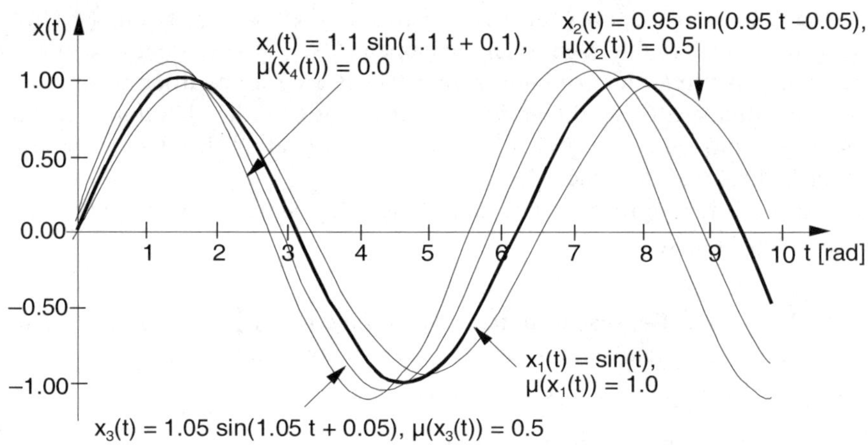

Fig. 2.25. Some trajectories of the fuzzy function $x(\tilde{\underline{s}}, t) = \tilde{a} \cdot \sin(\tilde{\omega}_0 t + \tilde{b})$

The application of fuzzy functions for describing and solving engineering problems requires the numerical treatment of these functions. For each crisp argument $\underline{t} \in \underline{\mathbf{T}}$ the fuzzy result $\tilde{x}(\underline{t})$ may be computed by applying the extension principle (Sect. 2.1.7) or α-level optimization (Sects. 5.2.2 and 5.2.3).

As the α-level optimization has proved to be advantageous for engineering applications, the idea of *α-discretization* (Sect. 2.1.9) is adopted and extended to fuzzy functions.

Based on the bunch parameter representation of fuzzy functions in accordance with Eqs. (2.67) and (2.68) α-level sets of fuzzy functions may be generated. For this purpose, the α-discretization is applied to the fuzzy bunch parameters $\tilde{s}_j$. This leads to the α-level sets S_{j,α_k} (Sect. 2.1.9). In general, n_s fuzzy bunch parameters $\tilde{s}_1, ..., \tilde{s}_j, ..., \tilde{s}_{n_s}$ exist and the α-level sets S_{j,α_k} form the crisp subspace $\underline{S}_{\alpha_k}$ for each α-level α_k. This is illustrated in Fig. 2.21, whereby the sketched fuzzy sets $\tilde{A}_i$ are to be interpreted as fuzzy bunch parameters $\tilde{s}_j$ and the generated subspace $\underline{X}_{\alpha_k}$ represents $\underline{S}_{\alpha_k}$.

As can already be recognized in Eq. (2.45), for the subspaces $\underline{S}_{\alpha_i}$ and $\underline{S}_{\alpha_k}$ the relationship

$$\underline{S}_{\alpha_i} \subseteq \underline{S}_{\alpha_k} \quad \forall\ \alpha_i, \alpha_k \mid \alpha_i, \alpha_k \in (0,1];\ \alpha_i \geq \alpha_k \tag{2.70}$$

holds, and the support subspace $\underline{S}_{\alpha\to+0}$ contains all other subspaces $\underline{S}_{\alpha_k}$.

For each given value $\alpha = \alpha_k$ the bunch parameter vector may now take the values $\underline{s} \in \underline{S}_\alpha$. Evaluating all these $\underline{s}$ leads to a crisp set of functions, which is referred to as *α-function set* $X_\alpha(\underline{t})$

$$X_\alpha(\underline{t}) = \left\{ x(\underline{s},\underline{t})\ \ \forall\, \underline{s} \mid \underline{s} \in \underline{S}_\alpha;\ \alpha \in (0,1] \right\}. \tag{2.71}$$

For $\alpha \to +0$ the *function support*

$$X_{\alpha\to+0}(\underline{t}) = \left\{ x(\underline{s},\underline{t})\ \ \forall\, \underline{s} \mid \underline{s} \in \underline{S}_{\alpha\to+0} \right\} \tag{2.72}$$

of the fuzzy function is obtained.

Any arbitrary function $x(\underline{t})$ contained in $X_\alpha(\underline{t})$ is a trajectory of the fuzzy function on the level α. In the case that all functional values of a fuzzy function are fuzzy numbers, the trajectory for $\alpha = 1$ is referred to as a *trend function*.

For a sufficiently high number of α-levels a fuzzy function may be represented by the assigned α-function sets $X_\alpha(\underline{t})$

$$\tilde{x}(\underline{t}) = x(\underline{\tilde{s}}, \underline{t}) = \left\{ \left(X_\alpha(\underline{t}), \mu(X_\alpha(\underline{t})) \right) \mid \mu(X_\alpha(\underline{t})) = \alpha\ \forall\, \alpha \in (0,1] \right\}, \tag{2.73}$$

with $X_\alpha(\underline{t})$ from Eq. (2.71). This denotes a *general α-level representation* of Eq. (2.68).

Based on α-discretization Eq. (2.73) describes the fuzzy function $\tilde{x}(\underline{t})$ as a set of the α-function sets $X_\alpha(\underline{t})$ with the assigned membership values $\mu(X_\alpha(\underline{t})) = \alpha$. According to Eq. (2.71) each of these α-function sets represents an assessed bunch of functions (function bunch) that possesses at least the assigned membership α indicating the assessment result.

If *all fuzzy functional values of* $\tilde{x}(\underline{t})$ *are convex fuzzy sets*, $X_\alpha(\underline{t})$ is then represented by an interval $x_{t\,\alpha}$ for each crisp $\underline{t}$. The functional values $\tilde{x}_t$ of the fuzzy function $\tilde{x}(\underline{t})$ at specified points $\underline{t}$ may then be characterized by

$$\tilde{x}_t = x_t(\underline{\tilde{s}}) = \left\{ \left(x_{t\alpha}, \mu(x_{t\alpha}) \right) \mid x_{t\alpha} = [x_{t\alpha l}, x_{t\alpha r}];\ \mu(x_{t\alpha}) = \alpha\ \forall\, \alpha \in (0,1] \right\}, \tag{2.74}$$

with

$$x_{t\alpha l} = \min\left[x_t(\underline{s}) \mid \underline{s} \in \underline{S}_\alpha \right], \tag{2.75}$$

$$x_{t\alpha r} = \max\left[x_t(\underline{s}) \mid \underline{s} \in \underline{S}_\alpha \right]. \tag{2.76}$$

This represents the basis for numerically determining fuzzy functional values with the aid of the α-level optimization.

The crisp functions

$$x_{\alpha l}(\underline{t}) = \left\{ x_{t\alpha l}\ \ \forall\, \underline{t} \mid \underline{t} \in \mathbf{\underline{T}} \right\} \tag{2.77}$$

and

$$x_{\alpha r}(\underline{t}) = \left\{ x_{t\alpha r}\ \ \forall\, \underline{t} \mid \underline{t} \in \mathbf{\underline{T}} \right\} \tag{2.78}$$

constituted by the interval bounds from Eqs. (2.75) and (2.76) are the lower and upper *bounding functions* of the fuzzy function on the level α. If all fuzzy bunch parameters are fuzzy numbers, the bounding functions for α = 1 then coincide and form the trend function $x_{\alpha=1}(\underline{t}) = x_{\alpha=1\,l}(\underline{t}) = x_{\alpha=1\,r}(\underline{t})$. In general, bounding functions do not necessarily represent trajectories of the fuzzy function. At different points $\underline{t}_i$ and $\underline{t}_{i+1}$, frequently, different bunch parameter vectors $\underline{s} \in \tilde{\underline{s}}$, i.e., different trajectories, lead to the interval bounds in Eqs. (2.75) and (2.76).

Example 2.4. The one-dimensional fuzzy function $x(\tilde{\underline{s}},t) = \tilde{a}\cdot\sin(\tilde{\omega}_0 t + \tilde{b})$ from example 2.3 is considered for two different variants of bunch parameters.

The first variant is characterized by $\tilde{a}$ = <0.9, 1.0, 1.1>, $\tilde{\omega}_0 = \omega_0 = 1.0$, and $\tilde{b} = b = 0$, for the second variant the bunch parameters coincide with these from example 2.3, i.e., $\tilde{a}$ = <0.9, 1.0, 1.1>, $\tilde{\omega}_0$ = <0.9, 1.0, 1.1>, and $\tilde{b}$ = <−0.1, 0.0, 0.1>. The obtained curves for the bounding functions on the α-level $\alpha \to +0$ and the trend functions are displayed in Figs. 2.26 and 2.27.

Additionally, Fig. 2.27 shows the trajectories from Fig. 2.25 and indicates their assignment to the appropriate α-function sets. For different α-levels various trajectories may be chosen:

- $\alpha = 1$ — $X_{\alpha=1}(t)$ is the function set that contains exactly one element describing the function $x(t) = \sin(t)$ as the mean function of $\tilde{x}(t)$.
- $\alpha = 0.5$ — $X_{\alpha=0.5}(t)$ represents the function set comprising all functions $x(t) = a\cdot\sin(\omega_0 t + b)$ with arbitrary bunch parameter values within the intervals $a \in [0.95, 1.05]$, $\omega_0 \in [0.95, 1.05]$, and $b \in [-0.05, 0.05]$.
- $\alpha \to +0$ — $X_{\alpha\to+0}(t)$ is the function set of all functions $x(t) = a\cdot\sin(\omega_0 t + b)$ with arbitrary bunch parameter values within the intervals $a \in [0.9, 1.1]$, $\omega_0 \in [0.9, 1.1]$, and $b \in [-0.1, 0.1]$.

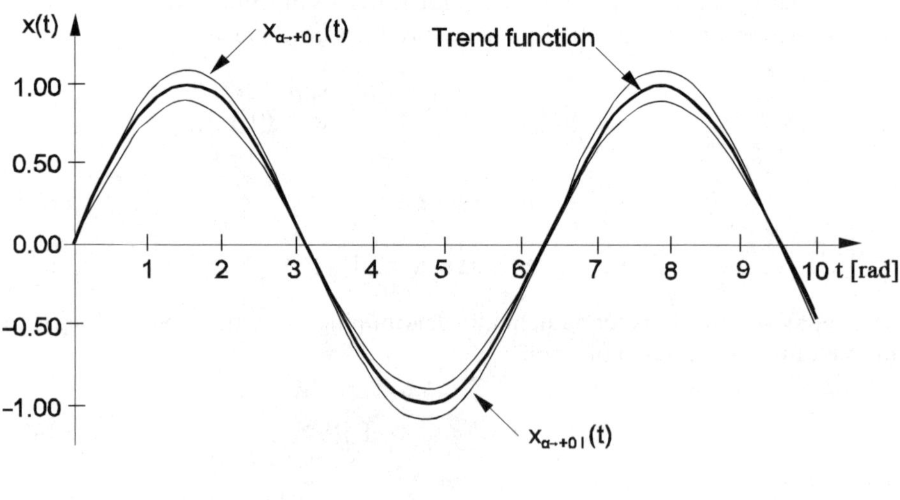

Fig. 2.26. Fuzzy sine function, first variant

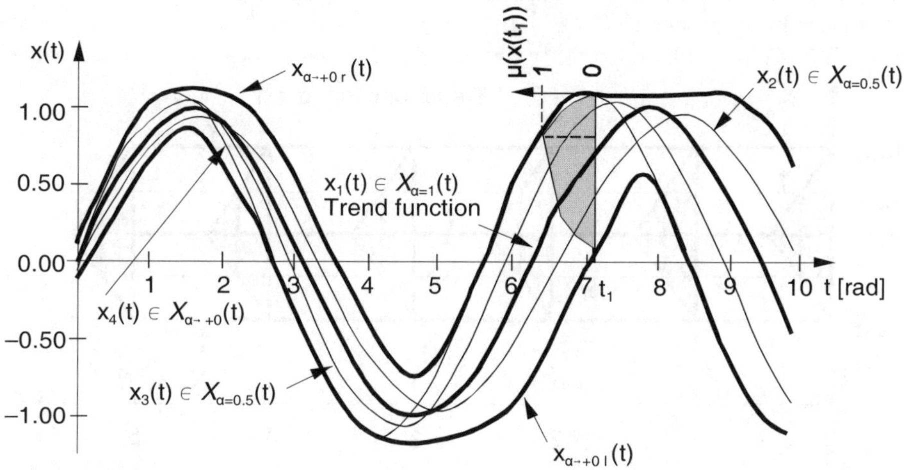

Fig. 2.27. Fuzzy sine function, second variant

Special Fuzzy Functions. Fuzzy functions possess several properties of particular interest, two of which are stated in the following.

A fuzzy function $\tilde{x}(\underline{t})$ with $\underline{t} \in \mathbf{\underline{T}}$ is said to be *stationary* if its fuzzy functional values $\tilde{x}_t = \tilde{x}(\underline{t}_i)$ are identical for all $\underline{t}_i \in \mathbf{\underline{T}}$

$$\tilde{x}(\underline{t}_i) = \tilde{x}(\underline{t}_i + \Delta\underline{t}) = \tilde{x}_t \quad \forall\, \underline{t}_i,\ \underline{t}_i + \Delta\underline{t} \in \mathbf{\underline{T}}. \tag{2.79}$$

The trend function as well as the lower and upper bounding functions for all α-levels of a stationary fuzzy function is a constant function, i.e., $x_{\alpha\,l}(\underline{t})$ = constant and $x_{\alpha\,r}(\underline{t})$ = constant.

A fuzzy function $\tilde{x}(\underline{t})$ with $\underline{t} \in \mathbf{\underline{T}}$ possesses *homogeneous or stationary increments* if the fuzzy functional value $\tilde{x}(\underline{t}_i + \underline{\varkappa}) + c$ is equal to the fuzzy functional value $\tilde{x}(\underline{t}_k + \underline{\varkappa})$, namely for all $\underline{\varkappa}$ with $\underline{\varkappa} + \underline{t}_i \in \mathbf{\underline{T}}$, $\underline{\varkappa} + \underline{t}_k \in \mathbf{\underline{T}}$, $\underline{t}_i \neq \underline{t}_k$, and c = constant. Thereby, $\underline{t}_1$ and $\underline{t}_2$ are arbitrary but invariant

$$\tilde{x}(\underline{t}_i + \underline{\varkappa}) + c = \tilde{x}(\underline{t}_k + \underline{\varkappa}) \quad \forall \underline{\varkappa} \mid \underline{t}_i \neq \underline{t}_k;\ \underline{\varkappa} + \underline{t}_i \in \mathbf{\underline{T}};\ \underline{\varkappa} + \underline{t}_k \in \mathbf{\underline{T}};\ c = \text{constant}. \tag{2.80}$$

The trend function and the lower and upper bounding functions for all α-levels of a fuzzy function with homogeneous or stationary increments represent linear functions possessing identical slopes.

Example 2.5. The fuzzy process $\tilde{x}(\tau)$ with $\tau \in \mathbf{T} = [0, 5]$ s is approximated with the aid of suitably distributed fuzzy interpolation nodes, as explained in more detail in the subsequent section. These fuzzy interpolation nodes are given by crisp points in time τ_i and associated fuzzy triangular numbers $< x_{\alpha\to+0\,l},\ x_{\alpha=1},\ x_{\alpha\to+0\,r} >$ describing identical fuzziness of the functional values $\tilde{x}(\tau_i)$ for every discrete τ_i. Six fuzzy interpolation nodes are plotted for τ_i = 0, 1, 2, 3, 4, 5 s in Fig. 2.28.

Between these τ_i the fuzzy process $\tilde{x}(\tau)$ is linearly approximated, it represents a stationary fuzzy process.

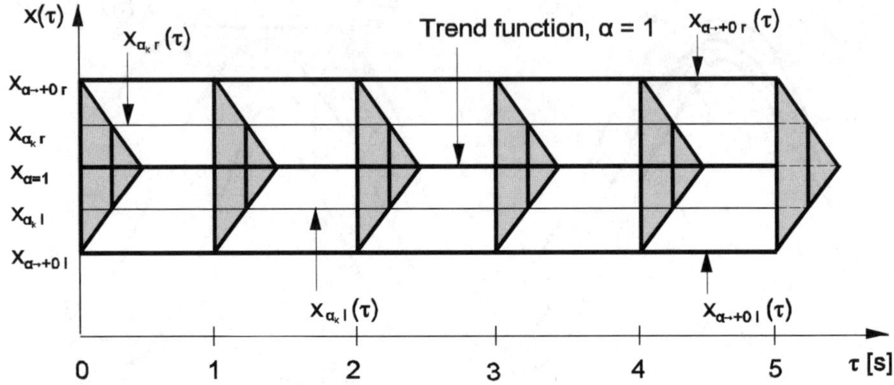

Fig. 2.28. Stationary fuzzy process

Example 2.6. For approximating the fuzzy process $\tilde{x}(\tau)$ with $\tau \in \mathbf{T} = [0, 5]$ s six equidistant fuzzy interpolation nodes are chosen (Fig. 2.29). Between the crisp points in time τ_i the fuzzy process $\tilde{x}(\tau)$ is linearly interpolated. Owing to the special fuzzy functional values at the τ_i the obtained fuzzy process possesses linear, homogeneous increments. The special condition

$$\tilde{x}(\tau_i) + c = \tilde{x}(\tau_{i+1}) \;\; ; \;\; i = 0, 1, 2, 3, 4 \;\; ; \;\; c = \text{constant} \tag{2.81}$$

is fulfilled.

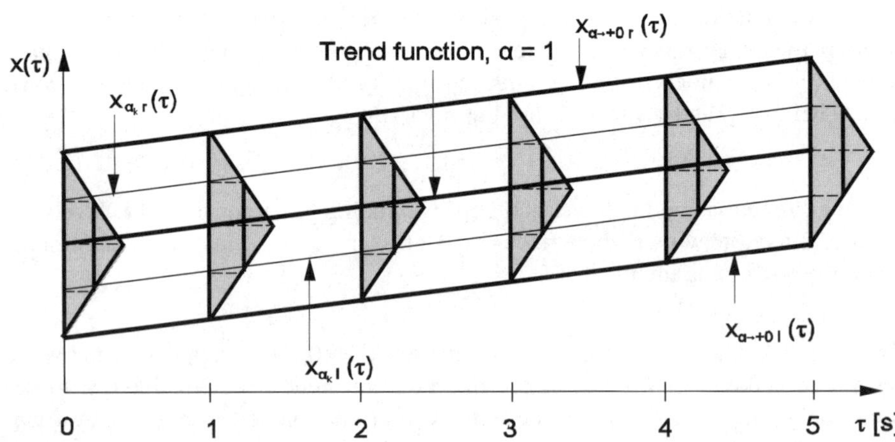

Fig. 2.29. Fuzzy process with linear, homogeneous or stationary increments

2.1.11.2 Point and Time Discretization of Fuzzy Functions

Numerical processing of fuzzy functions $\tilde{x}(\underline{t}) = x(\underline{\tilde{s}},\underline{t})$ demands not only their α-discretization but also the discretization of their argument $\underline{t}$, which constitutes the fundamental set $\underline{\mathbf{T}}$, if these are continuous. As in structural-engineering applications generally continuous fuzzy functions appear, *point and time discretization* are thus necessary in addition to α-discretization.

Point and time discretization means that the deterministic argument $\underline{t}$ is discretized by choosing k_0 interpolation nodes. Each discretized fuzzy function may then be represented by a sequence of fuzzy variables. If full interaction is presumed between the discrete fuzzy variables at the k_0 points, this leads to the sequence

$$\tilde{x}(\underline{t}) = \left\{x(\underline{\tilde{s}},\underline{t}_1),\ x(\underline{\tilde{s}},\underline{t}_2),\ \ldots,\ x(\underline{\tilde{s}},\underline{t}_{k_0})\ \middle|\ \underline{t}_1,\underline{t}_2,\ldots,\underline{t}_{k_0} \in \underline{\mathbf{T}}\right\}, \tag{2.82}$$

with the fuzzy bunch parameters $\underline{\tilde{s}}$. In the case that no interaction is assumed, the sequence

$$\tilde{x}(\underline{t}) = \left\{x(\underline{\tilde{s}}_1,\underline{t}_1),\ x(\underline{\tilde{s}}_2,\underline{t}_2),\ \ldots,\ x(\underline{\tilde{s}}_{k_0},\underline{t}_{k_0})\ \middle|\ \underline{t}_1,\underline{t}_2,\ldots,\underline{t}_{k_0} \in \underline{\mathbf{T}}\right\} \tag{2.83}$$

containing the fuzzy bunch parameters $\underline{\tilde{s}}_1,\ \underline{\tilde{s}}_2,\ldots,\ \underline{\tilde{s}}_{k_0}$ is obtained.

A fuzzy field $\tilde{x}(\underline{\theta})$ may be described by its values at the points in space $\underline{\theta}_j \in \underline{\boldsymbol{\Theta}}$

$$\tilde{x}(\underline{\theta}) = \left\{x(\underline{\tilde{s}},\underline{\theta}_1),\ x(\underline{\tilde{s}},\underline{\theta}_2),\ \ldots,\ x(\underline{\tilde{s}},\underline{\theta}_{k_0})\ \middle|\ \underline{\theta}_1,\underline{\theta}_2,\ldots,\underline{\theta}_{k_0} \in \underline{\boldsymbol{\Theta}}\right\}, \tag{2.84}$$

which characterizes the fuzzy field in the form of *point discretization*.

A fuzzy process $\tilde{x}(\tau)$ may be represented by specifying its values at specific points in time $\tau_i \in \mathbf{T}$

$$\tilde{x}(\tau) = \left\{x(\underline{\tilde{s}},\tau_1),\ x(\underline{\tilde{s}},\tau_2),\ \ldots,\ x(\underline{\tilde{s}},\tau_{k_0})\ \middle|\ \tau_1,\tau_2,\ldots,\tau_{k_0} \in \mathbf{T}\right\}. \tag{2.85}$$

This is referred to as *time discretization* of the fuzzy process.

In order to describe a fuzzy function $\tilde{x}(\underline{\theta},\tau)$, its functional values may be prescribed at both points in space $\underline{\theta}_j$ (j = 1, ..., k_P) and points in time τ_i (i = 1, ..., k_Z)

$$\tilde{x}(\underline{\theta},\tau) = \left\{\begin{matrix} x(\underline{\tilde{s}},\underline{\theta}_1,\tau_1), & \ldots, & x(\underline{\tilde{s}},\underline{\theta}_j,\tau_1), & \ldots, & x(\underline{\tilde{s}},\underline{\theta}_{k_P},\tau_1) \\ \ldots & & & & \\ x(\underline{\tilde{s}},\underline{\theta}_1,\tau_i), & \ldots, & x(\underline{\tilde{s}},\underline{\theta}_j,\tau_i), & \ldots, & x(\underline{\tilde{s}},\underline{\theta}_{k_P},\tau_i) \\ \ldots & & & & \\ x(\underline{\tilde{s}},\underline{\theta}_1,\tau_{k_Z}), & \ldots, & x(\underline{\tilde{s}},\underline{\theta}_j,\tau_{k_Z}), & \ldots, & x(\underline{\tilde{s}},\underline{\theta}_{k_P},\tau_{k_Z}) \end{matrix}\right\}. \tag{2.86}$$

Example 2.7. The one-dimensional fuzzy function $\tilde{x}(\theta) = \tilde{s}_1 \cdot \theta + \tilde{s}_2$ with the fuzzy bunch parameters $\tilde{s}_1$ = <0.8, 1, 1.2> and $\tilde{s}_2$ = <–0.2, 0, 0.2> is represented by point discretization (Fig. 2.30).

If this fuzzy function is additionally variable in time, the point discretization is combined with a time discretization. The fuzzy function $\tilde{x}(\theta,\tau) = \tilde{s}_1 \cdot \theta + \tilde{s}_2 + \tilde{s}_3 \cdot \tau$ with $\tilde{s}_3 = <-0.1, 0, 0.2>$ is then characterized using discrete points in space and time (Fig. 2.31).

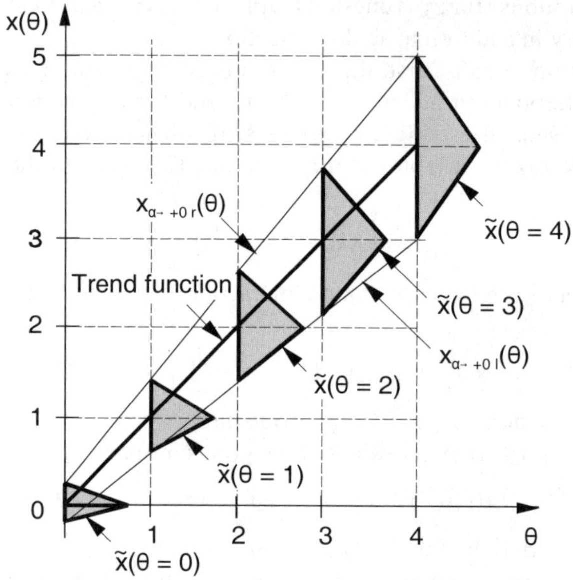

Fig. 2.30. Representation of the fuzzy field $\tilde{x}(\theta) = \tilde{s}_1 \cdot \theta + \tilde{s}_2$ by point discretization

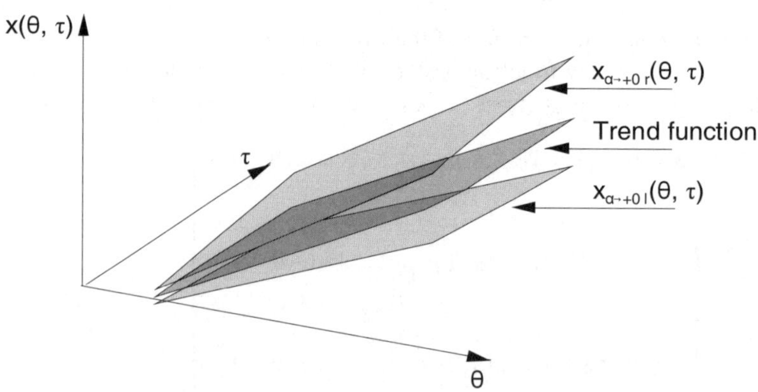

Fig. 2.31. Representation of the fuzzy function $\tilde{x}(\theta,\tau) = \tilde{s}_1 \cdot \theta + \tilde{s}_2 + \tilde{s}_3 \cdot \tau$ by point and time discretization

Consideration of Interaction. In the case of interaction the space of the fuzzy bunch parameters $\underline{\tilde{s}}_1$, $\underline{\tilde{s}}_2$, ..., $\underline{\tilde{s}}_{k_0}$ introduced in Eq. (2.83) is constrained by interactive dependencies $I_{i,\alpha}$ prescribed a priori, whereby i denotes the interaction counter. Each α-level set $\underline{S}_\alpha$ of the fuzzy bunch parameters $\underline{\tilde{s}}_1$, $\underline{\tilde{s}}_2$, ..., $\underline{\tilde{s}}_{k_0}$ forms a crisp n-dimensional parameter subspace. If interaction does not exist, this subspace is represented by an n-dimensional hypercuboid

$$\underline{S}_\alpha = \left\{\underline{S}_{1,\alpha}, \underline{S}_{2,\alpha}, \dots, \underline{S}_{k_0,\alpha}\right\}. \tag{2.87}$$

On the other hand, if $\underline{s}_1 = \underline{s}_2 = \dots = \underline{s}_{k_0} = \underline{s}$ holds, the fuzzy function is determined by the same fuzzy bunch parameter vector $\underline{\tilde{s}}$ for all points in space or time. This characterizes full interaction and Eq. (2.83) changes to Eq. (2.82).

Example 2.8. The effect of a priori prescribed interactive dependencies is demonstrated in the two-dimensional space of the fuzzy bunch parameters $\tilde{s}_1 = \langle 7, 8, 10 \rangle$ and $\tilde{s}_2 = \langle 1, 5, 6 \rangle$ (Fig. 2.32). The crisp bunch parameter subspace $\underline{S}_{\alpha \to +0}$ (support subspace) is constrained by the interactive dependencies $I_{1,\alpha \to +0} = \tilde{s}_1 - \tilde{s}_2 \leq 5$ and $I_{2,\alpha \to +0} = \tilde{s}_1 - 1.5 \cdot \tilde{s}_2 \geq 1$. Exclusively, the elements lying in the hatched domain of $\underline{S}_{\alpha \to +0}$ are to be considered.

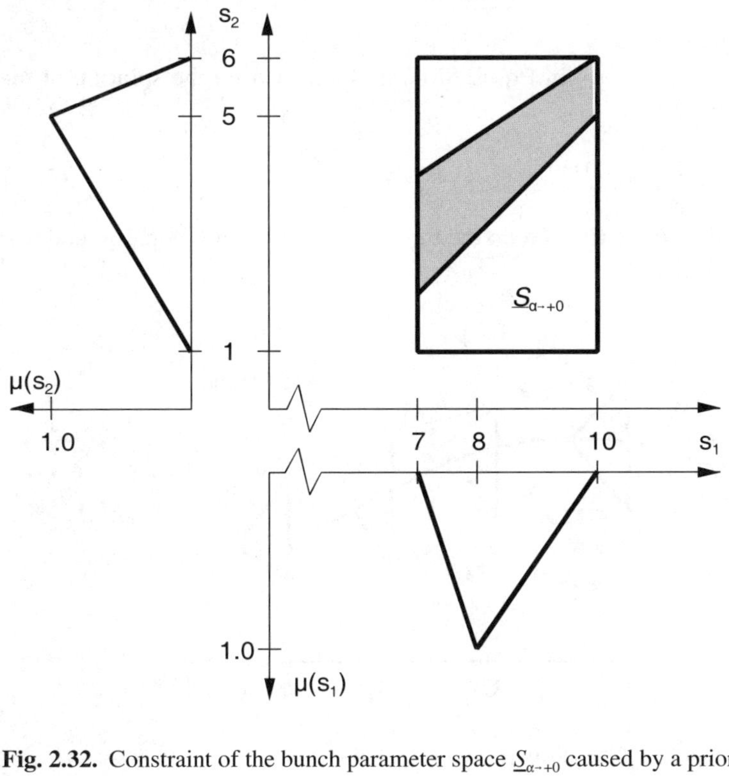

Fig. 2.32. Constraint of the bunch parameter space $\underline{S}_{\alpha \to +0}$ caused by a priori interaction

2.1.11.3 Approximative Description of Fuzzy Functions

In fuzzy structural analysis frequently physical input parameters, e.g., geometry, material, and load parameters, are only available at discrete points. If the discrete structural parameter values are uncertain, and if their uncertainty is described by fuzzy variables, these fuzzy variables may then be interpreted as fuzzy interpolation nodes of a fuzzy function.

With the aid of k_s specified fuzzy interpolation nodes $\tilde{x}_i = \tilde{x}(\underline{t}_i)$ with $i = 1, ..., k_s$ the fuzzy function $\tilde{x}(\underline{t})$ may be approximately determined. For this approximation common mathematical procedures may be applied. Owing to the fuzzy interpolation nodes, however, special solution techniques like α-level optimization (Sects. 5.2.2 and 5.2.3) are required. In the following examples the approximation procedure is demonstrated by means of fuzzy approximation polynomials.

Example 2.9. Provided that the fuzzy interpolation nodes $\tilde{x}_i = \tilde{x}(\underline{t}_i) = x(\underline{\tilde{s}}, \underline{t}_i)$ with $i = 1, ..., k_s$ are given on $\mathbb{R}^1$, a fuzzy approximation function based on the polynomial approach

$$x(\underline{\tilde{s}}, t) = \sum_{m=1}^{k_s} \tilde{s}_m \cdot t^{m-1} \tag{2.88}$$

may be drawn through the $\tilde{x}_i$.

The fuzzy free values $\tilde{s}_m$ in Eq. (2.88) are obtained from the solution of the fuzzy equation system

$$\sum_{m=1}^{k_s} \tilde{s}_m \cdot t_i^{m-1} = \tilde{x}_i \quad ; \quad i = 1, ..., k_s \,. \tag{2.89}$$

A fuzzy function approximated on the basis of six fuzzy interpolation nodes is shown in Fig. 2.33.

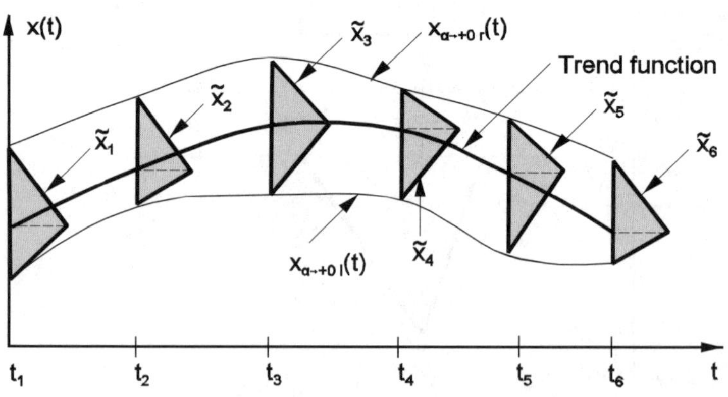

Fig. 2.33. Fuzzy function through six fuzzy interpolation nodes

Example 2.10. Starting from $k_s = k_1 \cdot k_2$ predefined fuzzy interpolation nodes $\tilde{x}_{i,j} = x(\tilde{\underline{s}}, \underline{t}_{i,j})$ with $i = 1, ..., k_1$ and $j = 1, ..., k_2$ on $\mathbb{R}^2$, the polynomial approach

$$x(\tilde{\underline{s}}, t_1, t_2) = \sum_{m=1}^{k_1} \sum_{n=1}^{k_2} \tilde{s}_{m,n} \cdot t_1^{m-1} \cdot t_2^{n-1} \tag{2.90}$$

may be chosen for approximately determining a fuzzy function.

The fuzzy free values in Eq. (2.90) are computed from the fuzzy equation system

$$\sum_{m=1}^{k_1} \sum_{n=1}^{k_2} \tilde{s}_{m,n} \cdot t_{1,i}^{m-1} \cdot t_{2,j}^{n-1} = \tilde{x}_{i,j} \; ; \quad i = 1, ..., k_1; \; j = 1, ..., k_2 . \tag{2.91}$$

A fuzzy function approximated on the basis of Eq. (2.90) is illustrated in Fig. 2.34.

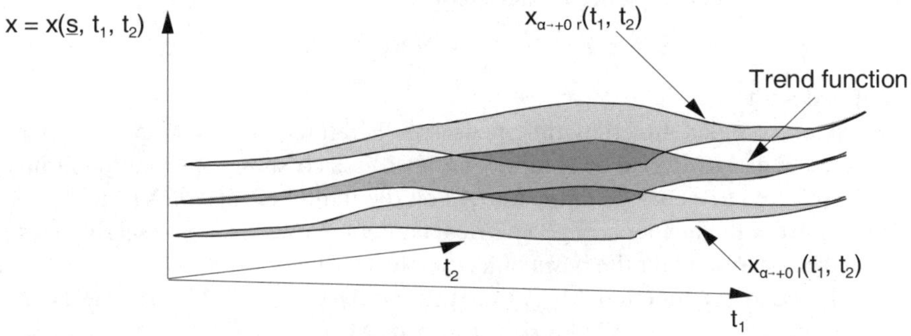

Fig. 2.34. Two-dimensional fuzzy function through 5·5 fuzzy interpolation nodes

2.2 Elements of Measure Theory

In order to establish a concept for assessing structural safety, an appropriate safety measure must be chosen. For this purpose, some measure theoretical basics are considered and discussed in view of their further application.

2.2.1 Measure and Uncertain Measure of Crisp Sets

The following explanations regarding measure theoretical foundations are consistent with [40, 60] as well as [10, 22, 212]. For defining the term *measure*, the concept of the elementary geometrical content represents the most common basis. This suggests that a measure serves for determining the content of geometrical formations and – in a generalized manner – of point sets. Well-known classical measures are the Riemann measure (associated with the Riemann integral), which utilizes the Jordan content, and the Lebesgue measure (associated with the Lebesgue integral), which is based on the Lebesgue content.

In set theory the measure concept is extended for assigning measure values to particular *crisp sets*, which belong to defined systems of crisp sets. A measure is thereby defined as being a special function, as explained on the subsequent basis.

For an arbitrary crisp set, e.g., the multidimensional fundamental set $\underline{\mathbf{X}}$, with the subsets $\underline{A}_i$, the set $\mathfrak{P}(\underline{\mathbf{X}})$ of *all* subsets $\underline{A}_i$ is referred to as the *power set* on $\underline{\mathbf{X}}$.

Subsets and systems of subsets of a power set $\mathfrak{P}(\underline{\mathbf{X}})$ are called *families of sets* $\mathfrak{M}(\underline{\mathbf{X}})$.

A family of sets $\mathfrak{M}(\underline{\mathbf{X}})$ is referred to as *σ-algebra* $\mathfrak{S}(\underline{\mathbf{X}})$, if

$$\underline{\mathbf{X}} \in \mathfrak{S}(\underline{\mathbf{X}}), \tag{2.92}$$

$$\underline{A}_i \in \mathfrak{S}(\underline{\mathbf{X}}) \Rightarrow \underline{A}_i^C \in \mathfrak{S}(\underline{\mathbf{X}}) \tag{2.93}$$

holds, and if for every sequence of subsets $\underline{A}_i$

$$\underline{A}_i \in \mathfrak{S}(\underline{\mathbf{X}});\ i = 1, 2, \ldots \Rightarrow \bigcup_i^{\infty} \underline{A}_i \in \mathfrak{S}(\underline{\mathbf{X}}) \tag{2.94}$$

is complied with.

A real or complex function that possesses a family of sets $\mathfrak{M}(\underline{\mathbf{X}})$ as set of definition and assigns a real or complex number to each subset $\underline{A}_i$ from the family of sets is referred to as *set function* $M_{\mathfrak{M}}(\underline{A}_i)$ on the family of sets $\mathfrak{M}(\underline{\mathbf{X}})$.

Starting from the elementary geometrical content, a measure for assessing crisp sets may be constituted on the basis of a suitable set function.

A real-valued set function $M_{\mathfrak{S}}(\underline{A}_i)$ on the family of sets $\mathfrak{S}(\underline{\mathbf{X}})$ that meets the following requirements is called (crisp) *measure* $M^s_{\mathfrak{S}}(\underline{A}_i)$ on $\mathfrak{S}(\underline{\mathbf{X}})$:

$$M^s_{\mathfrak{S}}(\varnothing) = 0, \tag{2.95}$$

$$\underline{A}_i \in \mathfrak{S}(\underline{\mathbf{X}}) \Rightarrow M^s_{\mathfrak{S}}(\underline{A}_i) \geq 0\,, \tag{2.96}$$

$$\underline{A}_i, \underline{A}_k \in \mathfrak{S}(\underline{\mathbf{X}});\ \underline{A}_i \cap \underline{A}_k = \varnothing \Rightarrow M^s_{\mathfrak{S}}(\underline{A}_i \cup \underline{A}_k) = M^s_{\mathfrak{S}}(\underline{A}_i) + M^s_{\mathfrak{S}}(\underline{A}_k)\,, \tag{2.97}$$

$$\underline{A}_i \in \mathfrak{S}(\underline{\mathbf{X}});\ i = 1, \ldots, n;\ \underline{A}_i \cap \underline{A}_k = \varnothing\ \forall i \neq k \Rightarrow M^s_{\mathfrak{S}}\Big(\bigcup_{i=1}^{\infty} \underline{A}_i\Big) = \sum_{i=1}^{\infty} M^s_{\mathfrak{S}}(\underline{A}_i)\,. \tag{2.98}$$

The measure $M^s_{\mathfrak{S}}(\underline{A}_i)$ is additive. If a maximum value exists for $M^s_{\mathfrak{S}}(\underline{A}_i)$, $M^s_{\mathfrak{S}}(\underline{A}_i)$ is then called finitely additive. The triple $[\underline{\mathbf{X}}, \mathfrak{S}, M^s_{\mathfrak{S}}]$ constitutes the (crisp) *measure space*.

When defining *uncertain measures* the idea of the elementary geometrical content no longer serves as the starting point, but the set theoretical approach associated with appropriate requirements represents the basis for uncertain measures.

In accordance with the properties of crisp measures, uncertain measures assign real measure values to the elements (crisp subsets) $\underline{A}_i$ of the family of sets $\mathfrak{S}(\underline{\mathbf{X}})$. The subsets $\underline{A}_i \in \mathfrak{S}(\underline{\mathbf{X}})$ may be interpreted as events $E_i = \underline{A}_i$. The assigned family of sets $\mathfrak{S}(\underline{\mathbf{X}})$ constitutes a space of events on the fundamental set $\underline{\mathbf{X}}$ (space of results or observations). The event E_i is considered to have occurred if at least one element from the fundamental set $\underline{x}_i \in \underline{\mathbf{X}}$ belongs to the subset $\underline{A}_i$. Thereby the $\underline{x}_i$ represent realizations, which are observed in consequence of elementary events, which are possible results of a trial and exclude each other. If the assignment of $\underline{x}_i$ to $\underline{A}_i$ cannot be specified in binary terms, i.e., $\underline{x}_i$ does neither completely belong to $\underline{A}_i$ nor to $\underline{A}_i^C$, the event $E_i = \underline{A}_i$ can then be regarded as partly occurred. The intensity of the occurrence of the event $E_i = \underline{A}_i$ is assessed with the aid of uncertain measures. Uncertain measures describe the degree of membership of the elements $\underline{x}_i$ of a fundamental set $\underline{\mathbf{X}}$ with which the $\underline{x}_i$ belong to the subset $\underline{A}_i$ from $\underline{\mathbf{X}}$. They map the elements of the fundamental set onto the interval [0, 1].

An *uncertain measure* is a special real-valued set function $M^u_{\mathfrak{S}}(\underline{A}_i)$ on the elements $\underline{A}_i$ (crisp subsets on the fundamental set $\underline{\mathbf{X}}$) of the family of sets $\mathfrak{S}(\underline{\mathbf{X}})$.

$M^u_{\mathfrak{S}}(\underline{A}_i)$ is normalized by means of

$$M^u_{\mathfrak{S}}(\varnothing) = 0 \,, \tag{2.99}$$

$$M^u_{\mathfrak{S}}(\underline{\mathbf{X}}) = 1 \,. \tag{2.100}$$

If the set $\underline{A}_i$ is enlarged, the measure value should not decrease but behave monotonically in accordance with the enlargement of the set to be measured

$$\underline{A}_i, \underline{A}_k \in \mathfrak{S}(\underline{\mathbf{X}});\ \underline{A}_i \subseteq \underline{A}_k \ \Rightarrow\ M^u_{\mathfrak{S}}(\underline{A}_i) \le M^u_{\mathfrak{S}}(\underline{A}_k) \,. \tag{2.101}$$

In the case of nonfinite fundamental sets $\underline{\mathbf{X}}$, additionally, continuity is demanded for sequences of the sets $\underline{A}_i$,

$$\underline{A}_i \in \mathfrak{S}(\underline{\mathbf{X}});\ i = 1, \ldots, n;\ \underline{A}_1 \subseteq \underline{A}_2 \subseteq \ldots \subseteq \underline{A}_n \Rightarrow \lim_{n\to\infty} M^u_{\mathfrak{S}}(\underline{A}_n) = M^u_{\mathfrak{S}}(\lim_{n\to\infty} \underline{A}_n) \,, \tag{2.102}$$

or

$$\underline{A}_i \in \mathfrak{S}(\underline{\mathbf{X}});\ i = 1, \ldots, n;\ \underline{A}_1 \supseteq \underline{A}_2 \supseteq \ldots \supseteq \underline{A}_n \Rightarrow \lim_{n\to\infty} M^u_{\mathfrak{S}}(\underline{A}_n) = M^u_{\mathfrak{S}}(\lim_{n\to\infty} \underline{A}_n) \,. \tag{2.103}$$

The uncertain measure $M^u_{\mathfrak{S}}(\underline{A}_i)$ is not necessarily additive. As also introduced for crisp measures, the triple $[\underline{\mathbf{X}}, \mathfrak{S}, M^u_{\mathfrak{S}}]$ is referred to as an *uncertain measure space*, likewise.

Every uncertain measure deduced from this definition complies with

$$\underline{A}_i, \underline{A}_k \in \mathfrak{S}(\underline{\mathbf{X}}) \ \Rightarrow\ M^u_{\mathfrak{S}}(\underline{A}_i \cup \underline{A}_k) \ge \max[M^u_{\mathfrak{S}}(\underline{A}_i), M^u_{\mathfrak{S}}(\underline{A}_k)] \,, \tag{2.104}$$

$$\underline{A}_i, \underline{A}_k \in \mathfrak{S}(\underline{\mathbf{X}}) \ \Rightarrow\ M^u_{\mathfrak{S}}(\underline{A}_i \cap \underline{A}_k) \le \min[M^u_{\mathfrak{S}}(\underline{A}_i), M^u_{\mathfrak{S}}(\underline{A}_k)] \,. \tag{2.105}$$

For conveniently describing uncertain measures, *distribution functions*, which possess the properties of set functions, may be established. For this purpose, the n-dimensional Euclidian space $\mathbb{R}^n$ is chosen to represent the fundamental set $\underline{\mathbf{X}}$. The consideration is focused on the system $\mathfrak{M}_0(\mathbb{R}^n)$ of the open subsets

$$\underline{A}_i = \left\{ \underline{t} = (t_1, \ldots, t_k, \ldots, t_n) \mid \underline{x} = \underline{x}_i;\ \underline{x}, \underline{t} \in \mathbb{R}^n;\ t_k < x_k;\ k = 1, \ldots, n \right\} \tag{2.106}$$

in $\mathbb{R}^n$. This family of sets $\mathfrak{M}_0(\mathbb{R}^n)$ constitutes the σ-algebra $\mathfrak{S}_0(\mathbb{R}^n)$ as *Borel σ-algebra of the* $\mathbb{R}^n$. $\mathfrak{S}_0(\mathbb{R}^n)$ is a *Boolean set algebra.*

A *distribution function* $V(\underline{A}_i)$ is a set function $M^u_{\mathfrak{S}}(\underline{A}_i)$ on the open subsets $\underline{A}_i$ of the family of sets $\mathfrak{S}_0(\mathbb{R}^n)$ according to Eq. (2.106). $V(\underline{A}_i)$ represents an *uncertain measure* with the properties

$$0 \le V(\underline{A}_i) \le 1 \quad \forall \underline{A}_i \in \mathfrak{S}_0(\mathbb{R}^n), \tag{2.107}$$

$$\underline{A}_i, \underline{A}_k \in \mathfrak{S}_0(\mathbb{R}^n);\ \underline{A}_i \subseteq \underline{A}_k \quad \Rightarrow \quad V(\underline{A}_i) \le V(\underline{A}_k), \tag{2.108}$$

$$\underline{A}_i \in \mathfrak{S}_0(\mathbb{R}^n);\ i = 1,\dots,n;\ \underline{A}_1 \subseteq \dots \subseteq \underline{A}_n \quad \Rightarrow \quad \lim_{n\to\infty} V(\underline{A}_n) = V(\lim_{n\to\infty} \underline{A}_n) = 1\,, \tag{2.109}$$

$$\underline{A}_i \in \mathfrak{S}_0(\mathbb{R}^n);\ i = 1,\dots,n;\ \underline{A}_1 \supseteq \dots \supseteq \underline{A}_n \quad \Rightarrow \quad \lim_{n\to\infty} V(\underline{A}_n) = V(\lim_{n\to\infty} \underline{A}_n) = 0\,. \tag{2.110}$$

Furthermore, $V(\underline{A}_i)$ must be left-continuous for all $\underline{A}_i$

$$\lim_{\underline{D}_i \to \emptyset} V(\underline{A}_i \setminus \underline{D}_i) = V(\underline{A}_i) \quad \forall \underline{A}_i, \underline{D}_i \in \mathfrak{S}_0(\mathbb{R}^n) \mid \underline{D}_i \subseteq \underline{A}_i;\ \underline{D}_i \ne \emptyset. \tag{2.111}$$

When interpreting the $\underline{A}_i$ as events E_i, $\underline{A}_i$ may be replaced by $\underline{x}_i$. The distribution function is then described by

$$V(\underline{x}) = V(\underline{A}) \tag{2.112}$$

for all i without indexing. The association between $\underline{A} = \underline{A}_i$ and $\underline{x} = \underline{x}_i$ results from Eq. (2.106).

Based on the general definitions given above particular uncertain measures are considered in the subsequent sections.

2.2.2 Probability

The description of the probability measure relates to [10, 22, 40, 47, 60, 155, 212] and is based on the axiomatic characterization of probability after Kolmogorov.

Regarding crisp sets the *probability* $P(\underline{A}_i)$ represents an uncertain measure $M^u_{\mathfrak{S}}(\underline{A}_i)$ on $\mathfrak{S}(\underline{\mathbf{X}})$. $P(\underline{A}_i)$ meets the conditions in Eqs. (2.99) to (2.103). The requirement of monotonicity according to Eq. (2.101) is replaced by the stronger demand of additivity

$$\underline{A}_i \in \mathfrak{S}(\underline{\mathbf{X}});\ i = 1,\dots,n;\ \underline{A}_i \cap \underline{A}_k = \emptyset \ \forall i \ne k \quad \Rightarrow \quad P(\bigcup_i \underline{A}_i) = \sum_i P(\underline{A}_i). \tag{2.113}$$

When restricting to finite fundamental sets $\underline{\mathbf{X}}$, Eq. (2.113) may be expressed in a weakened manner

$$\underline{A}_1, \underline{A}_2 \in \mathfrak{S}(\underline{\mathbf{X}});\ \underline{A}_1 \cap \underline{A}_2 = \emptyset \quad \Rightarrow \quad P(\underline{A}_1 \cup \underline{A}_2) = P(\underline{A}_1) + P(\underline{A}_2). \tag{2.114}$$

Probability is a special uncertain measure that, additionally, meets the requirements stated for a crisp measure, as $P(\underline{A}_i)$ is finitely additive. The associated uncertain measure space $[\underline{\mathbf{X}}, \mathfrak{S}, M^u_{\mathfrak{S}}]$ is referred to as probability space $[\underline{\mathbf{X}}, \mathfrak{S}, P]$.

Frequently, the probability space is also specified by $[\Omega, \mathfrak{S}, P]$. Thereby Ω denotes the set of all possible results ω (elementary events) of a random trial that exclude each other. For a convenient evaluation, i.e., measurement of the events observed, Ω is mapped onto a suitable fundamental set $\underline{\mathbf{X}}$, here chosen to be $\underline{\mathbf{X}} = \mathbb{R}^n$. The elementary events $\omega \in \Omega$ lead to the realizations $\underline{x} \in \underline{\mathbf{X}}$ in $\mathbb{R}^n$. The fundamental set $\underline{\mathbf{X}}$ comprises all possible realizations that might be observed in consequence of the elementary events ω.

The result of the mapping

$$\underline{X}:\quad \Omega \rightarrow \underline{\mathbf{X}} = \mathbb{R}^n\,, \tag{2.115}$$

characterized by the real-valued function $\underline{X} = g(\omega)$ on Ω, which is measurable on $\mathfrak{S}(\underline{\mathbf{X}})$, is referred to as *random variable* or *random vector*. The functional values $\underline{X}$ thereby represent random values (realizations) $\underline{x} \in \underline{\mathbf{X}}$, which are assigned to the elementary events ω.

The probability measure $P(\underline{A}_i)$ is to be understood as a gradual assessment of the proposition $\underline{X} \in \underline{A}_i$. It expresses the probability with which $\underline{X}$ takes a value $\underline{x} \in \underline{\mathbf{X}}$ belonging to $\underline{A}_i$. Interpreting the $\underline{A}_i$ as events the distribution function for describing the probability on $\underline{\mathbf{X}}$ is defined according to Eqs. (2.107) to (2.112).

The *probability distribution function* $F(\underline{x})$ on $\underline{\mathbf{X}} = \mathbb{R}^n$ is a distribution function $V(\underline{x})$ that makes use of the uncertain measure *probability* $P(\underline{A}_i)$ to assess the (possible) events $\underline{A}_i$ from Eq. (2.106)

$$F\big(\underline{x} = (x_1, ..., x_n)\big) = P\big(\underline{X} = \underline{t} = (t_1, ..., t_n) \mid \underline{x}, \underline{t} \in \underline{\mathbf{X}} = \mathbb{R}^n;\, t_k < x_k;\, k = 1, ..., n\big). \tag{2.116}$$

An integrable function $f(\underline{t})$ with $\underline{t} = (t_1, ..., t_n) \in \underline{\mathbf{X}}$ that belongs to $F(\underline{x})$ is referred to as a *probability density function* if

$$F(\underline{x}) = \int_{t_1=-\infty}^{t_1=x_1} \cdots \int_{t_k=-\infty}^{t_k=x_k} \cdots \int_{t_n=-\infty}^{t_n=x_n} f(\underline{t})\, d\underline{t} \tag{2.117}$$

holds. Thereby $\underline{X}$ is presumed to be a continuous random variable. In the discrete case the integration is replaced by summation. Subsequently, $f(\underline{x})$ is used instead of $f(\underline{t})$.

Probability represents a special *objective* uncertain measure. When analyzing real data, the quality of inferences determined by probability is conditioned by statistical laws, which presuppose constant reproduction conditions and preferably large samples from the universe.

In the approach within the scope of this book probability density and probability distribution functions $f(\underline{x})$ and $F(\underline{x})$ on $\underline{\mathbf{X}}$ are exclusively generated on the basis of objective information, i.e., observations or measurements of objective physical parameters, subjective assessments are not taken into account. For determining the open set $\underline{A}_i$ according to Eq. (2.106) a value $\underline{x}_i$ is specified. The random vector $\underline{X}$ with its random realizations $\underline{t} = \underline{x}_j \in \underline{\mathbf{X}}$ is observed with regard to the random event $\underline{X} \in \underline{A}_i$: *the random vector* $\underline{X}$ *takes the value* $\underline{x}_j$ *for which every coordinate* $x_{j,k}$ *is less than the assigned coordinate* $x_{i,k}$ *of the specified value* $\underline{x}_i \in \underline{\mathbf{X}}$. Thereby all constraints that may influence the results of the "trial" must remain unchanged. Influences from an arbitrarily altered process situation cannot be considered. After the evaluation of a quasi-infinite number of sample elements $\underline{x}_j$, the functional value of the probability distribution function $F(\underline{x}_i) = P(\underline{X} \in \underline{A}_i)$ is known at the specified point $\underline{x}_i \in \mathbf{\underline{X}}$. This functional value $F(\underline{x}_i)$ indicates the probability with which the observed event $\underline{X} \in \underline{A}_i$ occurs in consequence of a random realization of $\underline{X}$ that has not yet been observed. The set of the functional values $F(\underline{x}_i)$ comprising all possible values $\underline{x}_i \in \mathbf{\underline{X}}$ that might be specified constitutes the sought probability distribution function $F(\underline{x}) = P(\underline{X} \in \underline{A}_i)$. This function assesses the membership of the elements $\underline{x}$ from the fundamental set $\mathbf{\underline{X}}$ in relation to the crisp sets $\underline{A}_i$ representing the events $\underline{X} \in \underline{A}_i$; it maps the fundamental set onto the interval [0, 1].

The probability measure must be extended if the fuzzy sets $\tilde{\underline{A}}_1, \ldots, \tilde{\underline{A}}_n$ characterized by their membership functions $\mu_{A_1}(\underline{x}), \ldots, \mu_{A_n}(\underline{x})$ belong to the family of sets being under consideration. When extended by the fuzzy sets $\tilde{\underline{A}}_1, \ldots, \tilde{\underline{A}}_n$, the family of sets $\mathfrak{S}(\mathbf{\underline{X}} = \mathbb{R}^n)$ no longer constitutes a Boolean σ-algebra. The fuzzy sets $\tilde{\underline{A}}_1, \ldots, \tilde{\underline{A}}_n$ are interpreted as being fuzzy events $\underline{X} \in \tilde{\underline{A}}_i$. For this purpose, it is presupposed by definition that the membership functions $\mu_{A_1}(\underline{x}), \ldots, \mu_{A_n}(\underline{x})$ must be Borel measurable [155, 212]. For measuring the $\tilde{\underline{A}}_1, \ldots, \tilde{\underline{A}}_n$ by means of the probability measure, two basic approaches may be pursued; the probability with which a fuzzy event occurs may either be computed as a scalar or as a fuzzy set [212].

The interpretation of the probability of a fuzzy event as a scalar value traces back to [206]. The measure value searched for is defined with the aid of the Lebesgue-Stieltjes integral

$$P(\tilde{\underline{A}}_i) = \int_{\mathbb{R}^n} \mu_{A_i}(\underline{x})\, dP(\underline{x}). \tag{2.118}$$

Applying the probability density function $f(\underline{x})$ on $\mathbf{\underline{X}}$ Eq. (2.118) gives

$$P(\tilde{\underline{A}}_i) = \int_{\underline{x}} \cdots \int \mu_{A_i}(\underline{x}) \cdot f(\underline{x})\, d\underline{x}. \tag{2.119}$$

The probability $P(\tilde{\underline{A}}_i)$ for the occurrence of a fuzzy event $\tilde{\underline{A}}_i$ defined by Eq. (2.118) represents a crisp number in the interval [0, 1].

An alternative approach for calculating the probability of a fuzzy event as a fuzzy set is proposed in [203]. The measure value searched for is determined by means of α-discretization of the $\tilde{\underline{A}}_1, \ldots, \tilde{\underline{A}}_n$. The event $\underline{X} \in \tilde{\underline{A}}_i$ is investigated for all α-levels, i.e., the probabilities $P(\underline{X} \in \underline{A}_{i,\alpha})$ concerning the events $\underline{X} \in \underline{A}_{i,\alpha}$ are computed. Each α-level set $\underline{A}_{i,\alpha}$ yields a crisp value for $P(\underline{X} \in \underline{A}_{i,\alpha})$. The set of these measure values for all α-levels together with the membership values $\mu = \alpha$ leads to the fuzzy probability

$$\begin{aligned}\tilde{P}(\tilde{\underline{A}}_i) = \{(P(\underline{A}_{i,\alpha}), \mu(P(\underline{A}_{i,\alpha}))) \mid P(\underline{A}_{i,\alpha}) = P(\underline{x} \in \underline{A}_{i,\alpha});\\ \mu(P(\underline{A}_{i,\alpha})) = \alpha\ \forall\ \alpha \in (0, 1]\}\end{aligned} \tag{2.120}$$

searched for. With the aid of the probability density function $f(\underline{x})$ on $\mathbf{\underline{X}}$

$$P(\underline{X} \in \underline{A}_{i,\alpha}) = \int_{\underline{x} \mid \underline{x} \in \underline{A}_{i,\alpha}} \cdots \int f(\underline{x})\, d\underline{x} \tag{2.121}$$

holds. The basic idea of this approach is used to introduce the fuzzy probability in Sect. 2.3.1.

Further treatments of the probability measure - like conditional probabilities or Bayes theorem - going beyond the fundamentals described above and concerning, e.g., dependencies between the events to be assessed are not basically considered within the scope of this book.

2.2.3 Other Uncertain Measures

According to the definition of uncertain measures, not only does probability meet the requirements in Eqs. (2.99) to (2.103), but also other assumptions may be established within the margins of this set of conditions. Many other uncertain measures, which exploit these boundary conditions, have already been derived and exist besides probability. Some of these uncertain measures are the *possibility*, the *necessity*, and other approaches expressing *plausibility*, *credibility*, or *belief*.

On the basis of [178] various uncertain measures on finite fundamental sets may be deduced by means of the following measure theoretical approach, see also [10, 46, 47, 155]. On a fundamental set **X** all subsets A_i, i.e., all elements of the power set $\mathfrak{P}(\mathbf{X})$, are assessed by distributing the "weight" w = 1 over the A_i. The sum of the "weights" of all A_i is thereby determined by

$$\sum_{i} \mathrm{w}(A_i) = 1. \tag{2.122}$$

This assessment may, e.g., describe the membership of an element x from the fundamental set **X** in relation to A_i. The sets A_i possessing a "weight" $\mathrm{w}(A_i) > 0$ are referred as to *focal subsets*. All focal sets together with their "weights" $\mathrm{w}(A_i)$ constitute the *body of evidence* as being the "entirety of all clues". Due to the requirement in Eq. (2.122) the assessment of the A_i is called the *basic probability assignment*. The term *probability* is thereby not linked to its definition according to the previous section. The focal subsets do not need to be disjointed nor need their union yield the fundamental set **X**.

In order to establish an uncertain measure $\mathrm{M}^{\mathrm{u}}_{\mathfrak{S}}(B_j)$, a family of sets $\mathfrak{S}(\mathbf{X})$ is defined that comprises all sets B_j that are to be measured. Considering the events $A_i \cap B_j = \varnothing$ the degree of *plausibility* $\mathrm{Pl}(B_j)$ may be determined, whereas analyzing $A_i \subseteq B_j$ leads to the degree of *credibility* $\mathrm{Cr}(B_j)$. In each case the measure values are calculated adding the "weights" of all focal subsets A_i being struck

$$\mathrm{Pl}(B_j) = \sum_{i} \mathrm{w}(A_i) \;\mid\; A_i \cap B_j \neq \varnothing\,, \tag{2.123}$$

$$\mathrm{Cr}(B_j) = \sum_{i} \mathrm{w}(A_i) \;\mid\; A_i \subseteq B_j\,. \tag{2.124}$$

The degrees of plausibility and credibility are "extreme" assessments of the sets B_j, they are linked by

$$\mathrm{Pl}(B_j) = 1 - \mathrm{Cr}(B_j^{\mathrm{C}})\,. \tag{2.125}$$

All other uncertain measures $M^u_{\mathfrak{S}}(B_j)$ that may be deduced from a basic probability assignment are bounded by $Pl(B_j)$ and $Cr(B_j)$

$$Cr(B_j) \le M^u_{\mathfrak{S}}(B_j) \le Pl(B_j) . \tag{2.126}$$

Particularly determining the focal subsets, the uncertain measure *probability* may be derived, for example. If the focal subsets are disjoint singletons, i.e., in the *dissonant* case, the uncertain measure $M^u_{\mathfrak{S}}(B_j)$ characterizes the probability measure

$$P(B_j) = Cr(B_j) = Pl(B_j) . \tag{2.127}$$

In this approach the focal subsets represent the elementary events.

The *possibility* being another uncertain measure of particular interest is obtained if the focal subsets form a nested sequence of sets specifying the *consonant* case

$$A_1 \subseteq A_2 \subseteq \dots \subseteq A_i \subseteq \dots \subseteq A_n . \tag{2.128}$$

The possibility measure then characterizes a special case of the degree of plausibility

$$\Pi(B_j) = Pl(B_j) . \tag{2.129}$$

The class of *λ-uncertain measures* $M^u_{\mathfrak{S},\lambda}(B_j)$ introduced in [186] forms a subset of the uncertain measures that may be deduced from a basic probability assignment [10, 155]. The $M^u_{\mathfrak{S},\lambda}(B_j)$ are considered advantageous regarding both their flexibility when being adapted to basic requirements and the simple treatment of their formulas [10]. The requirement of monotonicity according to Eq. (2.101) is expressed as a "restricted additivity requirement" for λ-uncertain measures. Considering two disjoint elements B_1 and B_2 ($B_1 \cap B_2 = \varnothing$) of the family of sets $\mathfrak{S}(\mathbf{X})$ the rule

$$M^u_{\mathfrak{S},\lambda}(B_1 \cup B_2) = M^u_{\mathfrak{S},\lambda}(B_1) + M^u_{\mathfrak{S},\lambda}(B_2) + \lambda \cdot M^u_{\mathfrak{S},\lambda}(B_1) \cdot M^u_{\mathfrak{S},\lambda}(B_2) \tag{2.130}$$

is suggested with $\lambda > -1$. For $\lambda = 0$ $M^u_{\mathfrak{S},\lambda}(B_j)$ characterizes the probability measure and Eq. (2.130) then coincides with Eq. (2.114). In contrast to the probability, the possibility measure does not belong to the λ-uncertain measures.

Owing to the importance of the possibility measure and its close relationship to fuzzy set theory some pertinent basics are discussed in the subsequent paragraphs and compared to the probability measure.

Possibility. The *possibility measure* is introduced on the basis of the approaches in [10, 22, 47, 155, 212]. Its relationship to fuzzy set theory is thereby pointed out.

The possibility $\Pi(\underline{A}_i)$ is an uncertain measure $M^u_{\mathfrak{S}}(\underline{A}_i)$ on $\mathfrak{S}(\underline{\mathbf{X}})$ in accordance with Eqs. (2.99) to (2.103), which may be used to measure crisp sets. The monotonicity requirement in Eq. (2.101) is here substituted by the stronger demand for the supremum

$$\underline{A}_i \in \mathfrak{S}(\underline{\mathbf{X}});\ i = 1, \dots, n \quad \Rightarrow \quad \Pi(\bigcup_i \underline{A}_i) = \sup_i \Pi(\underline{A}_i) . \tag{2.131}$$

Regarding finite fundamental sets $\underline{\mathbf{X}}$ this condition may be weakened to

$$\underline{A}_1, \underline{A}_2 \in \mathfrak{S}(\underline{\mathbf{X}}) \;\Rightarrow\; \Pi(\underline{A}_1 \cup \underline{A}_2) = \max[\Pi(\underline{A}_1), \Pi(\underline{A}_2)]. \tag{2.132}$$

$\Pi(\underline{A}_i)$ is not additive. The continuity in the sense of Eqs. (2.102) and (2.103) is only for finite sets strictly complied with [149]. Equations (2.99) and (2.100) are always satisfied. The possibility thus represents a special uncertain measure, the assigned uncertain measure space is denoted by $[\underline{\mathbf{X}}, \mathfrak{S}, \Pi]$.

The possibility $\Pi(\underline{A}_i)$ characterizes the gradual assessment of the proposition $\tilde{\underline{x}} \subseteq \underline{A}_i$. $\Pi(\underline{A}_i)$ expresses the possibility with which $\tilde{\underline{x}}$ belongs to $\underline{A}_i$ or – according to the set theoretical interpretation – the possibility with which the values $\underline{x} \in \tilde{\underline{x}}$ simultaneously belong to $\underline{A}_i$. The possibility measure denotes the subjective assessment of the possibility of the occurrence of an event, i.e., the *epistemic* possibility, but not the objective physical possibility.

As already demonstrated for the probability measure, possibility density and possibility distribution functions may be defined. The crisp sets $\underline{A}_i$ according to Eq. (2.106) are interpreted as being events and the distribution function describing the possibility on $\underline{\mathbf{X}}$ is established starting from its general definition.

The *possibility distribution function* $\pi(\underline{x})$ on $\underline{\mathbf{X}} = \mathbb{R}^n$ is a distribution function $V(\underline{x})$ that assesses the possible events $\underline{A}_i$ according to Eq. (2.106) with the aid of the special uncertain measure *possibility* $\Pi(\underline{A}_i)$

$$\pi\big(\underline{x} = (x_1, \ldots, x_n)\big) = \Pi\big(\tilde{\underline{x}} = \underline{t} = (t_1, \ldots, t_n) \mid \underline{x}, \underline{t} \in \underline{\mathbf{X}} = \mathbb{R}^n;\, t_k < x_k;\, k = 1, \ldots, n\big). \tag{2.133}$$

The *possibility density function* assigned to $\pi(\underline{x})$ is denoted by $\bar{\mu}(\underline{x})$. With $\bar{\mu}(\underline{x}) = \bar{\mu}(\underline{t})$ the relationship

$$\pi(\underline{x}) = \sup_{t_k < x_k;\, k = 1, \ldots, n} \bar{\mu}(\underline{t}) \tag{2.134}$$

holds.

Both a possibility density function $\bar{\mu}(x)$ and a possibility distribution function $\pi(x)$ are shown in Fig. 2.35 for the one-dimensional case.

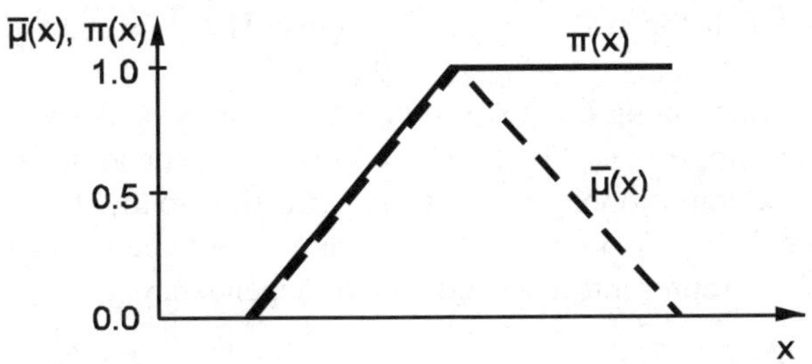

Fig. 2.35. Possibility density function $\bar{\mu}(x)$, assigned possibility distribution function $\pi(x)$

The definition of the normalized membership function $\mu(\underline{x})$ of a fuzzy set $\tilde{\underline{A}}_i$ (Sect. 2.2.1) permits its direct interpretation as a possibility density function

$$\bar{\mu}(\underline{x}) = \mu(\underline{x}). \tag{2.135}$$

For this purpose, the fuzzy set $\tilde{\underline{A}}_i$ is considered an uncertain constraint for the values $\underline{x}$ [10]. The membership value $\mu(\underline{x} = \underline{x}_j)$ indicates the degree of possibility with which the element $\underline{x}_j$ satisfies the uncertain proposition belonging to $\tilde{\underline{A}}_i$. The membership values "induce" the possibility values. The fuzzy sets $\tilde{\underline{A}}_i$ assessed by means of possibility are denoted by $\tilde{\underline{X}}$.

The possibility distribution function $\pi(\underline{x}) = \Pi(\tilde{\underline{X}} < \underline{x})$ assesses the fuzzy vector $\tilde{\underline{X}}$ regarding the event $\tilde{\underline{X}} < \underline{x}$. It indicates the possibility with which a value $\underline{x}_j$ that has not yet been observed occurs in such a way that every coordinate $x_{j,k}$ is less than the assigned coordinate x_k of the specified value $\underline{x} \in \underline{\mathbf{X}}$. The functional values $\pi(\underline{x})$ are a measure for the membership of the elements $\underline{x}$ from the fundamental set $\underline{\mathbf{X}}$ in relation to the crisp sets $\underline{A}_i$ interpreted as being events $\tilde{\underline{X}} < \underline{x}$, whereby $\underline{A}_i \in \mathbf{X}$ holds. The possibility distribution function $\pi(\underline{x})$ maps the fundamental set $\underline{\mathbf{X}}$ onto the interval [0, 1].

The possibility is a special uncertain measure that may not be linked to statistical approaches. Regarding reproduction conditions or repeatability of trials no demands are stated. Based on possibility theory, in addition to objective physical parameters, also subjective assessments by experts may be taken into account. Regarding samples of measured or observed parameter values – in contrast to the probability measure considered here – no requirements exist for establishing possibility density and possibility distribution functions. Objective information is evaluated with the aid of the possibility measure whereby expert knowledge is incorporated. Linguistic assessments, e.g., concerning the quality of a construction: *very good*, *good*, *satisfying*, *suboptimal*, *bad* may be included in the evaluation.

As pointed out for the probability, the possibility measure must also be extended if the fuzzy sets $\tilde{\underline{A}}_1, \ldots, \tilde{\underline{A}}_n$ belong to the family of sets considered. The fuzzy sets $\tilde{\underline{A}}_1, \ldots, \tilde{\underline{A}}_n$ are interpreted as being fuzzy events $\tilde{E}_i$: $\underline{x} \in \tilde{\underline{A}}_i$ with $\underline{x} \in \tilde{\underline{X}}$. Based on a given possibility density function $\bar{\mu}(\underline{x})$ on $\underline{\mathbf{X}}$,

$$\Pi(\tilde{\underline{A}}_i) = \sup_{\underline{x}} \min\left[\mu_{A_i}(\underline{x}), \bar{\mu}(\underline{x})\right] \tag{2.136}$$

denotes the possibility with which the fuzzy event $\tilde{E}_i$ occurs. $\Pi(\tilde{\underline{A}}_i)$ is a crisp number in the interval [0, 1].

Further relationships for characterizing the possibility measure may be derived as analogs to the probabilistic approach. For taking account of dependencies between events, for example, the conditional possibility and an analog to the Bayes theorem have been introduced [10]. Moreover, the interaction between fuzzy variables dealt with in Sect. 2.1.8 represents a special type of dependency.

Comparison of Probability and Possibility. The basic difference between probability and possibility arises from different concrete realizations of the monotonicity requirement according to Eq. (2.101). The logical operations $\wedge$ and $\vee$ for combining events E_i are realized on the basis of different arithmetic operations between the assigned measure values $P(E_i)$ and $\Pi(E_i)$. Whereas, concerning the probability measure the operation $\vee$ is accomplished by adding the measure values $P(E_i)$ for mutually disjoint events E_i, see Eq. (2.113), the operator *sup* is adopted for linking the possibilities $\Pi(E_i)$, see Eq. (2.131). In probabilistics the operation $\wedge$ is represented by the multiplication of the measure values $P(E_i)$ for independent events E_i, but in possibilistics

$$\Pi(E_1 \wedge E_2) = \min[\Pi(E_1), \Pi(E_2)] \tag{2.137}$$

holds for noninteractive events E_i and

$$\Pi(\underline{A}_1 \cap \underline{A}_2) = \min[\Pi(\underline{A}_1), \Pi(\underline{A}_2)] \tag{2.138}$$

applies accordingly to noninteractive sets $\underline{A}_i$ if these are interpreted as events. The operator *min* is thereby used in accordance with the Cartesian product between fuzzy sets.

For comparing the uncertain measures probability and possibility, the event E_i is considered: *the value* $\underline{x}$ *may belong to the set* $\underline{A}_i$. The measure values $P(\underline{A}_i)$ and $\Pi(\underline{A}_i)$ characterizing the membership of $\underline{x}$ in relation to $\underline{A}_i$ are compared by means of the demand in Eq. (2.104). Applying Eqs. (2.131) and (2.132) the smallest possible measure value $M^u_{\circledast}(\underline{A}_i) = \Pi(\underline{A}_i)$ within the bounds of Eq. (2.104) is determined. Every other measure value $M^u_{\circledast}(\underline{A}_i)$ satisfying Eq. (2.104) must be at least equal to $\Pi(\underline{A}_i)$

$$M^u_{\circledast}(\underline{A}_i) \geq \Pi(\underline{A}_i). \tag{2.139}$$

The possibility measure thus represents the weakest proposition regarding the realization of an event. The possibility concerning the occurrence of the event E_i represents the gradual assessment of the proposition: *it may happen that* E_i *occurs*. In contrast to this, the probability regarding the occurrence of the event E_i assesses the proposition: *the event* E_i *occurs* in the statistical mean. From this it follows that a probable event with $P(\underline{A}_i) > 0$ is also possible with $\Pi(\underline{A}_i) > 0$ in each case, but a possible event need not be necessarily probable too.

With the aid of the degrees of *credibility* and *plausibility* [10, 47, 155] it can be inferred that the probabilities $P(\underline{A}_i)$ may be considered lower bounds with regard to the possibilities $\Pi(\underline{A}_i)$

$$\Pi(\underline{A}_i) \geq P(\underline{A}_i); \quad \forall \underline{A}_i \in \mathfrak{S}(\mathbf{\underline{X}}). \tag{2.140}$$

The functional values of probability distribution functions $F(\underline{x})$ and possibility distribution functions $\pi(\underline{x})$ are thus associated via the relation

$$\pi(\underline{x}) \geq F(\underline{x}); \ \forall \underline{x} \in \mathbf{\underline{X}}. \tag{2.141}$$

From Eqs. (2.140) and (2.141) results a weak monotonicity relationship between the uncertain measures probability and possibility

$$\Pi(\underline{A}_1) \geq \Pi(\underline{A}_2) \Rightarrow P(\underline{A}_1) \geq P(\underline{A}_2); \quad \forall \underline{A}_1, \underline{A}_2 \in \mathfrak{S}(\mathbf{\underline{X}}). \tag{2.142}$$

Equation (2.140) together with the probabilistic complementary relationship

$$P(\underline{A}_i) + P(\underline{A}_i^C) = 1; \quad \forall \underline{A}_i, \underline{A}_i^C \in \mathfrak{S}(\mathbf{\underline{X}}) \tag{2.143}$$

yields the weak inequality

$$\Pi(\underline{A}_i) + \Pi(\underline{A}_i^C) \geq 1; \quad \forall \underline{A}_i, \underline{A}_i^C \in \mathfrak{S}(\mathbf{X}) \tag{2.144}$$

for the possibility measure.

Moreover, the characteristic of the possibility measure stated in Eq. (2.144) directly results from Eqs. (2.125), (2.126) and (2.129). Additionally, taking account of Eq. (2.127) also explains the relationship between the uncertain measures probability and possibility in Eq. (2.140).

2.2.4 Measures of Uncertainty

In fuzzy set theory, in probabilistics, in possibilistics, and in the theory of fuzzy random variables considered subsequently *local uncertainty* is assessed when specifying sets, i.e., the focus is set on the uncertainty concerning the assignment of particular elements from the fundamental set to the set of interest. How uncertain the set appears as a whole, i.e., how extensive the global uncertainty of the set is, has not yet been assessed with the approaches considered so far. For assessing this global uncertainty, however, numerous approaches are available, a variety of which is presented, e.g., in [139]. Further basic explanations may be found, e.g., in [10, 22].

A *measure of uncertainty* may be defined as a special set function that represents a crisp measure according to Sect. 2.2.1 that assesses the global uncertainty of subsets from a family of sets. It permits comparison of the measured subsets with regard to their uncertainty.

The various measures of uncertainty differ from each other according to

- The family of sets that represents the reference point for the assessment
- The specification of the most uncertain set, and
- The manner in which the sets are compared regarding their uncertainty

The most important measure systems concerning uncertainty are represented by entropy measures and energy measures. In this book, only the entropy measure after Shannon is applied to the one-dimensional case.

Entropy Measure after Shannon. Shannon's entropy measure has been derived to quantify information [60], it stems from information theory. Information comprises the elements x_i selected from a declared character set representing the fundamental set **X**. The interpretation of these elements as being events permits the assignment of probabilities $P(x_i)$ to the x_i and the generation of information may thus be described as a random process. The entropy after Shannon is stated as being an information measure based on the assessment of possible binary decisions.

The *Shannon entropy* H represents the average information content described by a number of binary decisions required for characterizing a state if the n objectively, physically possible elements of the declared character set possess – also different – probabilities. Considering discrete information

$$H = -\sum_{i=1}^{n} P(x_i) \cdot \mathrm{ld}\left(P(x_i)\right) \tag{2.145}$$

holds and in the continuous case

$$H = -\int_{x=-\infty}^{x=+\infty} f(x) \cdot \mathrm{ld}\left(f(x)\right) dx \tag{2.146}$$

applies. The unit of Shannon's entropy is the bit (binary digit).

A plausible explanation of the axiomatically determined entropy measure after Shannon, which addresses an intuitive imagination of the information content, is discussed in [139].

Modification of Shannon's Entropy Measure. On the basis of the Shannon entropy the uncertainty of a set can be measured. The elements of the declared character set are interpreted as elements of an uncertain set, their membership in relation to the considered set is indicated by an uncertain measure.

For assessing the fuzziness of the fuzzy set $\tilde{A}_i \subseteq \mathbf{X}$, the functional values of the membership function $\mu(x)$ of $\tilde{A}_i$ are applied as measure values of the elements. The dyadic logarithm in Eqs. (2.145) and (2.146) is transformed into the natural logarithm

$$\mathrm{ld}\left(\mu(x)\right) = k \cdot \ln\left(\mu(x)\right);\ k = \frac{1}{\ln(2)} \tag{2.147}$$

and the parameter k is released with $k > 0$. Additionally, the suitable requirement that the uncertainty of the fuzzy set $\tilde{A}_i$ must coincide with the uncertainty of its complementary set $\tilde{A}_i^C$ is introduced. Therewith, the definition of Shannon's entropy for measuring fuzzy sets is obtained, as presented in [10] and [22] for the discrete case.

Shannon's uncertainty measure of the discrete fuzzy set $\tilde{A}_i$ is defined by

$$H_u(\tilde{A}_i) = -k \cdot \sum_{i=1}^{n} \left[\mu(x_i) \cdot \ln\left(\mu(x_i)\right) + \left(1 - \mu(x_i)\right) \cdot \ln\left(1 - \mu(x_i)\right)\right], \tag{2.148}$$

and for the continuous fuzzy set $\tilde{A}_i$

$$H_u(\tilde{A}_i) = -k \cdot \int_{x=-\infty}^{x=+\infty} \left[\mu(x) \cdot \ln\left(\mu(x)\right) + \left(1 - \mu(x)\right) \cdot \ln\left(1 - \mu(x)\right)\right] dx \tag{2.149}$$

holds.

The Shannon entropy represents a measure characterizing the "steepness" of the membership function $\mu(x)$. When assessing the crisp set A_i the measure value

$H_u(A_i) = 0$ is obtained. The most uncertain set $\tilde{A}_i$ characterized by $\mu(x) = 0.50$ for all $x \in \tilde{A}_i$ except at the mean value yields the maximal measure value $H_u(\tilde{A}_i) = H_{max}$. With increasing conformity of the fuzzy set $\tilde{A}_i$ with the crisp set A_i the uncertainty measure $H_u(\tilde{A}_i)$ converges towards zero. The better the conformity of $\tilde{A}_i$ with the most uncertain fuzzy set is, the greater $H_u(\tilde{A}_i)$ becomes.

For measuring multidimensional fuzzy sets, Eqs. (2.148) and (2.149) may be easily extended to these cases.

2.3 Fuzzy Randomness

When considering major problems in engineering sciences, like damage processes, risk assessments, lifetime prognoses for structures, or loading in consequence of natural incidents, uncertain data are present. The classical modeling of these data using random variables or random processes is problematic if the requirements regarding this modeling are not complied with, e.g., if the reproduction conditions do not remain constant during the generation of sample elements. Such uncertain data possess further uncertainty in addition to their random properties. If this additional uncertainty is interpreted as being fuzziness and associated with the simultaneously existing randomness, fuzzy randomness arises.

Definitions and basic terms concerning fuzzy randomness have been introduced by Kwakernaak [96, 97] and enhanced, e.g., by Puri and Ralescu [150], Wang and Zhang [198], Bandemer and Näther [11], Viertl [194], Näther and Körner [134], and Krätschmer [90]. The formal characterization of fuzzy randomness chosen by these authors, however, only barely represents a suitable basis for numerical simulations in engineering sciences.

On the basis of α-discretization and with the aid of fuzzy probability distribution functions, a more suitable representation of fuzzy randomness may be gained in view of numerical simulations. In Sects. 2.3.1 and 2.3.2 the formal characterization of fuzzy random vectors and fuzzy random functions is addressed. Their application is demonstrated in Chap. 5 for fuzzy stochastic structural analysis and in Chap. 6 for fuzzy probabilistic safety assessment.

In contrast to the descriptions regarding fuzziness in Sect. 2.1, the uncertainty characteristic fuzzy randomness is directly presented for the multidimensional case.

2.3.1 Fuzzy Random Vectors

2.3.1.1 Definition of Fuzzy Random Vectors

In accordance with probability theory, the space of the random elementary events Ω is introduced. Instead of a real realization, a fuzzy realization of the form $\underline{\tilde{x}}(\omega) = (\tilde{x}_1, ..., \tilde{x}_n) \subseteq \mathbf{\underline{X}}$ is now assigned to each elementary event $\omega \in \Omega$. The n-tuple $\underline{\tilde{x}}(\omega)$ is constituted from the n fuzzy numbers $\tilde{x}_1, ..., \tilde{x}_n$ on the fundamental set $\underline{\mathbf{X}} = \mathbb{R}^n$. Each fuzzy number is defined as a convex, normalized fuzzy set

$$\tilde{x}_i = \{(x, \mu_{x_i}(x)) \mid x \in X\}, \tag{2.150}$$

whose membership function is at least segmentally continuous and possesses the functional value $\mu(x) = 1$ at precisely one value x. The fuzzy realizations $\tilde{x}_1 = \tilde{x}(\omega_1), ..., \tilde{x}_6 = \tilde{x}(\omega_6)$ on $\mathbb{R}^1$ resulting from six elementary events are illustrated in Fig. 2.36.

A *fuzzy random variable or fuzzy random vector* $\underline{\tilde{X}}$ is the fuzzy result of the uncertain mapping

$$\underline{\tilde{X}}: \Omega \tilde{\rightarrow} \mathbf{F}(\underline{\mathbf{X}}), \tag{2.151}$$

where $\mathbf{F}(\underline{\mathbf{X}})$ is the set of all fuzzy numbers on $\mathbb{R}^n$. With Eq. (2.151) a vector is defined whose elements possess the property fuzzy randomness.

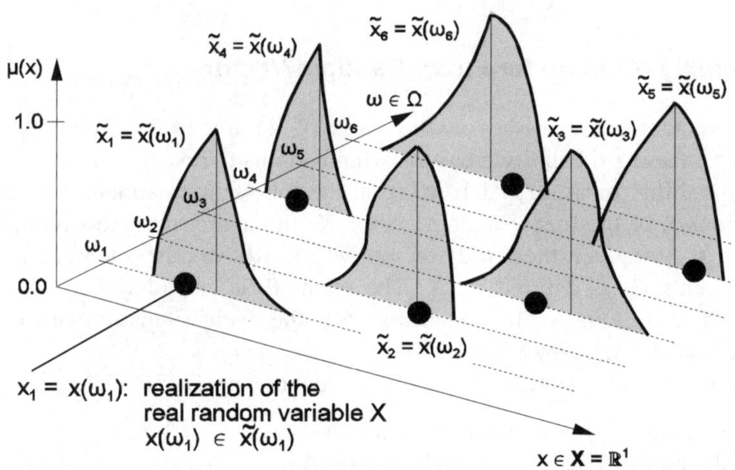

Fig. 2.36. Realizations of a one-dimensional fuzzy random variable

If the realization $\underline{x}$ of a real random vector $\underline{X}$ as well as the fuzzy realization $\underline{\tilde{x}}$ of a fuzzy random vector $\underline{\tilde{X}}$ may be assigned to an elementary event ω, and if $\underline{x} \in \underline{\tilde{x}}$ holds, this means that $\underline{x}$ is contained in $\underline{\tilde{x}}$. If, for all elementary events $\omega \in \Omega$, the $\underline{x}$ are contained in the $\underline{\tilde{x}}$, the $\underline{x}$ then constitute an *original* $\underline{X} = \underline{X}_j$ of the fuzzy random vector $\underline{\tilde{X}}$. The original $\underline{X}_j$ is referred to as *completely* contained in $\underline{\tilde{X}}$. Each real random vector $\underline{X}$ (without fuzziness) on $\underline{\mathbf{X}}$ that is completely contained in $\underline{\tilde{X}}$ is thus an original $\underline{X}_j$ of $\underline{\tilde{X}}$, i.e., in the Ω-direction each original $\underline{X}_j$ must be consistent with the fuzziness of $\underline{\tilde{X}}$. This means that the fuzzy random vector $\underline{\tilde{X}}$ is the *fuzzy set of all possible originals* $\underline{X}_j$ contained in $\underline{\tilde{X}}$. Fuzzy random vectors may be continuous or discrete with regard to both their randomness and fuzziness. The following considerations are primarily devoted to the continuous case as the others may easily be derived from this. Realizations of the real random variable X representing an original X_j of $\tilde{X}$ are displayed as black points in Fig. 2.36, for these $x(\omega_i) \in \tilde{x}(\omega_i)$ holds.

Each fuzzy random vector $\underline{\tilde{X}}$ contains at least one real random vector $\underline{X}$ as an original $\underline{X}_j$ of $\underline{\tilde{X}}$. It thus follows that each fuzzy random vector $\underline{\tilde{X}}$ that only possesses precisely one original is a real random vector $\underline{X}$. The description of fuzzy random vectors by means of their originals ensures that real random vectors are contained in fuzzy random vectors as a special case. As each realization $\underline{\tilde{x}}$ of the fuzzy random vector $\underline{\tilde{X}}$ according to Eq. (2.151) is a fuzzy number, the special case of a real random vector is uniquely defined by the mean values of the fuzzy realizations $\underline{\tilde{x}}$ (membership level $\mu = 1$). By this means it is possible to take account of random vectors and fuzzy random vectors simultaneously.

The fuzzy random vector is referred to as *normalized* if each realization $\underline{\tilde{x}}$ of $\underline{\tilde{X}}$ represents a normalized fuzzy variable.

As in Eq. (2.151) $\mathbf{F}(\underline{\mathbf{X}})$ represents the set of all fuzzy numbers on $\mathbb{R}^n$, it is ensured that the fuzzy random vector $\underline{\tilde{X}}$ is normalized.

2.3.1.2 Probability Measure for Fuzzy Random Vectors

Fuzzy random vectors are assessed on the basis of an uncertain probability measure, namely a fuzzy probability measure, which admits to reveal fuzziness.

The fuzzy probability measure is derived from the following treatment: Given are the realizations $\underline{\tilde{x}}$ of the fuzzy random vector $\underline{\tilde{X}}$; the $\underline{\tilde{x}}$ are fuzzy sets on the fundamental set $\underline{\mathbf{X}}$. Moreover, the family of sets $\mathfrak{M}(\underline{\mathbf{X}})$, whose crisp elements $\underline{A}_i$ are treated as events E_i, is defined on $\underline{\mathbf{X}}$. The event E_i is considered to have occurred when $\underline{\tilde{X}} \in \underline{A}_i$. Due to the fuzziness of $\underline{\tilde{x}}$ the event E_i also possesses fuzziness, this becomes the fuzzy event $\tilde{E}_i$.

Considering $\underline{\tilde{X}} \in \underline{A}_i$ three states are taken into account (Fig. 2.37):

- The fuzzy realization $\underline{\tilde{x}}_1$ lies completely inside the set $\underline{A}_i$.
- The fuzzy realization $\underline{\tilde{x}}_2$ lies only partially inside $\underline{A}_i$.
- The fuzzy realization $\underline{\tilde{x}}_3$ lies completely outside the set $\underline{A}_i$.

From this it may be inferred that the probability $P(\underline{\tilde{X}} \in \underline{A}_i)$ does not merely represent a crisp value, but rather a set of probabilities. Furthermore, it is obvious that the degree of the partial occurrence of $\underline{\tilde{X}} \in \underline{A}_i$ may be assessed by membership values induced by the fuzziness of $\underline{\tilde{X}}$.

The fuzzy probability $\tilde{P}(\underline{A}_i)$ *is the set of all probabilities* $P(\underline{\tilde{X}} \in \underline{A}_i)$ *with the corresponding membership values* $\mu(P(\underline{\tilde{X}} \in \underline{A}_i))$, *which takes into account all states of the occurrence of* $\underline{\tilde{X}} \in \underline{A}_i$.

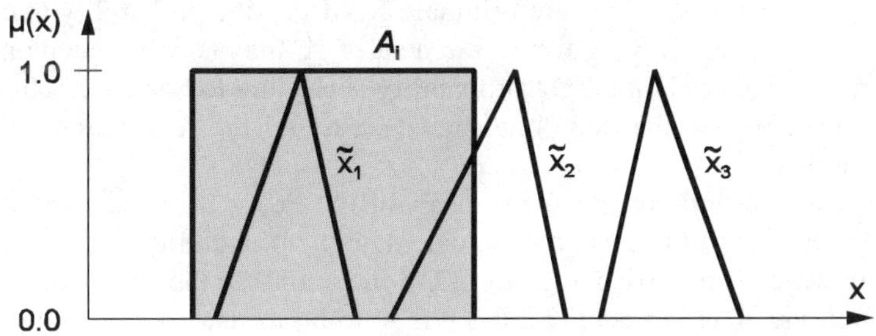

Fig 2.37. Interpretation of $\tilde{\underline{X}} \in \underline{A}_i$ for the one-dimensional case

The probability space $[\underline{X}, \mathfrak{S}, P]$ belonging to the real-valued probability is extended by the membership values $\mu(P(\tilde{\underline{X}} \in \underline{A}_i))$ to constitute the fuzzy probability space $[\underline{X}, \mathfrak{S}, P, \mu]$ or $[\underline{X}, \mathfrak{S}, \tilde{P}]$.

Based on α-discretization the determination of $\tilde{P}(\underline{A}_i)$ is accomplished by calculating a set of real-valued probabilities in the probability space $[\underline{X}, \mathfrak{S}, P]$ on each α-level. For this purpose the fuzzy random variable $\tilde{\underline{X}}$ is also subdivided into α-level sets

$$\underline{X}_\alpha = \left\{ \underline{X} = \underline{X}_j \mid \mu(\underline{X}_j) \geq \alpha \right\}. \tag{2.152}$$

The $\underline{X}_\alpha$ are crisp random sets. In the one-dimensional case a closed random interval $[X_{\alpha\,l}, X_{\alpha\,r}]$ is obtained.

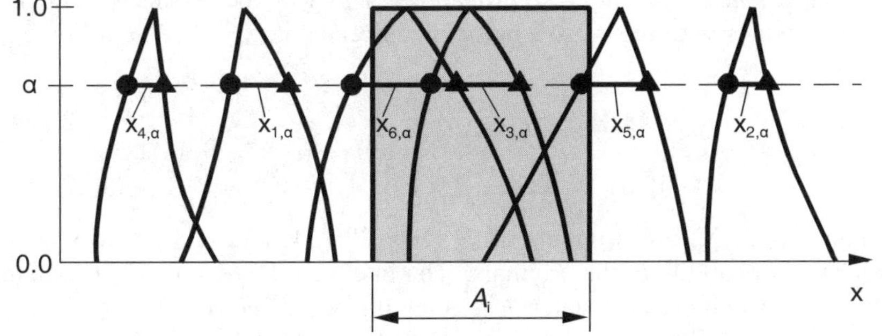

Fig. 2.38. Crisp set A_i and realizations $\tilde{x}$ of the fuzzy random variable $\tilde{\underline{X}}$ by applying α-discretization

As all elements of $\underline{X}_\alpha$ or $[X_{\alpha l}, X_{\alpha r}]$ are originals $\underline{X}_j$ of $\tilde{\underline{X}}$, the probability that the element $\underline{X}_j$ of the α-level set $\underline{X}_\alpha$ is also an element of $\underline{A}_i$ may now be stated on each α-level using the uncertain measure probability according to Sect. 2.2. This is illustrated in Fig. 2.38 for the one-dimensional case and the realizations of the originals $X_{\alpha l}$ and $X_{\alpha r}$.

The approach for determining the fuzzy probability $\tilde{P}(\underline{A}_i)$ by means of α-discretization corresponds to the basic idea to calculate the probability of a fuzzy event as fuzzy set, which is proposed, e.g., in [203]. In contrast to the extension of the probability measure stated in Sect. 2.2 the sets $\underline{A}_i$ to be measured now do not represent fuzzy sets, but the realizations $\tilde{\underline{x}}$ of the fuzzy random vector $\tilde{\underline{X}}$ are characterized by fuzziness.

The probability with which elements of a random, not yet observed, realization of the fuzzy random vector $\tilde{\underline{X}}$ and elements of a crisp set $\underline{A}_i$ – defined on the fundamental set $\mathbf{\underline{X}}$ – coincide is referred to as the *fuzzy probability* $\tilde{P}(\underline{A}_i)$. As the realizations of fuzzy random vectors are formed by fuzzy numbers, the fuzzy probability always represent convex fuzzy sets. It may thus be calculated by means of α-discretization without any requirements

$$\tilde{P}(\underline{A}_i) = \{(P_\alpha(\underline{A}_i), \mu(P_\alpha(\underline{A}_i))) \mid P_\alpha(\underline{A}_i) = [P_{\alpha l}(\underline{A}_i), P_{\alpha r}(\underline{A}_i)]; \; \mu(P_\alpha(\underline{A}_i)) = \alpha \,\forall\, \alpha \in (0, 1]\}. \tag{2.153}$$

The right-hand side of Eq. (2.153) is evaluated for each α-level with the aid of the uncertain measure probability according to Sect. 2.2. The bounds $P_{\alpha l}(\underline{A}_i)$ and $P_{\alpha r}(\underline{A}_i)$ of the α-level sets $P_\alpha(\underline{A}_i)$ are obtained by assessing the events

- E_1: "$\underline{X}_\alpha$ is contained in $\underline{A}_i$: $\underline{X}_\alpha \subseteq \underline{A}_i$", and
- E_2: "$\underline{X}_\alpha$ and $\underline{A}_i$ possess at least one common element: $\underline{X}_\alpha \cap \underline{A}_i \neq \emptyset$"

These two events are "extreme" interpretations of the proposition $\tilde{\underline{X}} \in \underline{A}_i$, the basic idea of which is also used for bounding probability by other uncertain measures, Eqs. (2.123) and (2.124). The event E_1 yields the least probability $P_{\alpha l}(\underline{A}_i)$, whereas the event E_2 yields the highest probability $P_{\alpha r}(\underline{A}_i)$. The events E_1 and E_2 characterize bounds with regard to a partial occurrence of $\tilde{\underline{X}} \in \underline{A}_i$. The following holds

$$P_{\alpha l}(\underline{A}_i) = P(\underline{X}_\alpha \subseteq \underline{A}_i), \tag{2.154}$$

$$P_{\alpha r}(\underline{A}_i) = P(\underline{X}_\alpha \cap \underline{A}_i \neq \emptyset). \tag{2.155}$$

As *all* elements of $\underline{X}_\alpha$ are originals of $\tilde{\underline{X}}$, Eqs. (2.154) and (2.155) also yield bounds for the probability of the originals. The intervals $[P_{\alpha l}(\underline{A}_i), P_{\alpha r}(\underline{A}_i)]$ contain the probabilities of all possible states describing the occurrence of $\tilde{\underline{X}} \in \underline{A}_i$.

The fuzzy probability $\tilde{P}(\underline{A}_i)$ thus represents the set of the uncertain measures $P_j(\underline{A}_i)$ for all originals $\underline{X}_j$ of $\tilde{\underline{X}}$ with the membership values $\mu(P_j(\underline{A}_i))$. As the dimension fuzziness of the fuzzy probability space $[\mathbf{\underline{X}}, \mathfrak{S}, \tilde{P}]$ is decomposed by α-discretization, for each original of $\tilde{\underline{X}}$ the probability measure in the probability space $[\mathbf{\underline{X}}, \mathfrak{S}, P]$ according to Sect. 2.2 may be computed.

For the special case of real random vectors $\underline{X}$, both $\underline{X}_\alpha \cap \underline{A}_i \neq \varnothing$ and $\underline{X}_\alpha \subseteq \underline{A}_i$ reduce to $\underline{X} \in \underline{A}_i$, i.e., $P_{\alpha\,l}(\underline{A}_i)$ and $P_{\alpha\,r}(\underline{A}_i)$ coincide; the event $\underline{X} \in \underline{A}_i$ cannot occur partially.

The partial occurrence of $\tilde{\underline{X}} \in \underline{A}_i$ is induced by the fuzziness of $\tilde{\underline{X}}$. The evaluation of all states, for which $\underline{X}_\alpha \cap \underline{A}_i \neq \varnothing$ is satisfied, but for which $\underline{X}_\alpha \subseteq \underline{A}_i$ does not apply, yields the probability shadow, which exclusively contains the fuzziness of the event $\tilde{\underline{X}} \in \underline{A}_i$.

The *probability shadow* $\tilde{P}_S(\underline{A}_i)$ is the probability with which the event $\tilde{\underline{X}} \in \underline{A}_i$ exclusively partially occurs, it characterizes the fuzziness of $\tilde{P}(\underline{A}_i)$ in Eq. (2.153)

$$\tilde{P}_S(\underline{A}_i) = \{(P_{S,\alpha}(\underline{A}_i), \mu(P_{S,\alpha}(\underline{A}_i))) \mid P_{S,\alpha}(\underline{A}_i) = [0, P_{S,\alpha\,r}(\underline{A}_i)]; \\ \mu(P_{S,\alpha}(\underline{A}_i)) = \alpha \;\forall\; \alpha \in (0,1]\}. \tag{2.156}$$

The α-level sets $P_{S,\alpha}(\underline{A}_i)$ are determined from

$$P_{S,\alpha\,r}(\underline{A}_i) = P(\underline{X}_\alpha \cap \underline{A}_i \neq \varnothing \wedge \underline{X}_\alpha \not\subseteq \underline{A}_i), \tag{2.157}$$

i.e., the following holds

$$P_{S,\alpha\,r}(\underline{A}_i) = P_{\alpha\,r}(\underline{A}_i) - P_{\alpha\,l}(\underline{A}_i). \tag{2.158}$$

Real random vectors $\underline{X}$ do not possess probability shadows $\tilde{P}_S(\underline{A}_i)$. A fuzzy random vector may thus be considered to be a "real random vector extended by the probability shadow".

The application of Eqs. (2.154) and (2.155) to the one-dimensional sets

$$A_i = \{x \mid x \in \mathbf{X};\, x_1 \le x \le x_2\} \tag{2.159}$$

leads to

$$P_{\alpha\,l}(A_i) = \max\big[0, P(X_{\alpha\,r} = t_r \mid x_2, t_r \in \mathbf{X};\; t_r \le x_2) \\ - P(X_{\alpha\,l} = t_l \mid x_1, t_l \in \mathbf{X};\; t_l < x_1)\big], \tag{2.160}$$

and

$$P_{\alpha\,r}(A_i) = P(X_{\alpha\,l} = t_l \mid x_2, t_l \in \mathbf{X};\; t_l \le x_2) - P(X_{\alpha\,r} = t_r \mid x_1, t_r \in \mathbf{X};\; t_r < x_1). \tag{2.161}$$

Thereby, the property that the bounds of the random α-level sets $[X_{\alpha\,l}, X_{\alpha\,r}]$ are originals of $\tilde{X}$ is used (Fig. 2.39).

Real random variables with $X = X_{\alpha\,l} = X_{\alpha\,r}$ lead to

$$P_{\alpha\,l}(A_i) = P_{\alpha\,r}(A_i) = P(X = t \mid x_1, x_2, t \in \mathbf{X};\; x_1 \le t \le x_2). \tag{2.162}$$

If the set $\underline{A}_i$ comprises only one element $\underline{A}_i = \underline{x}_i$, the fuzzy probability $\tilde{P}(\underline{A}_i)$ changes to $\tilde{P}(\underline{x}_i)$. The event $\underline{X}_\alpha \cap \underline{A}_i \neq \varnothing$ is then replaced by $\underline{x}_i \in \underline{X}_\alpha$ and $\underline{X}_\alpha \subseteq \underline{A}_i$ becomes $\underline{X}_\alpha = \underline{x}_i$. The probability $P_{\alpha\,l}(\underline{x}_i) = P(\underline{X}_\alpha = \underline{x}_i)$ may take values greater than zero only if a realization of $\underline{X}_\alpha$ exists that possesses exactly one element $\underline{X}_\alpha = \underline{t}$ with $\underline{t} = \underline{x}_i$ and if this element $\underline{t}$ represents a realization of a discrete original of $\underline{X}_\alpha$. Otherwise, $P_{\alpha\,l}(\underline{x}_i) = 0$ holds and the fuzziness of $\tilde{P}(\underline{x}_i)$ is exclusively specified by $P_{\alpha\,r}(\underline{x}_i)$. If $\underline{X}_\alpha$ contains only one original, $\underline{x}_i \in \underline{X}_\alpha$ also changes to $\underline{X}_\alpha = \underline{x}_i$ and $P_{\alpha\,r}(\underline{x}_i) = P_{\alpha\,l}(\underline{x}_i)$ holds.

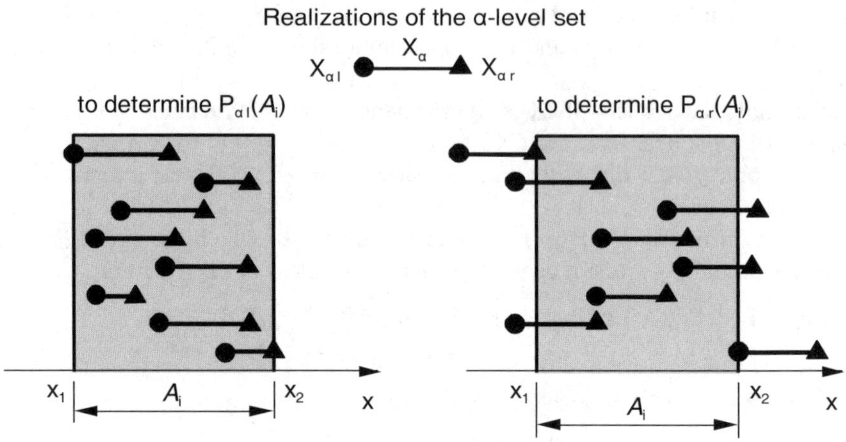

Fig. 2.39. Events for determining $P_{\alpha l}(A_i)$ and $P_{\alpha r}(A_i)$

In the one-dimensional case substituting $x_i = x_1 = x_2$ in Eq. (2.159) leads to

$$P_{\alpha l}(x_i) = P\left(X_{\alpha l} = X_{\alpha r} = t \mid x_i, t \in \mathbf{X};\ t = x_i\right) \ , \tag{2.163}$$

and

$$P_{\alpha r}(x_i) = P\left(X_{\alpha l} = t_l \mid x_i, t_l \in \mathbf{X};\ t_l \le x_i\right) - P\left(X_{\alpha r} = t_r \mid x_i, t_r \in \mathbf{X};\ t_r < x_i\right) \ . \tag{2.164}$$

The properties of the *fuzzy probability* $\tilde{P}(\underline{A}_i)$ result from the properties of the uncertain measure probability according to Sect. 2.2 under consideration of all α-levels at once. For example, a complementary relationship may be derived for $\tilde{P}(\underline{A}_i)$. From the equivalence

$$(\underline{\mathbf{X}}_\alpha \subseteq \underline{A}_i) \leftrightarrow (\underline{\mathbf{X}}_\alpha \cap \underline{A}_i^C = \varnothing) \tag{2.165}$$

for each α-level follows

$$P(\underline{\mathbf{X}}_\alpha \subseteq \underline{A}_i) = P(\underline{\mathbf{X}}_\alpha \cap \underline{A}_i^C = \varnothing) \ . \tag{2.166}$$

Since the complementary event belonging to $\underline{\mathbf{X}}_\alpha \cap \underline{A}_i^C = \varnothing$ reads $\underline{\mathbf{X}}_\alpha \cap \underline{A}_i^C \neq \varnothing$, for the probabilities of the assigned events it holds that

$$P(\underline{\mathbf{X}}_\alpha \cap \underline{A}_i^C = \varnothing) = 1 - P(\underline{\mathbf{X}}_\alpha \cap \underline{A}_i^C \neq \varnothing) \ . \tag{2.167}$$

Taking account of Eqs. (2.154), (2.155), and (2.165)

$$P_{\alpha l}(\underline{A}_i) = 1 - P_{\alpha r}(\underline{A}_i^C) \ , \tag{2.168}$$

and accordingly

$$P_{\alpha r}(\underline{A}_i) = 1 - P_{\alpha l}(\underline{A}_i^C) \tag{2.169}$$

is obtained. The combination of all α-levels in accordance with Eq. (2.153) yields

$$\tilde{P}(\underline{A}_i) = 1 - \tilde{P}(\underline{A}_i^C). \tag{2.170}$$

Further treatments of the fuzzy probability concerning, e.g., the consideration of dependencies between the events to be evaluated are not described here.

2.3.1.3 Fuzzy Probability Distributions

The fuzzy probability $\tilde{P}(\underline{A}_i)$ may be computed for each arbitrary set $\underline{A}_i \in \mathfrak{S}(\underline{\mathbf{X}})$. If – as a special family of sets $\mathfrak{S}(\underline{\mathbf{X}})$ – the system $\mathfrak{S}_0(\mathbb{R}^n)$ of the open sets $\underline{A}_i$ according to Eq. (2.106) is chosen, the concept of the probability distribution function may then be applied to fuzzy random vectors.

Fuzzy Probability Distribution Function. The *fuzzy probability distribution function* $\tilde{F}(\underline{x})$ of the fuzzy random vector $\tilde{\underline{X}}$ on $\underline{\mathbf{X}} = \mathbb{R}^n$ is the fuzzy probability $\tilde{P}(\underline{A}_i)$ of $\underline{A}_i$ – $\underline{A}_i$ according to Eq. (2.106) – for all $\underline{x}_i \in \underline{\mathbf{X}}$ (Fig 2.40). For each specified $\underline{x}$ the fuzzy functional value

$$\tilde{F}(\underline{x}) = \Big\{\big(F_\alpha(\underline{x}), \mu(F_\alpha(\underline{x}))\big) \mid F_\alpha(\underline{x}) = [F_{\alpha l}(\underline{x}), F_{\alpha r}(\underline{x})]; \; \mu(F_\alpha(\underline{x})) = \alpha \,\forall\, \alpha \in (0, 1]\Big\} \tag{2.171}$$

is defined for all α-levels by

$$F_{\alpha l}\big(\underline{x} = (x_1, \ldots, x_n)\big) = 1 - \max_{\underline{X}_j \in \underline{X}_\alpha} P\big(\underline{X}_j = \underline{t} = (t_1, \ldots, t_n) \mid \underline{x}, \underline{t} \in \underline{\mathbf{X}} = \mathbb{R}^n; \; \exists\, t_k \geq x_k;\, 1 \leq k \leq n\big), \tag{2.172}$$

and

$$F_{\alpha r}\big(\underline{x} = (x_1, \ldots, x_n)\big) = \max_{\underline{X}_j \in \underline{X}_\alpha} P\big(\underline{X}_j = \underline{t} = (t_1, \ldots, t_n) \mid \underline{x}, \underline{t} \in \underline{\mathbf{X}} = \mathbb{R}^n; \; t_k < x_k;\, k = 1, \ldots, n\big). \tag{2.173}$$

For determining $F_{\alpha l}(\underline{x})$ the relationship in Eq. (2.168) is used. All originals $\underline{X}_j$ of $\tilde{\underline{X}}$ that are contained in $\underline{X}_\alpha$ must be taken into account. *The fuzzy probability distribution function* $\tilde{F}(\underline{x})$ *of* $\tilde{\underline{X}}$ *may thus be interpreted as being the set of the probability distribution functions* $F_j(\underline{x})$ *of all originals* $\underline{X}_j$ *of* $\tilde{\underline{X}}$ *with the membership values* $\mu(F_j(\underline{x}))$. *It represents a fuzzy function in accordance with the definitions in Sect. 2.1.11.1 and each original* $\underline{X}_j$ *determines precisely one trajectory* $F_j(\underline{x}) \in \tilde{F}(\underline{x})$.

From the fuzzy probability distribution function $\tilde{F}(\underline{x})$ of the fuzzy random vector $\tilde{\underline{X}}$ the probability distribution shadow $\tilde{F}_S(\underline{x})$ characterizing exclusively the fuzziness of $\tilde{F}(\underline{x})$ may be computed.

The function

$$\tilde{F}_S(\underline{x}) = \Big\{\big(F_{S,\alpha}(\underline{x}), \mu(F_{S,\alpha}(\underline{x}))\big) \mid F_{S,\alpha}(\underline{x}) = [0, F_{S,\alpha r}(\underline{x})]; \; \mu(F_{S,\alpha}(\underline{x})) = \alpha \,\forall\, \alpha \in (0, 1]\Big\} \tag{2.174}$$

on $\underline{\mathbf{X}} = \mathbb{R}^n$ with

$$F_{S,\alpha r}(\underline{x}) = F_{\alpha r}(\underline{x}) - F_{\alpha l}(\underline{x}) \tag{2.175}$$

and $F_{\alpha r}(\underline{x})$ and $F_{\alpha l}(\underline{x})$ from Eqs. (2.172) and (2.173) is referred to as the *probability distribution shadow* of the fuzzy random vector $\tilde{\underline{X}}$.

In the one-dimensional case the fuzzy probability distribution function is determined by

$$F_{\alpha l}(x) = P\big(X_{\alpha r} = t_r \mid x, t_r \in \mathbf{X};\ t_r < x\big), \tag{2.176}$$

and

$$F_{\alpha r}(x) = P\big(X_{\alpha l} = t_l \mid x, t_l \in \mathbf{X};\ t_l < x\big). \tag{2.177}$$

The probability distribution shadow may also be specified with the aid of the functional value $P_{\alpha r}(x)$ according to Eq. (2.164)

$$F_{S,\alpha r}(x) = P_{\alpha r}(x) - P\big(X_{\alpha l} = t_l \mid x, t_l \in \mathbf{X};\ t_l = x\big). \tag{2.178}$$

Fuzzy Probability Density Function. The *fuzzy probability density function* $\tilde{f}(\underline{t})$ or $\tilde{f}(\underline{x})$ (Fig. 2.40) is a fuzzy function (Sect. 2.1.11.1) belonging to $\tilde{F}(\underline{x})$ which – in the continuous case in relation to randomness – is integrable for each original $\underline{X}_j$ of $\tilde{\underline{X}}$ and satisfies the relationship

$$F_j(\underline{x}) = \int_{t_1=-\infty}^{t_1=x_1} \ldots \int_{t_k=-\infty}^{t_k=x_k} \ldots \int_{t_n=-\infty}^{t_n=x_n} f_j(\underline{t})\, d\underline{t}\,, \tag{2.179}$$

with $\underline{t} = (t_1; \ldots; t_n) \in \mathbf{X}$. This means that for each original $\underline{X}_j$ the integration of the assigned trajectory $f_j(\underline{x}) \in \tilde{f}(\underline{x})$ leads to the trajectory $F_j(\underline{x}) \in \tilde{F}(\underline{x})$. For discrete fuzzy random variables – in relation to their randomness – the integral term reduces to a sum.

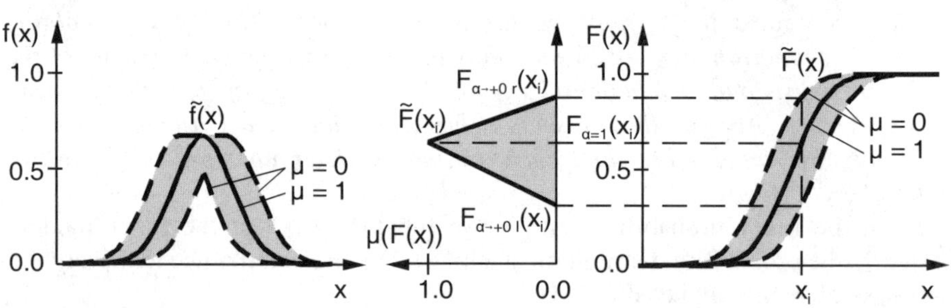

Fig. 2.40. Fuzzy probability density function $\tilde{f}(x)$ and fuzzy probability distribution function $\tilde{F}(x)$ of a continuous fuzzy random variable $\tilde{X}$

Bunch Parameter Representation. The fuzzy probability distribution function $\tilde{F}(\underline{x})$ according to Eq. (2.171) as well as the assigned fuzzy probability density function $\tilde{f}(\underline{x})$ represents an *assessed function bunch*, it may also be stated in the parametric form (Sect. 2.1.11.1)

$$\tilde{F}(\underline{x}) = F(\tilde{\underline{s}},\underline{x}). \qquad (2.180)$$

In Eq. (2.180) the assessment of the included crisp functions is realized by the fuzzy bunch parameter $\tilde{\underline{s}}$. As an example, the fuzzy probability distribution function

$$F(\tilde{\underline{s}},x) = \exp\left(-\exp(-\tilde{s}_1(x-\tilde{s}_2))\right) \qquad (2.181)$$

of Ex-Max Type I (Gumbel), with the fuzzy bunch parameters

$$\tilde{\underline{s}}_1 = \frac{\pi}{\tilde{\sigma}_X \cdot \sqrt{6}} \qquad (2.182)$$

and

$$\tilde{\underline{s}}_2 = \tilde{m}_X - 0.45 \cdot \tilde{\sigma}_X, \qquad (2.183)$$

is illustrated in Fig. 2.41.

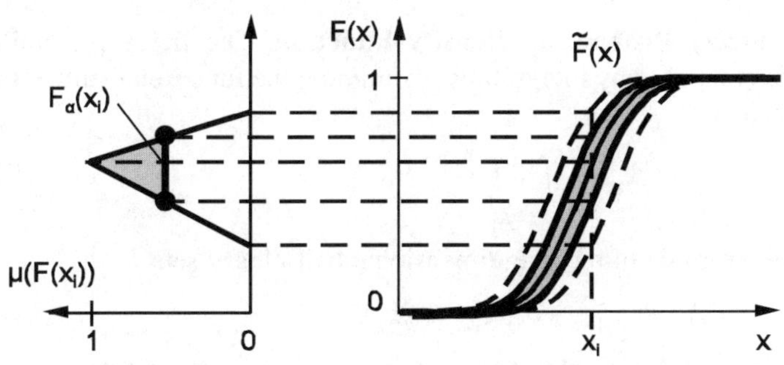

Fig. 2.41. Fuzzy bunch parameter representation

For introducing the $F(\tilde{\underline{s}},\underline{x})$ in numerical procedures the α-discretization is applied in the space of the fuzzy bunch parameters (Fig. 2.42). For each specified $\underline{x}$ this leads to the fuzzy functional value

$$\tilde{F}(\underline{x}) = F(\tilde{\underline{s}},\underline{x}) = \left\{\left(F_\alpha(\underline{x}), \mu(F_\alpha(\underline{x}))\right) \mid F_\alpha(\underline{x}) = [F_{\alpha l}(\underline{x}), F_{\alpha r}(\underline{x})]; \; \mu(F_\alpha(\underline{x})) = \alpha \,\forall\, \alpha \in (0, 1]\right\}, \qquad (2.184)$$

with

$$F_{\alpha l}(\underline{x}) = \min\left[F(\underline{s}, \underline{x}) \mid \underline{s} \in \underline{S}_\alpha\right], \qquad (2.185)$$

$$F_{\alpha r}(\underline{x}) = \max\left[F(\underline{s}, \underline{x}) \mid \underline{s} \in \underline{S}_\alpha\right]. \qquad (2.186)$$

The α-level set $\underline{S}_\alpha$ comprises the bunch parameters of all originals on the level α and all $\underline{S}_\alpha$ together form the fuzzy input set $\tilde{\underline{s}} = \tilde{\underline{S}}$ for the numerical treatment of $F(\tilde{\underline{s}},\underline{x})$. Thereby each element $\underline{s}_j$ from $\tilde{\underline{S}}$ determines one specific original $\underline{X}_j$ of $\tilde{\underline{X}}$ and the assigned trajectory $F_j(\underline{x}) = F(\underline{s}_j, \underline{x}) \in \tilde{F}(\underline{x})$ with the membership values $\mu(F_j(\underline{x})) = \mu(\underline{X}_j) = \mu(\underline{s}_j)$. For each $\underline{s}_j$ from $\underline{S}_\alpha$ the assigned original $\underline{X}_j$ belongs to the random α-level set $\underline{X}_\alpha$ and the functional values of the trajectory $F_j(\underline{x}) = F(\underline{s}_j, \underline{x})$ comply with $F_j(\underline{x}) \in F_\alpha(\underline{X})$ for each $\underline{x} \in \underline{\mathbf{X}}$.

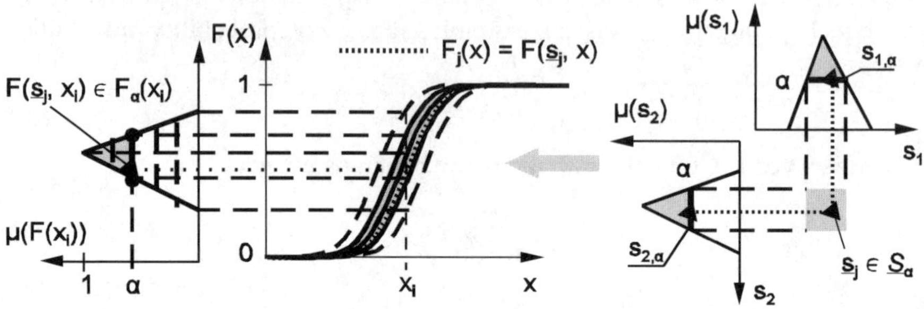

Fig. 2.42. α-discretization of $F(\tilde{\underline{s}}, x)$ from Eq. (2.181) in the space of the fuzzy bunch parameters $\tilde{s}_1$ and $\tilde{s}_2$

Integrating the Fuzzy Probability Density Function. The fuzzy probability $\tilde{P}(\underline{A}_i)$ may also be computed by integrating or summing the functional values $\tilde{f}(\underline{x})$ trajectory by trajectory

$$\tilde{P}(\underline{A}_i) = \int_{\underline{x} \mid \underline{x} \in \underline{A}_i} \cdots \int \tilde{f}(\underline{x})\, d\underline{x}. \tag{2.187}$$

Considering the example of the one-dimensional half-closed sets

$$A_i = \{x \mid x \in \mathbf{X};\, x_1 \le x < x_2\}, \tag{2.188}$$

the fuzzy probability $\tilde{P}(\underline{A}_i)$ is obtained as the set of all α-level sets $[P_{\alpha\,l}(A_i), P_{\alpha\,r}(A_i)]$ with

$$\begin{aligned} P_{\alpha l}(A_i) = \max\big[0,\, & P(X_{\alpha r} = t_r \mid x_2, t_r \in \mathbf{X};\ t_r < x_2) \\ & - P(X_{\alpha l} = t_l \mid x_1, t_l \in \mathbf{X};\ t_l < x_1)\big], \end{aligned} \tag{2.189}$$

and

$$P_{\alpha r}(A_i) = P(X_{\alpha l} = t_l \mid x_2, t_l \in \mathbf{X};\ t_l < x_2) - P(X_{\alpha r} = t_r \mid x_1, t_r \in \mathbf{X};\ t_r < x_1) \tag{2.190}$$

in the style of Eqs. (2.160) and (2.161). The substitution of the probability terms with the aid of Eqs. (2.176) and (2.177) yields

$$P_{\alpha l}(A_i) = \max\big[0,\, F_{\alpha l}(x_2) - F_{\alpha r}(x_1)\big], \tag{2.191}$$

and

$$P_{\alpha r}(A_i) = F_{\alpha r}(x_2) - F_{\alpha l}(x_1). \tag{2.192}$$

The combination of all α-levels results in

$$\tilde{P}(A_i) = \tilde{F}(x_2) - \tilde{F}(x_1) \ge 0, \tag{2.193}$$

in compliance with $P(A_i)$ concerning real random vectors.

In the case of fuzzy random variables that are continuous with regard to randomness, $P_{\alpha l}(A_i)$ and $P_{\alpha r}(A_i)$ may be determined by means of Eq. (2.179)

$$P_{\alpha l}(A_i) = \max\left[0,\ \int_{t=-\infty}^{t=x_2} f_1(t)\,dt - \int_{t=-\infty}^{t=x_1} f_2(t)\,dt \mid t \in \mathbf{X};\ X_1 = X_{\alpha r};\ X_2 = X_{\alpha l}\right], \quad (2.194)$$

$$P_{\alpha r}(A_i) = \int_{t=-\infty}^{t=x_2} f_2(t)\,dt - \int_{t=-\infty}^{t=x_1} f_1(t)\,dt \mid t \in \mathbf{X};\ X_1 = X_{\alpha r};\ X_2 = X_{\alpha l}\ . \quad (2.195)$$

By converting the left-hand integral terms the fuzzy probability $\tilde{P}(A_i)$ is obtained in a form that corresponds to the relationship in Eq. (2.170), i.e., $\tilde{P}(A_i)$ is gained by evaluating the complementary event $\tilde{X} \in A_i^C$

$$P_{\alpha l}(A_i) = \max\left[0,\ 1 - \left(\int_{t=-\infty}^{t=x_1} f_2(t)\,dt + \int_{t=x_2}^{t=+\infty} f_1(t)\,dt\right) \mid t \in \mathbf{X}; X_1 = X_{\alpha r}; X_2 = X_{\alpha l}\right], \quad (2.196)$$

$$P_{\alpha r}(A_i) = 1 - \left(\int_{t=-\infty}^{t=x_1} f_1(t)\,dt + \int_{t=x_2}^{t=+\infty} f_2(t)\,dt\right) \mid t \in \mathbf{X}; X_1 = X_{\alpha r}; X_2 = X_{\alpha l}\ . \quad (2.197)$$

The computation of $\tilde{P}(A_i)$ by integrating the trajectories $f_j(t)$ over A_i, as usually applied to real random vectors, requires taking account of the dependencies between $X_{\alpha l}$ and $X_{\alpha r}$ with the aid of conditional events or probabilities; the direct application of Eq. (2.170) is thus more advantageous.

2.3.1.4 Parameters of Fuzzy Random Vectors

Fuzzy probability is based on the special *objective* uncertain measure probability (Sect. 2.2.2), which must be applicable to each original $\underline{X}_j$ of a fuzzy random vector $\tilde{\underline{X}}$. The quality of inferences derived from this probabilistic approach is conditioned by statistical laws when the probability measure is exclusively applied to real data. The uncertainty described by one single original $\underline{X}_j$ only appears with the characteristic *randomness.* Uncertainty with the characteristic *fuzziness* is accounted for by considering different originals of $\tilde{\underline{X}}$. Based on observed, random realizations of $\tilde{\underline{X}}$ (or their subjective assessment) the originals $\underline{X}_j$ are determined. They are gained from the consideration: "Which $\underline{X}_j$ could have led to the observations made?" The detected originals $\underline{X}_j$ are assessed using membership values. The fuzzy random vector $\tilde{\underline{X}}$ is obtained as the fuzzy set of all originals $\underline{X}_j$. The corresponding fuzzy probability distribution function $\tilde{F}(\underline{x})$ maps the fundamental set $\underline{\mathbf{X}}$ onto the interval [0, 1]. In contrast to the probability distribution function $F(\underline{x})$ of real random vectors, the functional values of the fuzzy probability distribution function $\tilde{F}(\underline{x})$ are fuzzy numbers.

The distribution type and parameters of the fuzzy random vectors must thus be determined by evaluating all originals $\underline{X}_j$ of $\tilde{\underline{X}}$; these are obtained in the form of fuzzy parameters $\tilde{p}_t(\tilde{\underline{X}})$.

The *fuzzy parameter* $\tilde{p}_t(\tilde{\underline{X}})$ of the fuzzy random vector $\tilde{\underline{X}}$ comprises the fuzzy set of the parameters $p_t(\underline{X}_j)$ of all (real-valued) originals $\underline{X}_j$ with the membership values $\mu(p_t(\underline{X}_j))$

$$\tilde{p}_t(\tilde{\underline{X}}) = \left\{\left(p_t(\underline{X}_j), \mu(p_t(\underline{X}_j))\right) \mid \underline{X}_j \in \tilde{\underline{X}};\, \mu(p_t(\underline{X}_j)) = \mu(\underline{X}_j)\ \forall j\right\} . \tag{2.198}$$

By applying α-discretization, $\tilde{p}_t(\tilde{\underline{X}})$ is determined by

$$\tilde{p}_t(\tilde{\underline{X}}) = \left\{\left(p_{t,\alpha}(\tilde{\underline{X}}), \mu(p_{t,\alpha}(\tilde{\underline{X}}))\right) \mid \mu(p_{t,\alpha}(\tilde{\underline{X}})) = \alpha \,\forall\, \alpha \in (0,1]\right\} , \tag{2.199}$$

with

$$p_{t,\alpha}(\tilde{\underline{X}}) = [p_{t,\alpha l}(\tilde{\underline{X}}), p_{t,\alpha r}(\tilde{\underline{X}})] , \tag{2.200}$$

and

$$p_{t,\alpha l}(\tilde{\underline{X}}) = \min_j [p_t(\underline{X}_j) \mid \underline{X}_j \in \underline{X}_\alpha] , \tag{2.201}$$

$$p_{t,\alpha r}(\tilde{\underline{X}}) = \max_j [p_t(\underline{X}_j) \mid \underline{X}_j \in \underline{X}_\alpha] . \tag{2.202}$$

One-dimensional case. On this basis the moments of the fuzzy probability distribution of the one-dimensional fuzzy random variable $\tilde{X}$ may be stated. These are determined with the aid of α-discretization. The following definitions are formulated for continuous fuzzy random variables, in the discrete case the integrals change to sums.

The *k-th fuzzy initial moment* of the fuzzy probability distribution of $\tilde{X}$ is defined as

$$\tilde{m}_k = E\tilde{X}^k = \int_{x=-\infty}^{x=+\infty} x^k \cdot \tilde{f}(x)\, dx, \tag{2.203}$$

with $k \in \mathbb{N}$. The evaluation of Eq. (2.203) yields

$$\tilde{m}_k = \left\{\left(m_{k,\alpha}, \mu(m_{k,\alpha})\right) \mid m_{k,\alpha} = [m_{k,\alpha l}, m_{k,\alpha r}];\ \mu(m_{k,\alpha}) = \alpha \,\forall\, \alpha \in (0,1]\right\}, \tag{2.204}$$

in which

$$m_{k,\alpha l} = \min_j [\int_{x=-\infty}^{x=+\infty} x^k \cdot f_j(x)\, dx \mid X_j \in X_\alpha], \tag{2.205}$$

$$m_{k,\alpha r} = \max_j [\int_{x=-\infty}^{x=+\infty} x^k \cdot f_j(x)\, dx \mid X_j \in X_\alpha]. \tag{2.206}$$

For $k = 1$ Eq. (2.203) yields the *fuzzy expected value*

$$\tilde{m}_X = E\tilde{X} = \int_{x=-\infty}^{x=+\infty} x \cdot \tilde{f}(x)\, dx \tag{2.207}$$

of the fuzzy random variable $\tilde{X}$.

The *k-th fuzzy central moment* of the fuzzy probability distribution of $\tilde{X}$ reads

$$\tilde{\zeta}_k = E(\tilde{X} - \tilde{m}_X)^k = \int_{x=-\infty}^{x=+\infty} (x - \tilde{m}_X)^k \cdot \tilde{f}(x)\, dx \tag{2.208}$$

for $k \in \mathbb{N}$. The evaluation is carried out analogously using Eqs. (2.204), (2.205), and (2.206).

The *fuzzy variance* of the fuzzy random variable $\tilde{X}$ is given by Eq. (2.208) with $k = 2$ as

$$\tilde{\zeta}_2 = D^2\tilde{X} = \int_{x=-\infty}^{x=+\infty} (x - \tilde{m}_X)^2 \cdot \tilde{f}(x)\, dx. \tag{2.209}$$

The *fuzzy standard deviation* is obtained by extracting the square root of $D^2\tilde{X}$ in Eq. (2.209)

$$\tilde{\sigma}_X = \sqrt{D^2\tilde{X}} = \sqrt{\int_{x=-\infty}^{x=+\infty} (x - \tilde{m}_X)^2 \cdot \tilde{f}(x)\, dx}. \tag{2.210}$$

The laws governing the relationships between the moments of real random vectors may be applied in a similar way to the fuzzy moments of fuzzy random vectors. The interaction between the fuzzy variables involved must thereby be taken into account.

2.3.2 Fuzzy Random Functions

2.3.2.1 Definition of Fuzzy Random Functions

A fuzzy random function is a function whose functional values are fuzzy random vectors. These functional values may depend on the spatial coordinates $\underline{\theta} = (\theta_1, \theta_2, \theta_3)$, the time τ, and occasionally further parameters $\underline{\varphi} = (\varphi_1, \varphi_2, \ldots)$ from the parameter space $\underline{\mathbf{T}} \subseteq \mathbb{R}^m$. In general these parameters may represent fuzzy variables. With the fuzzy parameter vector $\underline{\tilde{t}} = (\tilde{\tau}, \underline{\tilde{\theta}}, \underline{\tilde{\varphi}}) \mid \underline{\tilde{t}} \in \mathbf{F}(\underline{\mathbf{T}})$, $\underline{\tilde{X}}(\underline{\tilde{t}})$ characterizes a fuzzy random function defined on the space $\mathbf{F}(\underline{\mathbf{T}}) \times \Omega$. Thereby Ω denotes the space of the random elementary events and $\mathbf{F}(\underline{\mathbf{T}})$ represents the set of all fuzzy vectors on $\underline{\mathbf{T}} \subseteq \mathbb{R}^m$. Extending the definition of fuzzy random vectors according to Eq. (2.151) with the aid of the basic characterization of fuzzy functions from Eq. (2.60) the *fuzzy random function* $\underline{\tilde{X}}(\underline{\tilde{t}})$ is defined as the fuzzy result of the uncertain mapping

$$\underline{\tilde{X}}(\underline{\tilde{t}}): \mathbf{F}(\underline{\mathbf{T}}) \times \Omega \ \tilde{\to}\ \mathbf{F}(\underline{\mathbf{X}}), \tag{2.211}$$

in which $\mathbf{F}(\underline{\mathbf{X}})$ characterizes the set of all fuzzy numbers on $\mathbb{R}^n$.

In accordance with the discussion of fuzzy functions (Sect. 2.1.11.1) the following explanations focus on crisp parameter vectors $\underline{t} = (\tau, \underline{\theta}, \underline{\varphi}) \mid \underline{t} \in \underline{\mathbf{T}}$, as these appear in common engineering applications. The definition of a fuzzy random function $\tilde{\underline{X}}(\underline{t})$ then reads

$$\tilde{\underline{X}}(\underline{t}):\ \underline{\mathbf{T}} \times \Omega \ \tilde{\to}\ \mathbf{F}(\underline{\mathbf{X}})\,. \tag{2.212}$$

For each specified point $\underline{t} \in \underline{\mathbf{T}}$ a fuzzy random function represents a fuzzy random vector $\tilde{\underline{X}}_t = \tilde{\underline{X}}(\underline{t})$ in accordance with Eq. (2.151). A fuzzy random function may thus also be defined as a set of fuzzy random vectors on the parameter space $\underline{\mathbf{T}}$

$$\tilde{\underline{X}}(\underline{t}) = \left\{ \tilde{\underline{X}}_t = \tilde{\underline{X}}(\underline{t}) \ \forall \underline{t} \mid \underline{t} \in \underline{\mathbf{T}} \right\}. \tag{2.213}$$

According to Eqs. (2.212) and (2.213), for all $\underline{t} \in \underline{\mathbf{T}}$ together a fuzzy function $\tilde{\underline{x}}(\underline{t})$ is assigned as a realization to each elementary event $\omega \in \Omega$. Two realizations $\tilde{x}_1(t)$ and $\tilde{x}_2(t)$ of a one-dimensional fuzzy random function are plotted in Fig. 2.43.

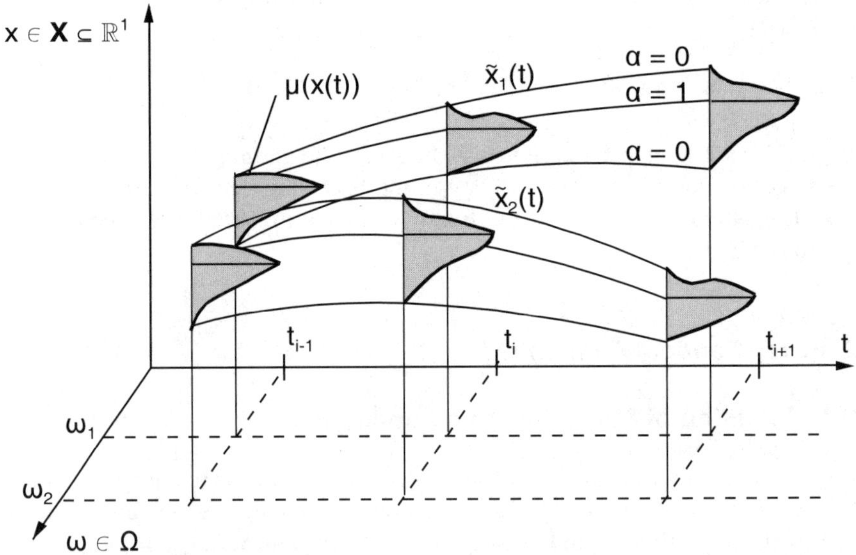

Fig. 2.43. Realizations of a one-dimensional fuzzy random function

If the realization $\underline{x}(\underline{t})$ of a real random function $\underline{X}(\underline{t})$ as well as the fuzzy realization $\tilde{\underline{x}}(\underline{t})$ of a fuzzy random function $\tilde{\underline{X}}(\underline{t})$ may be assigned to the same elementary event ω, and $\underline{x}_t = \underline{x}(\underline{t}) \in \tilde{\underline{x}}_t = \tilde{\underline{x}}(\underline{t})$ holds for all $\underline{t} \in \underline{\mathbf{T}}$, then $\underline{x}(\underline{t})$ is considered to be contained in $\tilde{\underline{x}}(\underline{t})$, i.e., $\underline{x}(\underline{t})$ represents a trajectory of $\tilde{\underline{x}}(\underline{t})$ (Sect. 2.1.11.1). If $\underline{x}(\underline{t})$ is contained in $\tilde{\underline{x}}(\underline{t})$ for every elementary event $\omega \in \Omega$, the $\underline{x}(\underline{t})$ constitute an original function $\underline{X}_j(\underline{t})$ of the fuzzy random function $\tilde{\underline{X}}(\underline{t})$. Each original function $\underline{X}_j(\underline{t})$ possesses the properties of a real random function $\underline{X}(\underline{t})$

(without fuzziness) on the parameter space $\underline{\mathbf{T}}$. This means that the fuzzy random function $\tilde{\underline{X}}(\underline{t})$ is the fuzzy set of all possible original functions $\underline{X}_j(\underline{t})$ contained in $\tilde{\underline{X}}(\underline{t})$. Thereby the original functions possess membership values $\mu(\underline{X}_j(\underline{t}))$ indicating the degree to which $\underline{X}_j(\underline{t})$ belongs to $\tilde{\underline{X}}(\underline{t})$.

In accordance with the treatment of fuzzy random vectors (Sect. 2.3.1), the *α-discretization* is adopted. Referring to Eq. (2.152) the set of original functions on the level α may be denoted by

$$\underline{X}_\alpha(\underline{t}) = \left\{\underline{X}(\underline{t}) = \underline{X}_j(\underline{t}) \mid \mu(\underline{X}_j(\underline{t})) \geq \alpha\right\}, \tag{2.214}$$

and is referred to as a *random α-function set*.

By combining the random α-function sets for all α-levels the α-level representation of a fuzzy random function

$$\tilde{\underline{X}}(\underline{t}) = \left\{\left(\underline{X}_\alpha(\underline{t}), \mu(\underline{X}_\alpha(\underline{t}))\right) \mid \mu(\underline{X}_\alpha(\underline{t})) = \alpha \;\forall\, \alpha \in (0,1]\right\} \tag{2.215}$$

is accomplished.

In the one-dimensional case for each specified $\underline{t}$ closed random intervals $[X_{t\alpha l}, X_{t\alpha r}]$ are obtained (Sect. 2.3.1) and the random α-function set according to Eq. (2.214) may be characterized by the set of these random intervals for all $\underline{t} \in \underline{\mathbf{T}}$

$$X_\alpha(\underline{t}) = \left\{[X_{t\alpha l}, X_{t\alpha r}] \;\forall\; \underline{t} \in \underline{\mathbf{T}}\right\}. \tag{2.216}$$

The random functions

$$X_{\alpha l}(\underline{t}) = \left\{X_{t\alpha l} \;\forall\; \underline{t} \in \underline{\mathbf{T}}\right\}, \tag{2.217}$$

and

$$X_{\alpha r}(\underline{t}) = \left\{X_{t\alpha r} \;\forall\; \underline{t} \in \underline{\mathbf{T}}\right\} \tag{2.218}$$

are the lower and upper *random bounding functions* of the fuzzy random function on the level α. In general, these random bounding functions do not necessarily represent original functions of the fuzzy random function, they merely envelope the random α-function sets $X_\alpha(\underline{t})$.

Bunch Parameter Representation. In accordance with the bunch parameter representation of fuzzy functions (Sect. 2.1.11.1), the fuzzy random function from Eq. (2.212) may be characterized by an assessed random function bunch denoted by

$$\tilde{\underline{X}}(\underline{t}) = \underline{X}(\tilde{\underline{s}}, \underline{t}). \tag{2.219}$$

The assessment of the individual real random functions included in $\underline{X}(\tilde{\underline{s}}, \underline{t})$ from Eq. (2.219) is again realized with the aid of fuzzy bunch parameters $\tilde{\underline{s}}$. The fuzzy random function according to Eq. (2.212) may now be defined in the space of the fuzzy bunch parameters

$$\tilde{\underline{X}}(\underline{t}) = \underline{X}(\tilde{\underline{s}}, \underline{t}) = \left\{\left(\underline{X}(\underline{t}), \mu(\underline{X}(\underline{t}))\right) \mid \underline{X}(\underline{t}) = \underline{X}(\underline{s}, \underline{t});\; \mu(\underline{X}(\underline{t})) = \mu(\underline{s}) \;\forall\, \underline{s} \in \tilde{\underline{s}}\right\}. \tag{2.220}$$

In accordance with Eq. (2.215) the fuzzy random function $\tilde{\underline{X}}(\underline{t})$ may also be characterized with the aid of random α-function sets described on the basis of bunch parameters

$$\underline{X}_{\alpha}(\underline{t}) = \left\{\underline{X}(\underline{s},\underline{t}) \;\forall\, \underline{s} \mid \underline{s} \in \underline{S}_{\alpha}\,;\; \alpha \in (0,1]\right\}. \tag{2.221}$$

The set $\underline{S}_\alpha$ addresses all original functions $\underline{X}_j(\underline{t}) = \underline{X}(\underline{s}_j, \underline{t})$ with $\underline{s}_j \in \underline{S}_\alpha$ on the level α. For each α it represents a subspace of $\tilde{\underline{s}}$. Each element $\underline{s}_j$ from $\underline{S}_\alpha$ determines precisely one original function, i.e., one real-valued random function on the level α.

Each fuzzy random function $\tilde{\underline{X}}(\underline{t})$ comprises at least one real random function $\underline{X}(\underline{t})$ as an original function $\underline{X}_j(\underline{t})$. Thus each fuzzy random function that possesses precisely one original function is a real random function $\underline{X}(\underline{t})$.

Real random functions accordingly represent a special case of fuzzy random functions. This special case appears for $\mu(\underline{X}_\alpha(\underline{t})) = 1$. Both real random functions and fuzzy random functions may thus be considered simultaneously in fuzzy stochastic structural analysis and fuzzy probabilistic safety assessment. This means that fuzzy random functions represent a generalized uncertainty model.

2.3.2.2 Parameters and Properties of Fuzzy Random Functions

As the fuzzy random functions according to Eq. (2.213) are defined as a set of fuzzy random vectors on the parameter space $\mathbf{\underline{T}}$, for each specified $\underline{t} \in \mathbf{\underline{T}}$ the fuzzy probability distribution function $\tilde{F}_t(\underline{x}) = \tilde{F}(\underline{x},\underline{t})$ characterizing the fuzzy random vector $\tilde{\underline{X}}_t = \tilde{\underline{X}}(\underline{t})$ is known. In accordance with Sect. 2.3.1.3, these fuzzy functions may be described with the aid of the original functions $\underline{X}_j(\underline{t})$ of $\tilde{\underline{X}}(\underline{t})$. For each crisp point $\underline{t}$ these original functions reduce to originals $\underline{X}_{t\,j} = \underline{X}_j(\underline{t})$ representing real-valued random vectors belonging to $\underline{t}$. The functional values $\tilde{F}_t(\underline{x})$ of the $\tilde{F}(\underline{x},\underline{t})$ at the specified point $\underline{t}$ may be determined in accordance with Eq. (2.171),

$$\tilde{F}_t(\underline{x}) = \left\{\left(F_{t\alpha}(\underline{x}), \mu(F_{t\alpha}(\underline{x}))\right) \mid F_{t\alpha}(\underline{x}) = [F_{t\alpha\,l}(\underline{x}), F_{t\alpha\,r}(\underline{x})]\,;\; \mu(F_{t\alpha}(\underline{x})) = \alpha \,\forall\, \alpha \in (0,1]\right\}, \tag{2.222}$$

with

$$F_{t\alpha\,l}(\underline{x}) = 1 - \max_{\underline{X}_{tj} \in \underline{X}_{t\alpha}} P\left(\underline{X}_{tj} = \underline{u} \mid \underline{x}, \underline{u} \in \mathbf{\underline{X}} = \mathbb{R}^n;\, \exists\, u_k \geq x_k;\, 1 \leq k \leq n\right), \tag{2.223}$$

$$F_{t\alpha\,r}(\underline{x}) = \max_{\underline{X}_{tj} \in \underline{X}_{t\alpha}} P\left(\underline{X}_{tj} = \underline{u} \mid \underline{x}, \underline{u} \in \mathbf{\underline{X}} = \mathbb{R}^n;\, u_k < x_k;\, k = 1,\ldots,n\right), \tag{2.224}$$

and $\underline{x} = (x_1, \ldots, x_n)$, $\underline{u} = (u_1, \ldots, u_n)$.

The fuzzy probability distribution function $\tilde{F}_t(\underline{x})$ of the fuzzy random vector $\tilde{\underline{X}}_t = \tilde{\underline{X}}(\underline{t})$ at the point $\underline{t}$ also represents the set of the probability distributions $F_{t\,j}(\underline{x})$ of all originals $\underline{X}_{t\,j}$ (not of the original functions!) with the membership values $\mu(F_{t\,j}(\underline{x}))$. In compliance with Sect. 2.3.1.3 the $F_{t\,j}(\underline{x})$ are trajectories of $\tilde{F}_t(\underline{x})$.

If for all $\underline{t} \in \mathbf{\underline{T}}$ the $\tilde{F}_t(\underline{x})$ represent fuzzy Gaussian normal distributions (Gaussian normal distributions with fuzzy functional parameters), the assigned fuzzy random function is referred to as a *fuzzy Gaussian process*.

Owing to the generally existing dependencies between the $\tilde{\underline{X}}_t$ at different points $\underline{t}$, the specification of the fuzzy probability distribution functions $\tilde{F}_t(\underline{x})$ for several $\underline{t}$ is not sufficient. To completely characterize a fuzzy random function the dependencies in the $\tilde{F}_t(\underline{x})$ must be taken into account by introducing the multidimensional fuzzy probability distribution function

$$\tilde{F}\left(\underline{x}_1 = \underline{x}(\underline{t}_1), \ldots, \underline{x}_n = \underline{x}(\underline{t}_n)\right) = \tilde{P}\left(\underline{A}_1 = \underline{A}_i(\underline{t}_1) \wedge \ldots \wedge \underline{A}_n = \underline{A}_i(\underline{t}_n)\right), \tag{2.225}$$

in which the $\underline{A}_i$ are to be defined according to Eq. (2.106) and n denotes the number of points $\underline{t}$ in the parameter space $\underline{\mathbf{T}}$. For fuzzy random functions multidimensional fuzzy probability density functions are also defined. The fuzzy probability density function $\tilde{f}(\underline{x},\underline{t})$ is a function belonging to $\tilde{F}(\underline{x},\underline{t})$, whereby – in the continuous case – each $f_j(\underline{x}, \underline{t})$ belonging to the original function $\underline{X}_j(\underline{t})$ is integrable at $\underline{t}$ and complies with the requirement

$$F_j\left(\underline{x}_1 = \underline{x}(\underline{t}_1), \ldots, \underline{x}_n = \underline{x}(\underline{t}_n)\right) = \int_{u_{11}=-\infty}^{u_{11}=x_{11}} \cdots \int_{u_{1r_1}=-\infty}^{u_{1r_1}=x_{1r_1}} \cdots \int_{u_{nr_n}=-\infty}^{u_{nr_n}=x_{nr_n}} f_j\left(\underline{u}_1 = \underline{u}(\underline{t}_1), \ldots, \underline{u}_n = \underline{u}(\underline{t}_n)\right) d\underline{u}_n \ldots d\underline{u}_1 . \tag{2.226}$$

In the discrete case the integral is to be replaced by the associated sum.

In this sense the fuzzy random function $\tilde{\underline{X}}(\underline{t})$ is characterized by an r-dimensional fuzzy random vector with the coordinates $x_{11}, \ldots, x_{1r_1}, x_{21} \ldots, x_{2r_2}, \ldots, x_{n1}, \ldots, x_{nr_n}$ and

$$r = \sum_{k=1}^{k=n} r_k . \tag{2.227}$$

Enhancing the concept of stochastic processes, the parameters and properties of fuzzy random functions may be defined. This enhancement relies on the parameters and properties of the original functions and on the α-level representation according to Eq. (2.215). In compliance with real stochastic processes each original function $\underline{X}(\underline{t})$, i.e., each real random function, may be evaluated with the aid of parameters. These parameters or characterizing terms like, e.g., the trend function, the covariance function, and the correlation function are defined as a bunch of functions assessed by membership values and thus represent fuzzy functions.

One-dimensional case. For the one-dimensional fuzzy random function $\tilde{X}(\underline{t})$ the *k-th fuzzy initial moment function* $\tilde{m}_k(\underline{t})$ may be defined as

$$\tilde{m}_k(\underline{t}) = E[\tilde{X}^k(\underline{t})] = \int_{x=-\infty}^{x=+\infty} x^k \cdot \tilde{f}(x,\underline{t})\, dx . \tag{2.228}$$

This integral has to be solved as a sequence of integrations of all $f_j(x, \underline{t})$, which are assigned to the original functions $X_j(\underline{t})$ included in $\tilde{X}(\underline{t})$. Applying α-discretization, for each specified $\underline{t}$

$$\tilde{m}_{k,t} = \{(m_{k,t\alpha}, \mu(m_{k,t\alpha})) \mid m_{k,t\alpha} = [m_{k,t\alpha l}, m_{k,t\alpha r}]; \; \mu(m_{k,t\alpha}) = \alpha \, \forall \, \alpha \in (0, 1]\}, \tag{2.229}$$

with

$$m_{k,t\alpha l} = \min_j [\int_{x=-\infty}^{x=+\infty} x^k \cdot f_{tj}(x)\, dx \mid X_{tj} \in X_{t\alpha}], \tag{2.230}$$

$$m_{k,t\alpha r} = \max_j [\int_{x=-\infty}^{x=+\infty} x^k \cdot f_{tj}(x)\, dx \mid X_{tj} \in X_{t\alpha}] \tag{2.231}$$

holds, in which $X_{t\alpha}$ denotes the random intervals $[X_{t\alpha\, l}, X_{t\alpha\, r}]$ referred to in Eq. (2.216).

The fuzzy moment functions $\tilde{m}_k(\underline{t})$ are determined by evaluating all original functions $X_j(\underline{t})$. They are fuzzy functions that explicitly depend on the parameter $\underline{t}$ but not on the $X_j(\underline{t})$.

For $k = 1$ Eq. (2.228) yields the *fuzzy expected value function*

$$\tilde{m}_X(\underline{t}) = E[\tilde{X}(\underline{t})] = \int_{x=-\infty}^{x=+\infty} x \cdot \tilde{f}(x, \underline{t})\, dx, \tag{2.232}$$

which characterizes the trend of the fuzzy random function. As each functional value represents a fuzzy number, with $\mu = 1$ the transition to the special case of a trend function of a real random function may be realized. With

$$\tilde{\zeta}_k = \tilde{E}[(\tilde{X}(\underline{t}) - \tilde{m}_X(\underline{t}))^k] = \int_{x=-\infty}^{x=+\infty} (x - \tilde{m}_X(\underline{t}))^k \cdot \tilde{f}(x, \underline{t})\, dx, \tag{2.233}$$

the *k-th fuzzy central moment function* of the one-dimensional fuzzy random function $\tilde{X}(\underline{t})$ is obtained. The evaluation of Eq. (2.233) is carried out corresponding to Eq. (2.229).

For $k = 2$ this leads to the *fuzzy variance function*

$$\mathrm{Var}[\tilde{X}(\underline{t})] = \tilde{\sigma}_X^2(\underline{t}) = \int_{x=-\infty}^{x=+\infty} (x - \tilde{m}_X(\underline{t}))^2 \cdot \tilde{f}(x, \underline{t})\, dx. \tag{2.234}$$

In view of the applications, moreover, the fuzzy covariance function and the fuzzy correlation function are required. The *fuzzy covariance function* regarding the parameter vectors $\underline{t}_1$ and $\underline{t}_2$ reads

$$\tilde{K}(\underline{t}_1, \underline{t}_2) = \mathrm{Cov}[\tilde{X}(\underline{t}_1), \tilde{X}(\underline{t}_2)] = E[(\tilde{X}(\underline{t}_1) - \tilde{m}_X(\underline{t}_1)) \cdot (\tilde{X}(\underline{t}_2) - \tilde{m}_X(\underline{t}_2))]. \tag{2.235}$$

Normalizing Eq. (2.235) with the aid of the fuzzy standard deviation yields the *fuzzy correlation function*

$$\tilde{\varrho}(\underline{t}_1, \underline{t}_2) = \frac{\tilde{K}(\underline{t}_1, \underline{t}_2)}{\sqrt{\mathrm{Var}(\tilde{X}(\underline{t}_1))} \cdot \sqrt{\mathrm{Var}(\tilde{X}(\underline{t}_2))}}. \tag{2.236}$$

The fuzzy covariance function and the fuzzy correlation function are symmetric functions, i.e., the following holds

$$\tilde{K}(\underline{t}_1,\underline{t}_2) = \tilde{K}(\underline{t}_2,\underline{t}_1), \tag{2.237}$$

$$\tilde{\varrho}(\underline{t}_1,\underline{t}_2) = \tilde{\varrho}(\underline{t}_2,\underline{t}_1). \tag{2.238}$$

In extension to real stationary processes, *stationary fuzzy random functions* are defined. A fuzzy random function is referred to as stationary in strong sense if the fuzzy random variable $\tilde{X}_t$ is independent of the parameter vector $\underline{t}$

$$\tilde{X}(\underline{t}) = \tilde{X}_t = \tilde{X} \ \forall \underline{t} \in \mathbf{T}. \tag{2.239}$$

This leads to a fuzzy probability distribution function $\tilde{F}_t(x)$ of $\tilde{X}_t$ that is also independent of $\underline{t}$, i.e., $\tilde{F}_t(x)$ does not vary for different $\underline{t} \in \mathbf{T}$

$$\tilde{F}(x,\underline{t}) = \tilde{F}_t(x) = \tilde{F}(x) \ \forall \underline{t} \in \mathbf{T}. \tag{2.240}$$

This definition also holds for the multidimensional case.

If the fuzzy random function represents a fuzzy second order process (as a formally extended real-valued second order process), the previous definition may be modified. A fuzzy second order process is considered stationary in wide sense if its fuzzy expected value function as well as its fuzzy variance function represents a stationary fuzzy function (Sect. 2.1.11.1), i.e., both fuzzy functions are independent of the parameter vector $\underline{t}$

$$\tilde{m}_X(\underline{t}) = E[\tilde{X}(\underline{t})] = \tilde{m}_X = \text{constant} \ \forall \underline{t} \in \mathbf{T}, \tag{2.241}$$

$$\tilde{\sigma}_X^2(\underline{t}) = \text{Var}[\tilde{X}(\underline{t})] = \tilde{\sigma}_X^2 = \text{constant} \ \forall \underline{t} \in \mathbf{T}. \tag{2.242}$$

From this it follows that the fuzzy correlation function and the fuzzy covariance function at all points $\underline{t}$ in the parameter space only depend on the difference $\underline{\delta} = \underline{t}_2 - \underline{t}_1$ in each particular direction.

As a fuzzy Gaussian process represents a second order process and is, furthermore, completely determined by its first two fuzzy moments, a fuzzy Gaussian process that is stationary in wide sense is also stationary in strong sense.

2.3.2.3 Fuzzy Random Field and Fuzzy Random Process as Special Cases of Fuzzy Random Functions

As fuzzy random functions combine both randomness and fuzziness and comprise arbitrary parameter vectors $\underline{t}$, the following special cases may be introduced:

- If the parameter vector $\underline{t}$ comprises only the time coordinate τ, i.e., $\underline{t} = \tau$, a special fuzzy random function arises as a *fuzzy random process*.
- If the parameter vector $\underline{t}$ is constituted only by the spatial coordinates $\underline{\theta} = (\theta_1, \theta_2, \theta_3)$, a *fuzzy random field* is obtained as a special case of the fuzzy random function.

- If the uncertainty to be specified exhibits only randomness but no fuzziness, the fuzzy bunch parameter vector $\underline{\tilde{s}}$ in the fuzzy bunch parameter representation according to Eq. (2.219) changes to the deterministic, i.e., crisp, parameter $\underline{s}$. Hence the special case of a real random function is stated.
- The special case of a fuzzy function is obtained if the uncertainty to be specified is characterized only by fuzziness. The uncertain mapping according to Eqs. (2.211) and (2.212) is then exclusively defined on the parameter space $\underline{\mathbf{T}}$.

The uncertainty characteristics fuzzy randomness, randomness, and fuzziness may be simultaneously considered in the structural analysis, in the safety assessment, and in the structural design, since both real random functions and fuzzy functions represent special cases of fuzzy random functions. Regarding applications in structural engineering, fuzzy random processes and fuzzy random fields are of particular interest.

Fuzzy Random Fields. With the aid of fuzzy random fields, fuzzy randomness that is time constant but varying subject to the spatial coordinates $\underline{\theta} = (\theta_1, \theta_2, \theta_3)$ may be described on a domain $\underline{B} \subseteq \mathbb{R}^n$. From Eq. (2.213) the special case

$$\underline{\tilde{X}}(\underline{\theta}) = \left\{\underline{\tilde{X}}_\theta = \underline{\tilde{X}}(\underline{\theta}) \;\; \forall\, \underline{\theta} \mid \underline{\theta} \in \underline{B} \subseteq \mathbb{R}^n\right\} \tag{2.243}$$

may be derived.

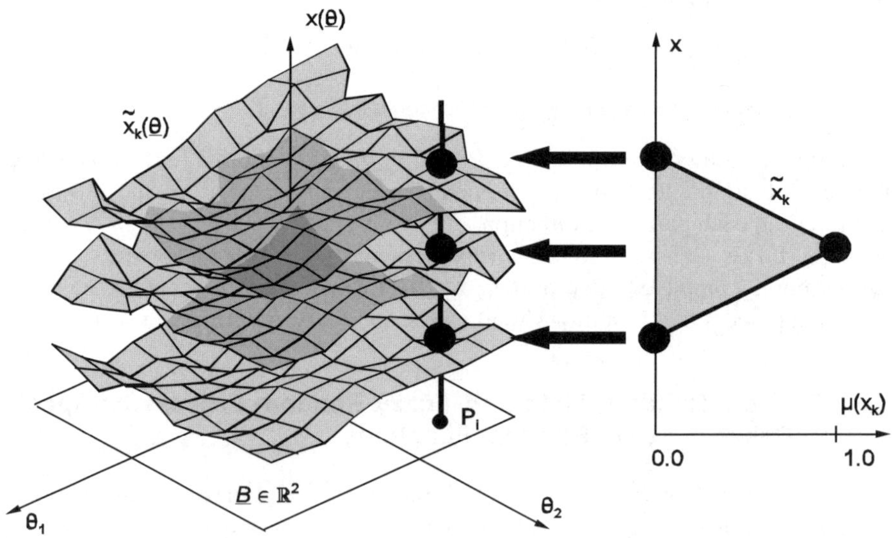

Fig. 2.44. Fuzzy realization of a fuzzy random field on a two-dimensional domain

With Eq. (2.243) the fuzzy random field is defined as a set of fuzzy random vectors on the domain $\underline{B}$. For a two-dimensional domain a fuzzy realization of a fuzzy random field is illustrated in Fig. 2.44. This realization represents a fuzzy function. With the realization $\omega_k \Rightarrow \tilde{x}_k$ the fuzzy variable $\tilde{x}_k$ is assigned to the point P_i.

For the numerical analysis the *bunch parameter representation*

$$\tilde{\underline{X}}(\underline{\theta}) = \underline{X}(\tilde{\underline{s}},\underline{\theta}) \tag{2.244}$$

is advantageous. From Eq. (2.220) follows

$$\tilde{\underline{X}}(\underline{\theta}) = \underline{X}(\tilde{\underline{s}},\underline{\theta}) = \Big\{\big(\underline{X}(\underline{\theta}),\, \mu(\underline{X}(\underline{\theta}))\big) \mid \underline{X}(\underline{\theta}) = \underline{X}(\underline{s},\, \underline{\theta});\; \mu(\underline{X}(\underline{\theta})) = \mu(\underline{s}) \;\; \forall \underline{s} \in \tilde{\underline{s}}\Big\}. \tag{2.245}$$

With the aid of *α-discretization* this may be described in accordance with Eq. (2.215)

$$\tilde{\underline{X}}(\underline{\theta}) = \Big\{\big(\underline{X}_{\alpha}(\underline{\theta}),\mu(\underline{X}_{\alpha}(\underline{\theta}))\big) \mid \mu(\underline{X}_{\alpha}(\underline{\theta})) = \alpha \;\forall\, \alpha \in (0,1]\Big\}, \tag{2.246}$$

whereby $\underline{X}_{\alpha}(\underline{\theta})$ denotes the random α-function sets

$$\underline{X}_{\alpha}(\underline{\theta}) = \Big\{\underline{X}(\underline{s},\underline{\theta}) \;\; \forall\, \underline{s} \mid \underline{s} \in \underline{S}_{\alpha};\;\; \alpha \in (0,1]\Big\}, \tag{2.247}$$

comprising all original functions on the level α.

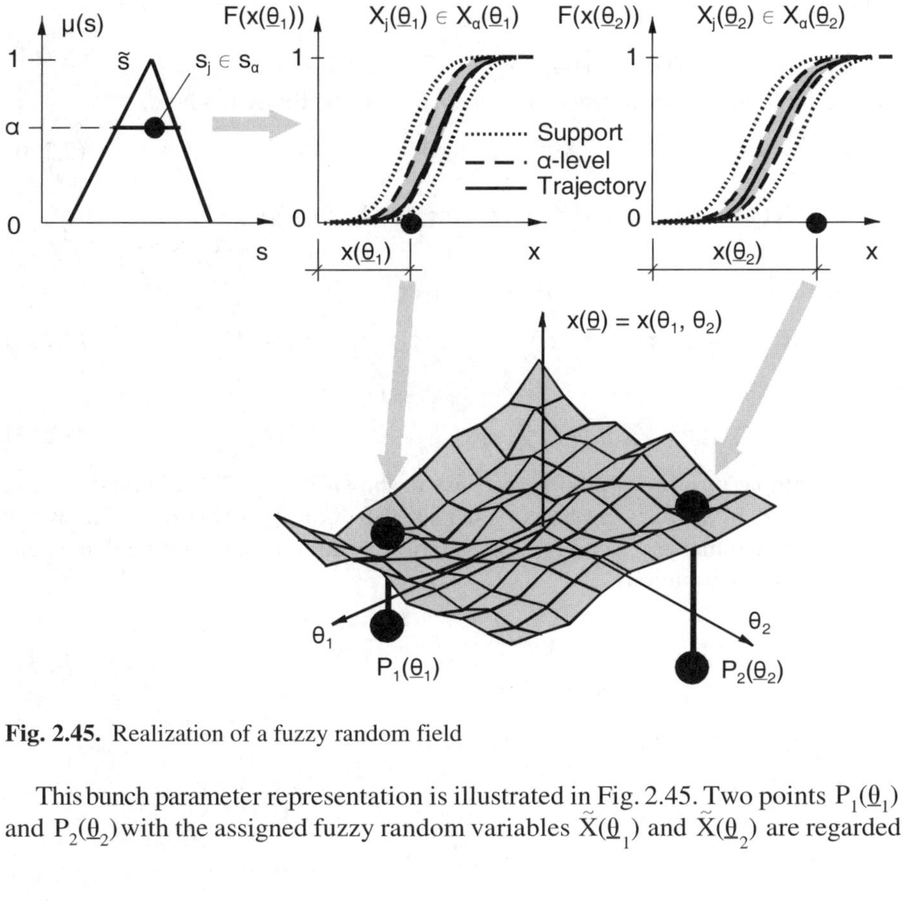

Fig. 2.45. Realization of a fuzzy random field

This bunch parameter representation is illustrated in Fig. 2.45. Two points $P_1(\underline{\theta}_1)$ and $P_2(\underline{\theta}_2)$ with the assigned fuzzy random variables $\tilde{X}(\underline{\theta}_1)$ and $\tilde{X}(\underline{\theta}_2)$ are regarded and the associated fuzzy probability distribution functions $\tilde{F}(x(\underline{\theta}_1))$ and $\tilde{F}(x(\underline{\theta}_2))$ are plotted. Both fuzzy probability distribution functions depend on the fuzzy bunch parameter $\underline{\tilde{s}}$. All originals that are accounted for by $\underline{s}_j \in \underline{s}_\alpha$ belong to the level α. With a specific value $\underline{s}_j$ precisely one original of each fuzzy random variable $\tilde{X}(\underline{\theta}_1)$ and $\tilde{X}(\underline{\theta}_2)$ is chosen. These originals are characterized by the fuzzy realizations $\tilde{x}(\theta_1)$ and $\tilde{x}(\theta_2)$. Considering the realizations of the originals for all points $P_i(\theta_i)$ then leads to one realization of one original function of the fuzzy random field.

A stationary fuzzy random field is called *isotropic* if the fuzzy correlation function and the fuzzy covariance function at all points $\underline{\theta} \in \underline{\mathbf{\Theta}}$ only depend on the absolute distance $\|\underline{\delta}\| = \|\underline{\theta}_2 - \underline{\theta}_1\|$

$$\tilde{K}(\underline{\theta}, \underline{\theta} + \underline{\delta}) = \tilde{K}(\|\underline{\delta}\|) \; \forall \, \underline{\theta} \in \underline{\mathbf{\Theta}} \; . \tag{2.248}$$

Fuzzy Random Processes. With the aid of fuzzy random processes time-dependent influences, like, e.g., a water pressure depending on the water level, may be accounted for. In the case of fuzzy random processes Eq. (2.213) takes the special form

$$\underline{\tilde{X}}(\tau) = \left\{ \underline{\tilde{X}}_\tau = \underline{\tilde{X}}(\tau) \;\; \forall \, \tau \mid \tau \in \mathbf{T} \right\} . \tag{2.249}$$

According to the *bunch parameter representation* the following holds

$$\underline{\tilde{X}}(\tau) = \underline{X}(\underline{\tilde{s}}, \tau), \tag{2.250}$$

$$\begin{aligned} \underline{\tilde{X}}(\tau) = \underline{X}(\underline{\tilde{s}}, \tau) = \{ (\underline{X}(\tau), \mu(\underline{X}(\tau))) \mid \; & \underline{X}(\tau) = \underline{X}(\underline{s}, \tau); \\ & \mu(\underline{X}(\tau)) = \mu(\underline{s}) \;\; \forall \, \underline{s} \in \underline{\tilde{s}} \} \; . \end{aligned} \tag{2.251}$$

The application of the *α-discretization* leads to

$$\underline{\tilde{X}}(\tau) = \left\{ (\underline{X}_\alpha(\tau), \mu(\underline{X}_\alpha(\tau))) \mid \mu(\underline{X}_\alpha(\tau)) = \alpha \; \forall \, \alpha \in (0,1] \right\} , \tag{2.252}$$

with

$$\underline{X}_\alpha(\tau) = \left\{ \underline{X}(\underline{s}, \tau) \;\; \forall \, \underline{s} \mid \underline{s} \in \underline{S}_\alpha \; ; \;\; \alpha \in (0,1] \right\} . \tag{2.253}$$

An example of a fuzzy random process is shown in Fig. 2.46. This process is characterized by a fuzzy probability density function $\tilde{f}(x)$ that depends on the fuzzy bunch parameters $\tilde{s}_1 = \tilde{\sigma}_x(\tau)$ and $\tilde{s}_2 = \tilde{m}_x(\tau)$ and represents a time-dependent fuzzy normal distribution

$$\tilde{f}(x, \tau) = \frac{1}{\tilde{\sigma}_X(\tau) \cdot \sqrt{2\pi}} \cdot e^{-0.5 \left(\frac{x - \tilde{m}_X(\tau)}{\tilde{\sigma}_X(\tau)} \right)} . \tag{2.254}$$

This is illustrated for three time points τ_1, τ_2, and τ_3.

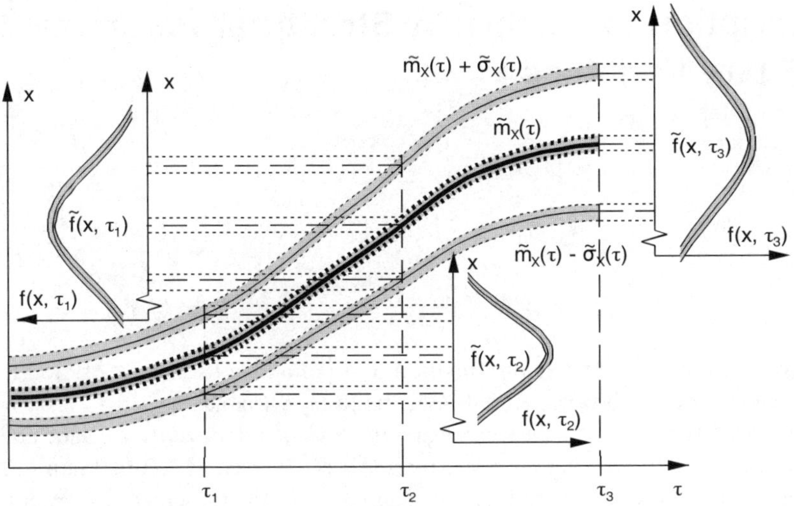

Fig. 2.46. Fuzzy random process

3 Description of Uncertain Structural Parameters as Fuzzy Variables

The specification of the membership function $\mu_A(\underline{x})$ of a fuzzy set $\tilde{\underline{A}}$ is referred to as fuzzification. The result of fuzzification is the fuzzy variable $\tilde{\underline{x}}$.

By means of fuzzification the *informal and lexical uncertainty* of data and model parameters with the characteristic *fuzziness* is assessed as an uncertain set. Uncertainty of this type may occur in physical structural parameters, the parameters of probability distributions, as well as in the processes of design and planning, construction, utilization, damage, preservation, and strengthening. These processes are characterized by highly subjective factors. These include, e.g., human mistakes and errors as well as uncertain boundary conditions. Uncertainty in the above-mentioned processes influences the physical structural parameters; the effects of uncertainty may be described mathematically by fuzzification [118].

The fuzzified variables may be fuzzy input parameters (data uncertainty) as well as fuzzy model parameters (model uncertainty). The criterion for distinguishing between data uncertainty and model uncertainty is the assignment of the respective uncertainty to the input variables or the model, as introduced in Sect. 1.1. In a specific case this assignment is linked to the definition of the model. In a safety assessment the respective mechanical model and hence the fuzzy model parameters exclusively enter the limit state surface. In the fuzzy probabilistic safety concept (Chap. 6) fuzzified physical structural parameters are always treated as fuzzy model parameters. The fuzzified parameters of the probability distributions of stochastic structural parameters form a part of the fuzzy probabilistic basic variables and hence a part of the data uncertainty.

The membership function of a fuzzy variable (Sect. 2.1) is independent of both the model and the type of variable to be fuzzified (physical structural parameter or parameter of a probability distribution).

The fuzzification of uncertain parameters is restricted here to the one-dimensional case, and only elementary fuzzification methods are considered. The fuzzification of physical structural parameters is demonstrated in the following. The application of the algorithms for quantifying the uncertain parameters of fuzzy probability distributions is dealt with in Chap. 4.

3.1 Data Uncertainty – Specification of Membership Functions

The specification of a membership function depends on the information available. Information on structural parameters may be available in the form of

- A sample containing only a small number of elements (information type I)
- A linguistic assessment (information type II)
- A singular, uncertain result of measurement (information type III)
- Knowledge based on experience (information type IV)

The information available generally consists of various types and must be processed by means of a combination of the fuzzification methods described in the following.

The starting point for fuzzification is the definition of the *fuzzy set* (Sect. 2.1). In order to describe the fuzzy set $\tilde{A}$ on the fundamental set **X** it is necessary to formulate a criterion that is more or less satisfactorily fulfilled by the elements $x \in \mathbf{X}$. This criterion may either represent an uncertain proposition or an event. In order to specify the membership function $\mu_A(x)$, all $x \in \mathbf{X}$ are gradually assessed in relation to the formulated criterion. The complementation of fuzzy sets according to Eq. (2.17) must thereby be observed, as the assessment of the membership of the elements $x \in \mathbf{X}$ to the fuzzy set $\tilde{A}$ also includes an assessment of their membership to the complementary set $\tilde{A}^C$. The fundamental set **X** is constructed using physical structural parameters or parameters of probability distributions.

The generated membership function reflects a subjective assessment of objective conditions. It is appropriate to choose simple functional formulations such as linear or polygonal types for the $\mu_A(x)$; complicated or involved descriptions are less suitable for this purpose.

Fuzzification for Information Type I. The membership function is specified on the basis of existing data comprising elements of a sample. The assessment criterion for the elements x is directly related to numerical values derived from **X**. If more than one sample element is available for the uncertain parameter, the fuzzification process may be backed up by the application of simple mathematical algorithms. For this purpose the objective information, i.e., the sample elements, are represented in a histogram, which provides a starting point for the fuzzification suggestion. In this context histograms serve as design aids for the membership function. A fuzzification suggestion obtained from the latter is referred to as an *initial draft* of the membership function. This forms a basis for the inclusion of subjective aspects for correcting or adapting the initial draft.

Preparation of the available data (sample elements) is necessary for the *initial draft* of the membership function. The parameter to be fuzzified has a definition domain representing the fundamental set **X** of the sample elements x_i. The fundamental set **X** is subdivided into disjoint crisp subsets X_k, in which the following holds

$$\bigcup_k X_k = \mathbf{X} . \tag{3.1}$$

The sample elements are assigned to the subsets X_k. The result of the assignment is plotted in the form of a histogram. The width of the subsets is chosen in such a way that an appropriate histogram construction is obtained. The use of different subset widths leads to variants of the membership function.

By way of example, four different types of function are proposed for specifying the membership function:

- A linear function
- A polygonal function
- A quadratic parabola, and
- A function corresponding to the probability distribution function of a normal distribution

The applied functions may be used to represent a fuzzy number or a fuzzy interval. For this purpose the functional formulation is matched to the data with the aid of the method of least squares. From the previously constructed histogram the subset X_k containing the largest number of sample elements is selected. All subsets containing smaller values of x are used for computing the left-hand branch of $\mu_A(x)$ while the remaining subsets are used for computing the right-hand branch. The subset X_k is taken into consideration in both branches. The branches of $\mu_A(x)$ are formulated as functions of x in such a way that their functional values yield the number of sample elements contained in the respective subsets as closely as possible. A requirement in this respect is that the sum of the squared differences between the actual number of sample elements in the respective subset $n(X_k)$ and the functional value $\mu_A(x_{k,m})$ in the middle of the subset (at $x_{k,m}$) is a minimum

$$\sum_k \left[n(X_k) - \mu_A(x_{k,m})\right]^2 \rightarrow \min . \tag{3.2}$$

The two functions obtained generally have a point of intersection. In the case of a fuzzy number this point of intersection represents the mean value of the fuzzy set. The two neighboring zeroes of the functions mark the interval bounds of the support. The obtained membership function is finally normalized so that the functional value at the mean value point is unity. The initial draft is completed when the membership function satisfies all the properties formulated in Sect. 2.1. Where necessary, the membership function must be matched to further requirements or additional constraints by subsequent correction (e.g., variation of the subset width). Fuzzy intervals arise when the membership function of the computed fuzzy number is multiplied by a factor larger than unity, and finally "truncated" for $\mu = 1.0$, i.e., when the α-level set for $\alpha = 1.0$ is taken to be the mean interval.

As a supplement or an alternative, additional conditions may be prescribed for the shape of the membership function. In the case of fuzzy numbers, for example, the mean value may be placed in the middle of the subset containing the largest number of sample elements. In the case of fuzzy intervals the mean interval may be defined in advance. It is also possible to prescribe the interval bounds of the support or the additional consideration of subsets that lie outside the sample domain and that contain no elements.

Example 3.1. The formulation of the initial draft of a membership function is demonstrated by considering the determination of the compressive strength of sandstone comprising a historical facade. In order to integrate a natural stone facade made of sandstone into a new structure it is necessary to check whether the facade also complies with safety standards under modified loading conditions. In order to determine the load-bearing capacity of the facade it is necessary to assess, among other factors, the compressive strength of the existing sandstone. For this purpose core drillings were extracted at 16 positions in the sandstone masonry. The drillings were performed at right angles to the plane of the wall. The direction of loading of the specimens in compressive tests and the in situ direction of loading are hence perpendicular to one another. The measured compressive strengths are listed in Table 3.1.

The following proposition holds for the model values: "At right angles to the plane of sedimentation an increase in the compressive strength could be possible to a certain degree, i.e., the downward scatter of the characteristic values of the compressive strength decreases". An inspection of the construction site revealed that the state of the historical ashlar stone work was *good* to *very good*.

Table 3.1. Measured compressive strength of sandstone specimens

Specimens labels	Compressive strength [N/mm²]	Specimens labels	Compressive strength [N/mm²]
1	22.3	9	27.5
2	23.4	10	19.8
3	28.8	11	45.7
4	11.4	12	32.6
5	17.5	13	32.1
6	22.5	14	44.7
7	26.5	15	41.8
8	25.2	16	42.1

Based on the objective information obtained in the compressive tests and the subjective opinion of experts it is necessary to specify the membership function for the fuzzy variable *compressive strength of the sandstone* $\tilde{\beta}_D$. For this purpose the measured values were lumped together in groups and plotted in a histogram (Fig. 3.1).

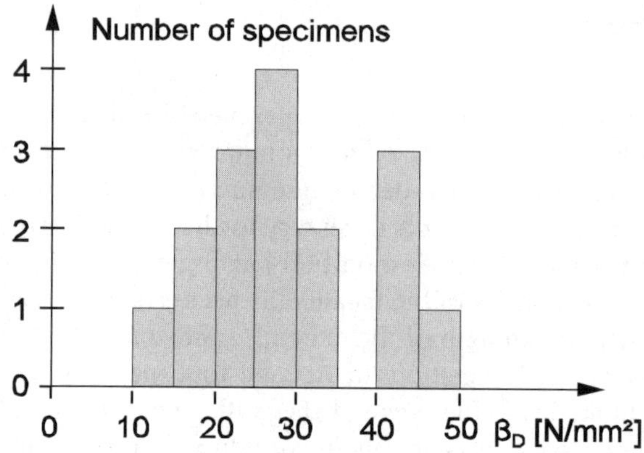

Fig. 3.1. Histogram of the compressive strength of sandstone specimens

With the aid of different functional formulations different suggestions were developed for the membership function $\mu(\beta_D)$. A fuzzy number and a fuzzy interval with linear branches of the membership function are shown in Fig. 3.2. Due to the fact that the fuzzy number significantly weights the lower compressive strength values it is appropriate to conduct a subsequent subjective modification. The fuzzification suggestions shown in Fig. 3.3 were generated using quadratic functions. Initial drafts obtained with the aid of a polygonal membership function are shown in Fig. 3.4. The suggestions shown in Fig. 3.5 are based on functional formulations corresponding to the probability distribution function of a normal distribution. The dashed curves shown in Figs. 3.3 and 3.4 do not fulfill the requirements imposed on a membership function. The initial drafts were modified by additional constraints and corrections.

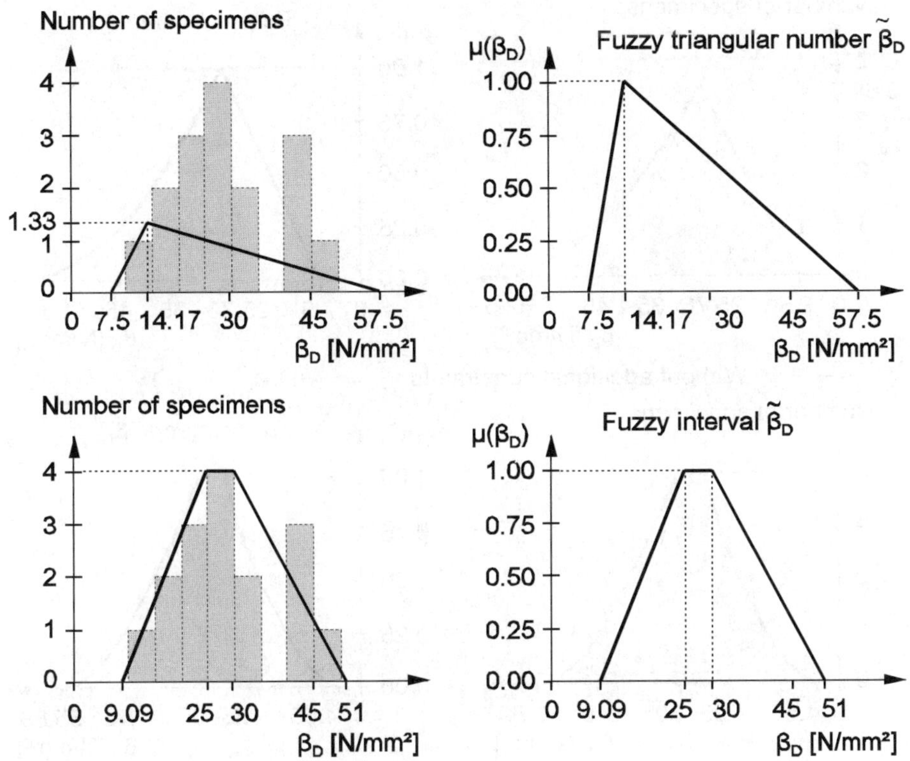

Fig. 3.2. Histogram and fuzzification of the compressive strength of sandstone; initial drafts for a fuzzy number and a fuzzy interval with linear branches; non-normalized and normalized membership functions

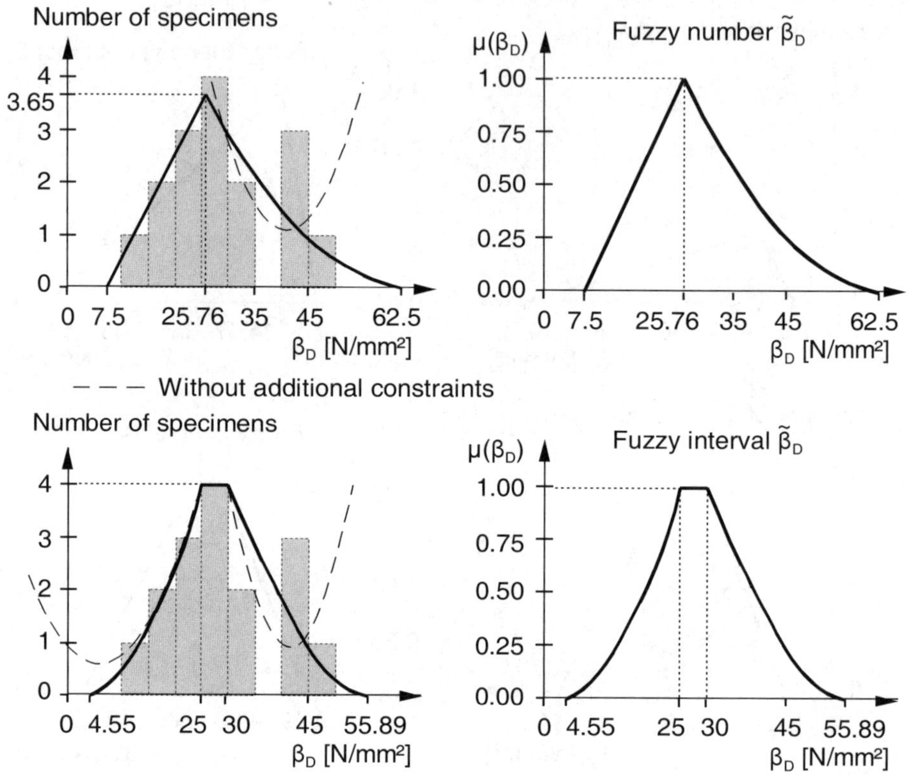

Fig. 3.3. Histogram and fuzzification of the compressive strength of sandstone; initial drafts for a fuzzy number and a fuzzy interval with quadratic branches; non-normalized and normalized membership functions

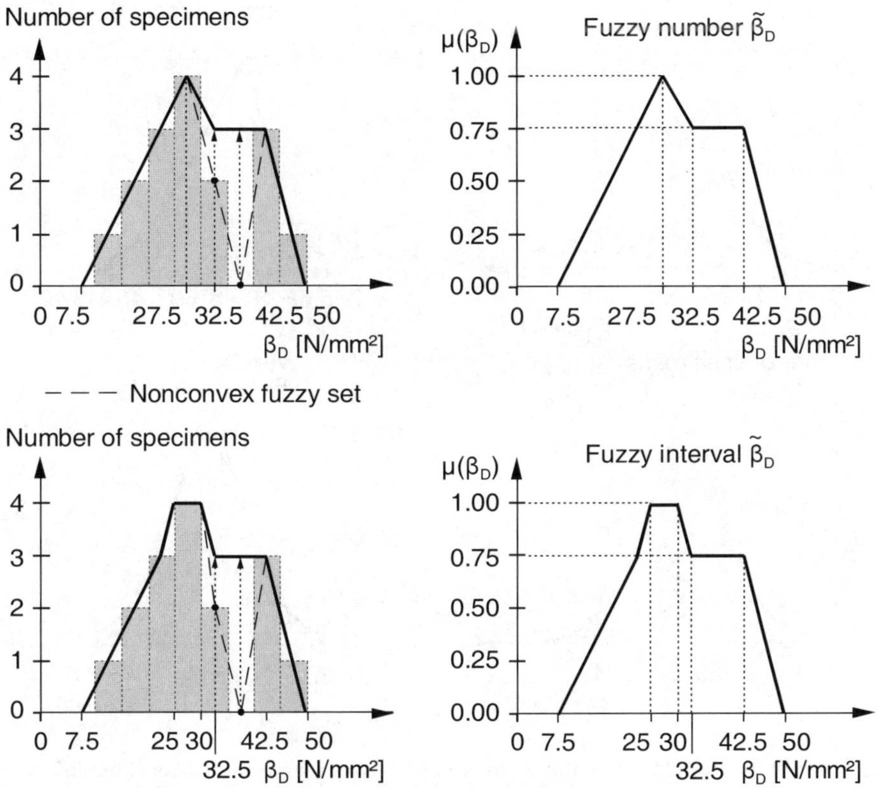

Fig. 3.4. Histogram and fuzzification of the compressive strength of sandstone; initial drafts for a fuzzy number and a fuzzy interval with polygonal branches; non-normalized and normalized membership functions

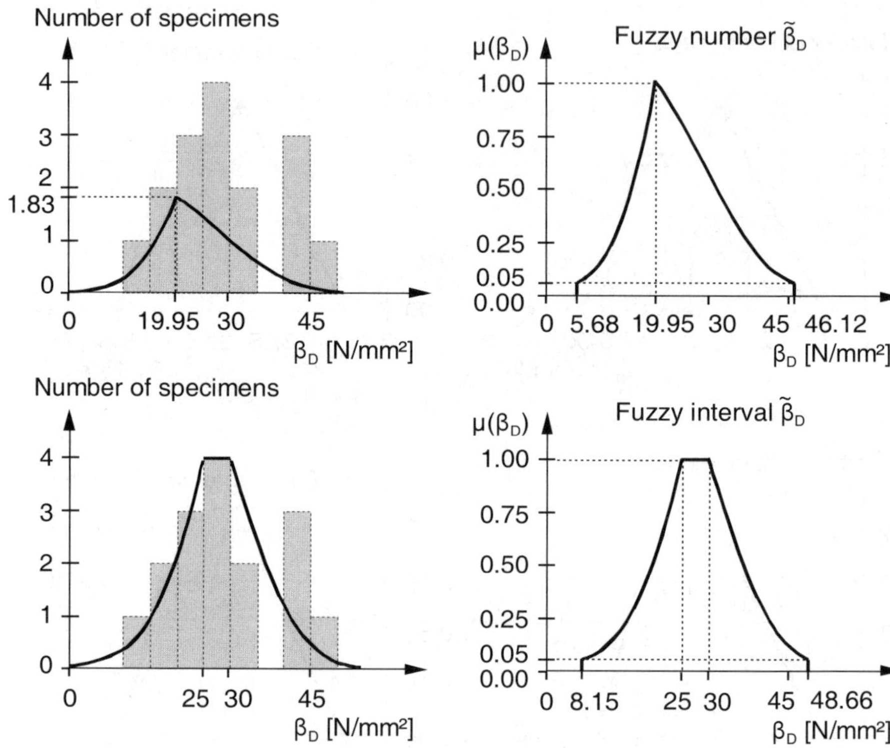

Fig. 3.5. Histogram and fuzzification of the compressive strength of sandstone; initial drafts for a fuzzy number and a fuzzy interval with membership functions in the form of the probability distribution function of a normal distribution; non-normalized and normalized membership functions

Fuzzification for Information Type II. The assessment criterion for the elements x of **X** may be expressed using linguistic variables (Sect. 2.1.4). As numerical values are required in order to carry out a structural analysis and safety assessment, it is necessary to transform the linguistic variables to a numerical scale. By combining the linguistic values with the aid of modifiers a wide spectrum is available for the purpose of assessment.

Example 3.2. In order to assess the foundation soil it is possible, e.g., to apply the linguistic variable *load-bearing capacity.* An assessment of the *load-bearing capacity* is carried out by assigning the linguistic values *high, medium,* and *low* in combination with modifiers such as *extremely* and *very.* The combinations of linguistic values and modifiers are mapped onto a numerical scale representing the sustainable soil pressure. This approach is demonstrated by way of example in Fig. 3.6.

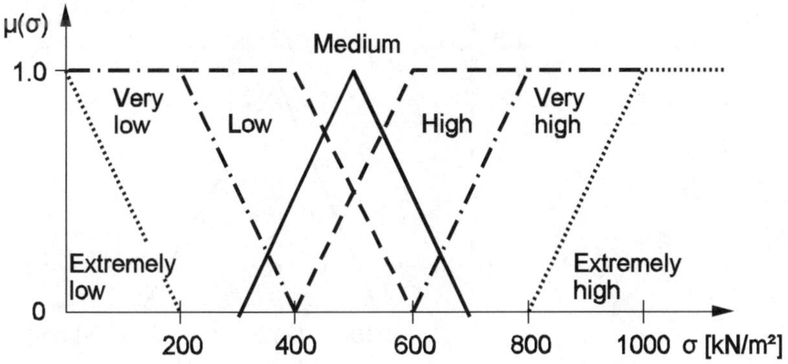

Fig. 3.6. Assessment of the linguistic variable *load-bearing capacity*

Fuzzification for Information Type III. If only a single numerical value from **X** is available as an uncertain result of measurement $\tilde{x}_m$, the assessment criterion for the elements x may be derived from the uncertainty of the measurement. The uncertainty observed in the measurement is quantified on the assigned numerical scale. A distinction is thereby made between the following two cases:

1. The measured result marks the uncertain position of the measured value on the scale.
2. The measured result describes the uncertain transition between two complementary states, somewhat similar to the gray zone of a black-white transition.

In case 1 the experimenter evaluates the uncertain observation for different membership levels. For the level $\mu_A(x) = 1$ a single measurement or a measurement interval is specified in such a way that the observation may be considered to be "as crisp as possible". For the level of the support ($\mu_A(x) = 0$) a measurement interval is determined that contains all possible measurements within the scope of the observation. An assessment of the uncertain measurements for intermediate levels is left up to the experimenter. The membership function is generated by interpolation or by connecting the determined points (x, μ_A(x)).

Example 3.3. An uncertain measurement in the sense of case 1 arises, e.g., when the thickness of a structural member with an uneven or rough surface is to be determined (Fig. 3.7). For measurements made using an analog measuring device a single scale value cannot be assigned without doubt to the thickness of the structural member. Uncertainty is attached to each measurement recorded. As shown in Fig. 3.7, all values between 210 mm and 215 mm are possible results of the measurement; these constitute the support of the fuzzy thickness $\tilde{d}$. The thickness d = 212 mm is chosen as the "best possible crisp" measured value. By connecting the three points the membership function for $\tilde{d}$ is obtained as a fuzzy triangular number.

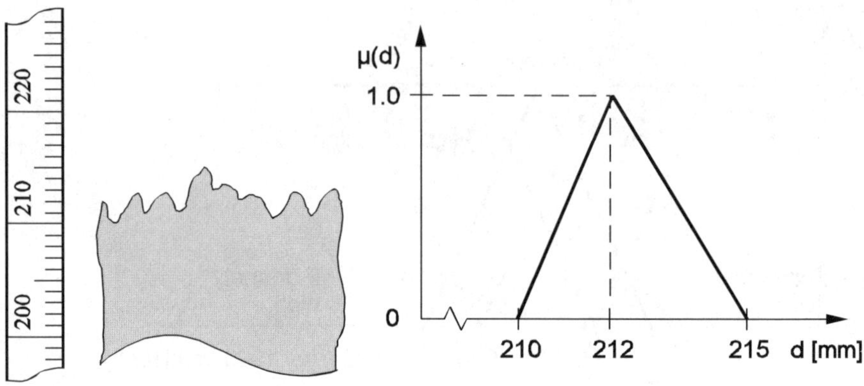

Fig. 3.7. Fuzzification of the uncertain thickness $\tilde{d}$ of a structural member

In case 2 the experimenter describes the uncertain observation by means of a continuous, monotonically increasing function B(x), which maps the possible measurements onto the interval [0, 1]. A suitable numerical scale is assigned to the observation. All scale values that may be assigned without doubt to state one are assessed with B(x) = 0. For measured values that belong without doubt to the complementary state two the functional value B(x) = 1 is adopted. For the uncertain transition from state one to state two the function B(x) is specified by the experimenter. The membership function $\mu_A(x)$ is obtained by differentiation of B(x) followed by normalization.

Example 3.4. For determining the water level w in a river a measuring staff is lowered onto the river bed for a short period of time. The water level is determined by noting the scale reading w on the staff separating the wet portion of the staff from the dry portion. The lower part of the staff may be considered without doubt to be "wet" (state one) and the upper part to be "not wet" (state two). As discussed in [194], a smooth transition is observed between these two states (Fig. 3.8). The function B(w) describes this transition. The membership function μ(w) developed from the latter describes the fuzzy water level $\tilde{w}$.

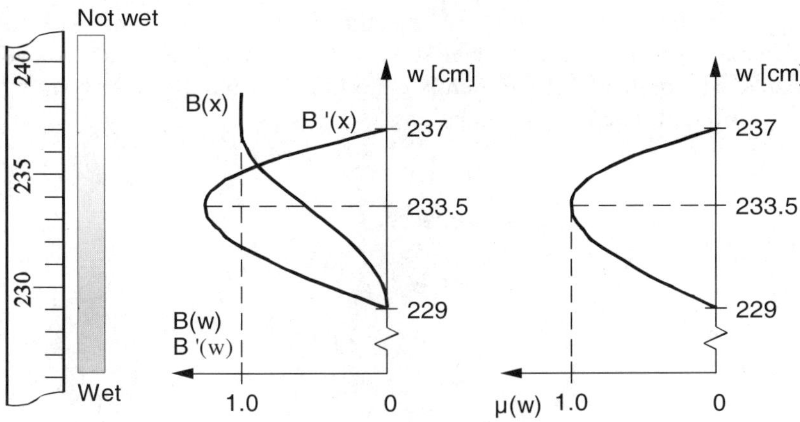

Fig. 3.8. Fuzzification of an uncertain water level $\tilde{w}$

Fuzzification for Information Type IV. When specifying the membership function it is always necessary to take into consideration the opinions of experts or expert groups, experience gained from comparable problems, and additional information where necessary. These subjective aspects generally supplement the initial draft of a membership function generated from measured values or linguistic variables. This also permits a consideration of the effects of the sample size, possible errors in measurement, and other inaccuracies attached to the fuzzification process. If neither sample elements nor linguistic variables are available, fuzzification depends entirely on estimates by experts.

When considering knowledge based on experience two alternative approaches may be adopted.

1. The fundamental set **X** is a discrete set containing known elements, or is available in a discretized form, e.g., as a set of subsets. A membership value must be directly specified for each element x of **X**. In order to generate the membership function of a continuous fuzzy set predefined points $(x, \mu_A(x))$ are connected either by a polygon or by means of interpolation functions. The suggested fuzzification algorithms for information type I may be applied for this purpose.
2. For a continuous, i.e., nondiscretized fundamental set **X**, it is necessary to assess a quasi-infinite number of elements. For this purpose a crisp set may initially be specified as a kernel set of the fuzzy set. The boundary regions of this crisp kernel set are finally "smeared" by assigned membership values $\mu_A(x) < 1$ to elements close to the boundary and leading the branches of $\mu_A(x)$ beyond the boundaries of the crisp kernel set monotonically to $\mu_A(x) = 0$ (Fig. 3.9). By this means elements that do not belong to the crisp kernel set, but are located "in the proximity" of the latter, are also assessed with membership values of $\mu_A(x) > 0$.

This approach may be extended by selecting several crisp kernel sets for different membership levels (α-level sets) and specifying the $\mu_A(x)$ in level increments. The branches of the membership function may be described with the aid of different functional formulations.

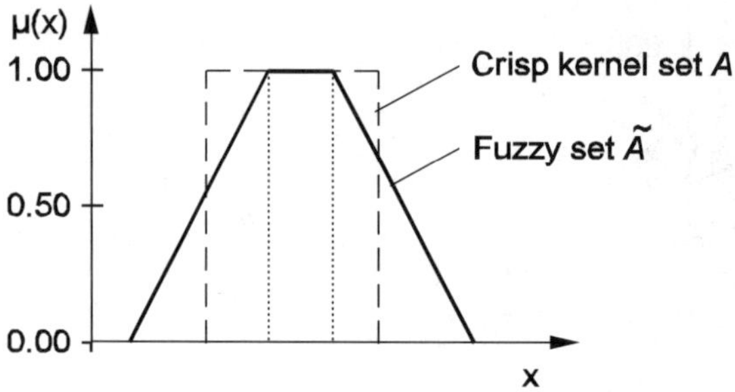

Fig. 3.9. Fuzzification with the aid of a crisp kernel set

Example 3.5. The initial draft of the membership function $\mu_A(\beta_D)$ for the compressive strength of sandstone (Example 3.1) is modified with the aid of expert knowledge. The starting point is the fuzzy interval according to Fig. 3.2.

A requirement is that all sample elements are accounted for by $\tilde{\beta}_D$. In other words the crisp kernel set should contain all test results; the crisp boundaries of this set then take on values of $\beta_{D\,l} = 11.4$ N/mm² and $\beta_{D\,r} = 45.7$ N/mm² (Fig. 3.10). For comparison: the value of the support of the initial draft according to Fig. 3.2 is $S(\tilde{\beta}_D) = [9.09, 51]$ N/mm². As the crisp kernel set is a subset of the support the above requirement is fulfilled. The left-hand boundary of the support $S(\tilde{\beta}_D)$ is obtained only slightly below the smallest measured compressive strength in the initial draft; a reduction in this value appears appropriate. According to the subjective opinions of several experts the sandstone compressive strength is considered to be good in overall terms. As a compromise, a strength below $\beta_D = 5$ N/mm² is thus ruled out. The right-hand boundary of the support is raised slightly to 52 N/mm². As one requirement imposed by the expert group on the mean interval has already been realized in the construction of the initial draft, this remains unaltered (Fig. 3.10).

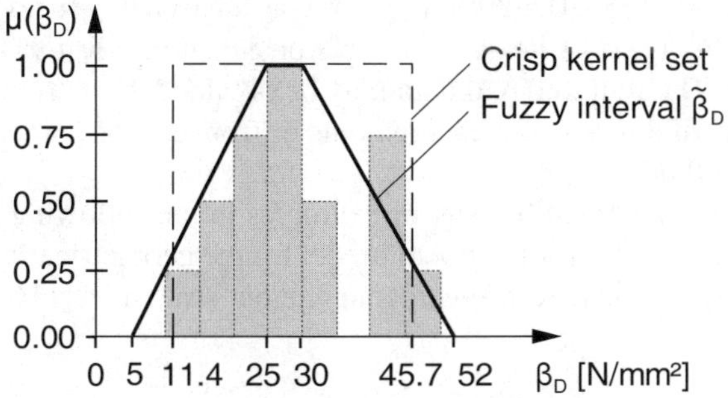

Fig. 3.10. Fuzzification suggestion for the compressive strength of sandstone

3.2 Model Uncertainty - Construction of Fuzzy Models

Fuzzy models are uncertainty models possessing the uncertainty characteristic fuzziness. Model uncertainty is generated during the abstraction process of model construction (Sect. 1.1). A distinction must be made between two essentially different approaches regarding model construction:

1. Formulation of an explicit algorithm with "free parameters". The parameters are linked to the formulated algorithm; the specific assignment of parameter values is problem dependent. These rank as model parameters. If the parameters are treated as fuzzy variables, the model becomes a fuzzy model. The fuzzy model parameters must comply with the definitions for fuzzy variables. Within the framework of the fuzzy probabilistic safety assessment (Chap. 6) model uncertainty enters the limit state surfaces, thus resulting in uncertain system behavior.
2. Formulation of rules for representing a physical process. The physical process to be modeled may only be characterized by a real "data set" comprised of observed input parameters and assigned result values. Based on this information a complex structure following certain rules is generated incrementally with the aid of simple, e.g., logical, combinations. This structure represents a model. After evaluating all of the available real data a certain amount of "leeway" remains in the established structure for mapping input parameters onto result values. This "leeway" represents model uncertainty. It is not possible to explicitly state the model parameters responsible for this model uncertainty. If the model uncertainty is identified as fuzziness, the structure is then a fuzzy model.

For the development of fuzzy models following the second approach, in which fuzzy model parameters are not explicitly known, information processing and control engineering methods may be applied. For example, the method of *approximate reasoning* [10, 22, 68, 155, 207, 208] may serve as a basis, or a fuzzy model may be formulated on the basis of *neural networks* or *genetic algorithms* [27, 35, 38, 75, 103, 109, 216]. The required real data may be obtained, e.g., from test results, from the evaluation of a series of measurements or from sets of points determined by an alternative method.

Only the first approach is demonstrated by way of examples in the following. The fuzzy models presented are submodels of a global model for the geometrically and physically nonlinear analysis of plane reinforced concrete bar structures [116, 133].

Fuzzy Model for Crack Formation and Tension Stiffening in Reinforced Concrete. In the abstraction process for constructing the model it is assumed that reinforced concrete is a heterogeneous material. Specific physical processes such as crack formation are dependent on a variety of factors that generally exhibit uncertainty. Factors that influence crack formation include, e.g., the granular shape and granular size of aggregates, the strength of the cement paste, and the bond between the aggregates and the cement paste. Exact prognoses concerning the location and spacing of anticipated cracks as well as crack widths and crack distribution over the cross section are not possible. Moreover, information regarding the exact behavior of the crack strain ε_{ctm} and crack stress f_{ctm} would also be necessary. Crack widths and crack spacing are thus treated as being uncertain.

The bond between reinforcement steel and the surrounding concrete results in tension stiffening, described more simply as the contributory action of the concrete between the cracks. Also, in the case of this physical phenomenon the inhomogeneous properties of the reinforced concrete lead to local uncertainty.

The frequently adopted phenomenological description of crack formation and tension stiffening results in the deterministic model presented in Fig. 3.11 for the stress-strain dependency in the tensile zone.

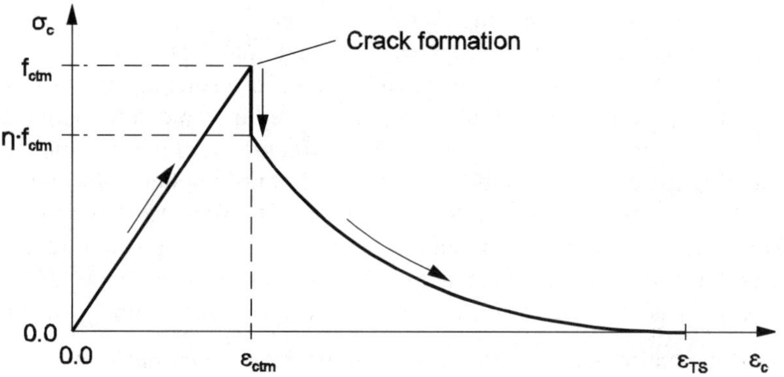

Fig. 3.11. Deterministic model for crack formation and tension stiffening

The falling branch of this curve is described by

$$\sigma_c = \eta \cdot f_{ctm} \cdot \left(\frac{\varepsilon_{TS} - \varepsilon_c}{\varepsilon_{TS} - \varepsilon_{ctm}} \right)^p . \tag{3.3}$$

All variables in Eq. (3.3) are deterministic model parameters. If the uncertainty that actually exists is described by the fuzzy variables $\tilde{\eta}$, $\tilde{\varepsilon}_{TS}$, $\tilde{\varepsilon}_{ctm}$, $\tilde{f}_{ctm}$, and $\tilde{p}$, Eq. (3.3) becomes

$$\tilde{\sigma}_c = \tilde{\eta} \cdot \tilde{f}_{ctm} \cdot \left(\frac{\tilde{\varepsilon}_{TS} - \varepsilon_c}{\tilde{\varepsilon}_{TS} - \tilde{\varepsilon}_{ctm}} \right)^{\tilde{p}} . \tag{3.4}$$

This equation is a fuzzy function that describes a fuzzy model for crack behavior and tension stiffening. The fuzzy model is shown in Fig. 3.12.

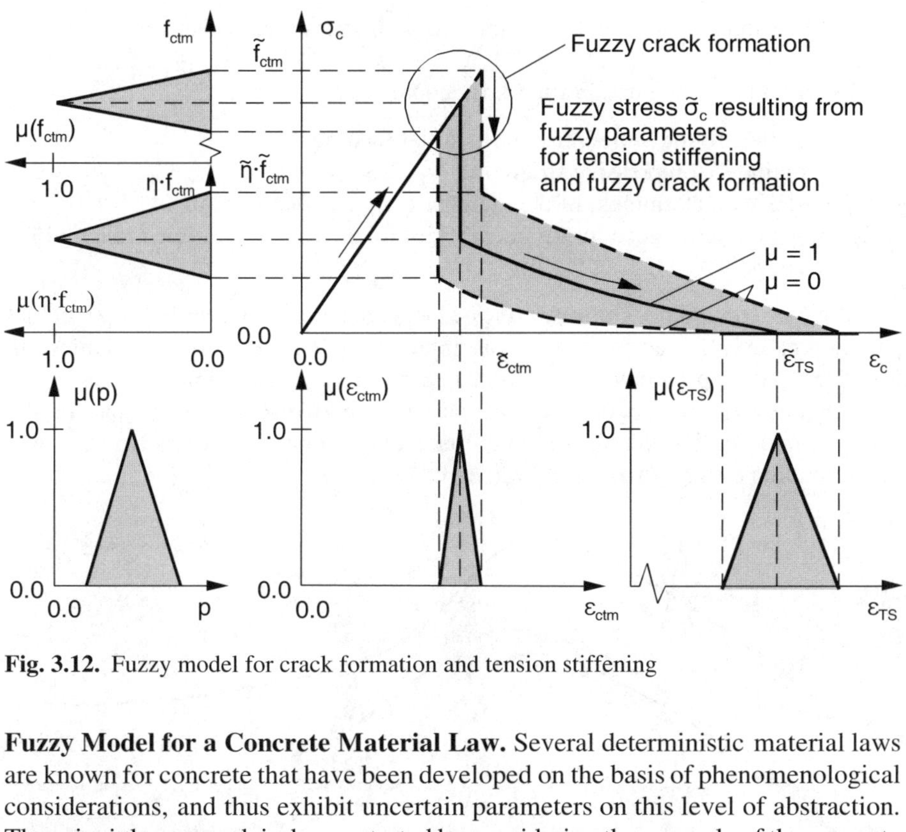

Fig. 3.12. Fuzzy model for crack formation and tension stiffening

Fuzzy Model for a Concrete Material Law. Several deterministic material laws are known for concrete that have been developed on the basis of phenomenological considerations, and thus exhibit uncertain parameters on this level of abstraction. The principle approach is demonstrated by considering the example of the concrete material law after Ma/Bertero [106] and Meskouris/Krätzig [113].

Using this material law it is possible to simulate monotonic stress-strain behavior as well as hysteretical material behavior. The functional curves are thereby dependent on several parameters that are able to account for special effects such as, e.g., the effects of the strapping action of stirrup reinforcement, the effects of local contact forces during crack closure, or strain softening.

In [106, 113, 133] deterministic values are assigned to these parameters, even though they exhibit uncertainty on closer examination. Empirical values from phenomenological observations, which are often small in number, are generally adopted as parameters. The limited number of observed results may still be used, however, to describe selected parameters as fuzzy model parameters.

In order to demonstrate this principle, attention is restricted to the initial loading curve. This is approximated by the continuous function

$$\sigma_c(\varepsilon) = \frac{\varepsilon\, E_c}{1 + \left(\dfrac{\varepsilon_{c0} E_c}{f_{c0}} - \dfrac{f_{cn}}{f_{cn}-1}\right)\dfrac{\varepsilon}{\varepsilon_{c0}} + \dfrac{1}{f_{cn}-1}\left(\dfrac{\varepsilon}{\varepsilon_{c0}}\right)^{f_{cn}}} \,. \tag{3.5}$$

This also permits a consideration of the action of stirrup reinforcement.

The parameters included in Eq. (3.5) denote the following:

- E_c: initial Young's modulus under initial loading
- f_{c0}: cylinder compressive strength
- ε_{c0}: strain on attainment of the cylinder compressive strength
- f_{cn}: parameter to account for lateral reinforcement in the falling branch of the curve

All parameters may exhibit fuzziness. Dependencies between the fuzzy model parameters may be accounted for as interaction between fuzzy variables. In Eq. (3.5) the crisp parameters are replaced by the fuzzy variables $\tilde{E}_c, \tilde{f}_{c0}, \tilde{\varepsilon}_{c0}, \tilde{f}_{cn}$. The obtained fuzzy function $\tilde{\sigma}_c(\tilde{\varepsilon})$is part of the overall fuzzy model for the concrete material law. The fuzzy function $\tilde{\sigma}_c(\tilde{\varepsilon})$, which corresponds to the fuzzy initial loading curve, is shown in Fig. 3.13.

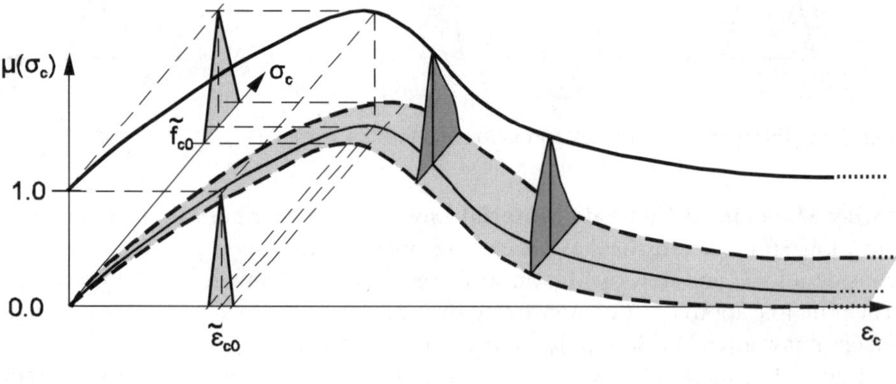

Fig. 3.13. Fuzzy model for a concrete material law; fuzzy initial loading curve

Fuzzy Model for the Arrangement of Reinforcement. Compliance with the prescribed arrangement of slack reinforcement for RC structures depends particularly on the care taken when laying the reinforcement steel. In the case of existing structures subject to postinspection the actual position of the reinforcement is often unknown.

As a result of human mistakes deviations from the prescribed reinforcement positioning often occur, which may exceed the permissible tolerances many times over. Uncertainty in the position, length, and shape of reinforcement steel may be described with the aid of continuous fuzzy variables. Uncertainty in the quality of reinforcement steel already installed as well as the number of placed steel bars and their diameters may be quantified using discrete fuzzy variables. Coarse inaccuracies or errors, accounted for in the past in a lumped manner as part of the accepted risk in safety assessments, may hence be directly introduced into the structural analysis.

The fuzzy position of a reinforcement bar, as defined by a function of fuzzy variables for the geometrical position of its beginning and end point $\tilde{x}_2(x_1 = 0)$ and $\tilde{x}_2(x_1 = L)$ is shown in Fig. 3.14.

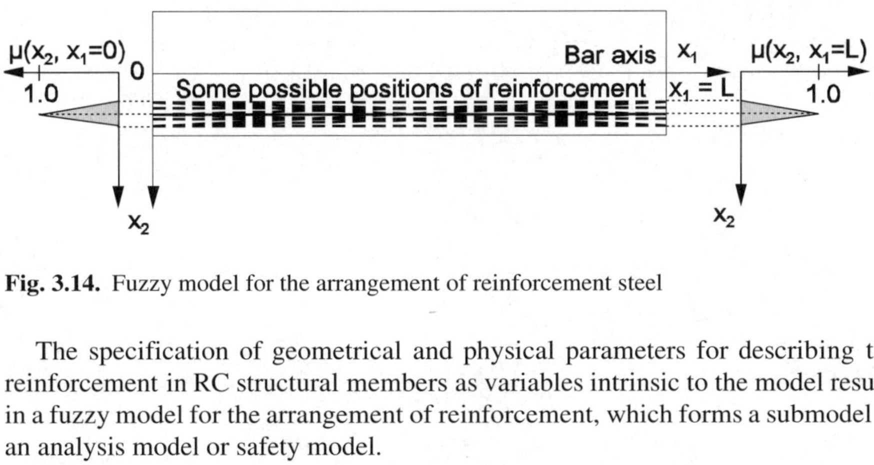

Fig. 3.14. Fuzzy model for the arrangement of reinforcement steel

The specification of geometrical and physical parameters for describing the reinforcement in RC structural members as variables intrinsic to the model results in a fuzzy model for the arrangement of reinforcement, which forms a submodel of an analysis model or safety model.

Fuzzy Model for Geometrical Imperfections. If geometrical imperfections are accounted for in the form of fuzzy model parameters, these enter the superordinate structural model (in a similar way to the arrangement of reinforcement) as a submodel.

The imperfections may be systematized, e.g., as imperfections in cross-sectional geometry, imperfections in the bar axis, imperfect bar connection eccentricities, or the imperfect position of nodes.

Besides imperfections that are introduced in the "normal case", e.g., production-related tolerances, considerably larger deviations from the design geometry may arise as a result of human mistakes. The fuzzy imperfections in the bar axis of a straight bar are represented in Fig. 3.15 in the form of a quadratic fuzzy parabola, which is dependent on the fuzzy model parameter $\tilde{f}_m = \tilde{f}(x_1 = 0.5 \cdot L)$ for the imperfection in the middle of the bar. Additional fuzzy model parameters may be similarly constructed to describe uncertain geometrical relationships. By this means it is possible to describe the overall structural geometry with the aid of a fuzzy model.

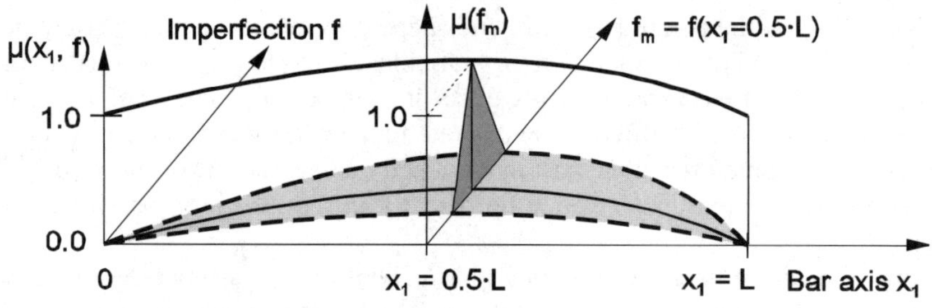

Fig. 3.15. Fuzzy model for imperfections in the bar axis

4 Description of Uncertain Structural Parameters as Fuzzy Random Variables

Structural parameters that possess the uncertainty characteristic *fuzzy randomness* are modeled as fuzzy random variables [119, 124]. For each fuzzy random variable it is necessary to generate the fuzzy probability distribution function, the fuzzy probability density function, and the associated fuzzy parameters on the basis of limited statistical data material. The data material possesses stochastic and non-stochastic properties, the latter resulting in particular from informal uncertainty or uncertainty due to fluctuating reproduction conditions.

The reasons for the fuzzy randomness of structural parameters are, e.g., manufacturing-related variations in the quality of construction materials, statistically insufficient data material for the quantification of applied loads and system resistances, or varying reproduction conditions during the period of observation.

The description of uncertain structural parameters as fuzzy random variables is presented in the following for the one-dimensional case. Analogous to random vectors in probability theory, multidimensional fuzzy random vectors may be constructed from a combination of one-dimensional fuzzy random variables, also under consideration of mutual dependencies.

First, a general technique is developed for modeling fuzzy random variables. The modeling of a fuzzy random variable is finally demonstrated for three typical data situations:

1. A data sample is of limited size, and no information is available on the statistical properties of the universe (Sect. 4.2.1).
2. Owing to unknown, nonconstant reproduction conditions the statistical data material possesses informal uncertainty (Sect. 4.2.2).
3. The fuzziness of the sample is due to nonconstant reproduction conditions that are known in detail (Sect. 4.2.3).

4.1 Techniques for Modeling Fuzzy Random Variables

A fuzzy random variable was defined in Sect. 2.3.1 as a fuzzy set of its originals. Its originals are real random variables. If all originals of a fuzzy random variable as well as the membership values are known, the fuzzy random variable is completely described by the fuzzy set of the originals. The problem consists in the determination of the originals from the available data material and their lumped description as a fuzzy set. Especially for fuzzy random variables that are continuous with regard to their fuzziness, i.e., which possess an infinite number of originals, an appropriate representation is necessary. Techniques for modeling continuous fuzzy random variables are presented in the following. These permit the formulation of fuzzy probability distribution functions including their parameters.

The fuzzy probability distribution function $\tilde{F}(x)$ of a fuzzy random variable $\tilde{X}$ can be described with the aid of the probability distribution functions $F_j(x)$ of their originals X_j in accordance with Eq. (2.171). A systematic computation of the $F_j(x)$ is possible by introducing α-levels. For each α-level the originals determine the interval $F_\alpha(x) = [F_{\alpha l}(x), F_{\alpha r}(x)]$. The computation of $F_{\alpha l}(x)$ and $F_{\alpha r}(x)$ is carried out in accordance with Eqs. (2.172) and (2.173), and for the one-dimensional case, in accordance with Eqs. (2.176) and (2.177).

The entirety of the originals yields a *bunch of functions*, which, together with the membership values, represent the fuzzy probability density function and the fuzzy probability distribution function.

In order to describe the probability density function and the probability distribution function of each original the *distribution parameters* and the *distribution type* must be known. The set of parameters forms the fuzzy bunch parameters of the bunch of functions sought. The fuzzy bunch parameters and the distribution type define the originals of the fuzzy random variable.

Because, by definition, the realizations of a fuzzy random variable have already been defined as fuzzy numbers (Sect. 2.3.1), the sought parameters of the distribution and the numerical parameters for describing the distribution type as well as the parameters occurring in the functional equations must also be fuzzy numbers. Precisely one original of the fuzzy random variable then exists for the membership level $\mu = \alpha = 1.0$. This original represents the special case of a real random variable.

4.1.1 Fuzzy Probability Distribution Function for Known Fuzzy Parameters

If fuzzy parameters such as the fuzzy expected value or the fuzzy standard deviation are known and the type of the probability distribution of the originals is also specified (e.g., as a logarithmic normal distribution), it is then possible to directly compute the unknown parameters of the fuzzy probability distribution function. These fuzzy functional parameters are fuzzy input variables of the sought fuzzy probability density function $\tilde{f}(x)$ and the fuzzy probability distribution function $\tilde{F}(x)$. The functional values of $\tilde{f}(x)$ and $\tilde{F}(x)$ may be computed with the aid of the extension principle or by α-level optimization (Sect. 5.2). The respective functional equation of the $f_j(x)$ and $F_j(x)$ of the originals is thereby applied as the mapping model. Interaction between the fuzzy functional parameters must be taken into account.

Example 4.1. The fuzzy expected value $\tilde{m}_x = <5.5, 6.0, 6.8>$ and the fuzzy standard deviation $\tilde{\sigma}_x = <0.8, 1.0, 1.1>$ are given. The probability distribution of the originals is assumed to be a normal distribution. All originals are then specified by

$$\tilde{f}(x) = \frac{1}{\tilde{\sigma}_x \cdot \sqrt{2 \cdot \pi}} \cdot e^{-0.5 \cdot \left(\frac{x - \tilde{m}_x}{\tilde{\sigma}_x} \right)^2}, \tag{4.1}$$

and

$$\tilde{F}(x) = \frac{1}{\tilde{\sigma}_x \cdot \sqrt{2 \cdot \pi}} \cdot \int_{-\infty}^{x} e^{-0.5 \cdot \left(\frac{t - \tilde{m}_x}{\tilde{\sigma}_x} \right)^2} dt. \tag{4.2}$$

The fuzzy expected value $\tilde{m}_x$ and the fuzzy standard deviation $\tilde{\sigma}_x$ occur directly in the fuzzy equations (4.1) and (4.2). These are at the same time fuzzy functional parameters. If $\tilde{m}_x$ is replaced by m_x and $\tilde{\sigma}_x$ by σ_x in Eqs. (4.1) and (4.2), the mapping model for the α-level optimization are obtained. The fuzzy probability density function and the fuzzy probability distribution function are presented in Figs. 4.1 and 4.2.

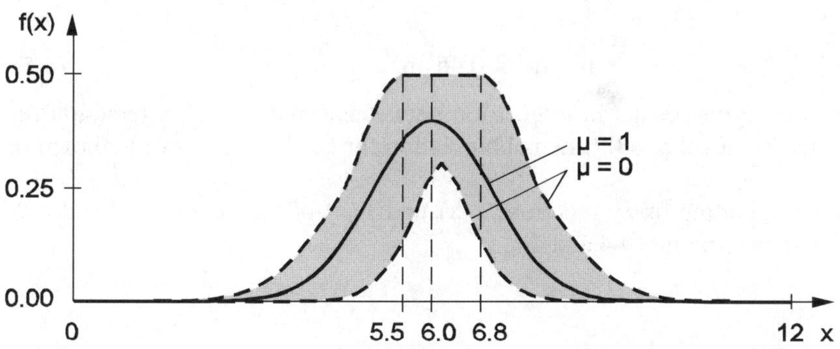

Fig. 4.1. Fuzzy probability density function $\tilde{f}(x)$ of a fuzzy normal distribution

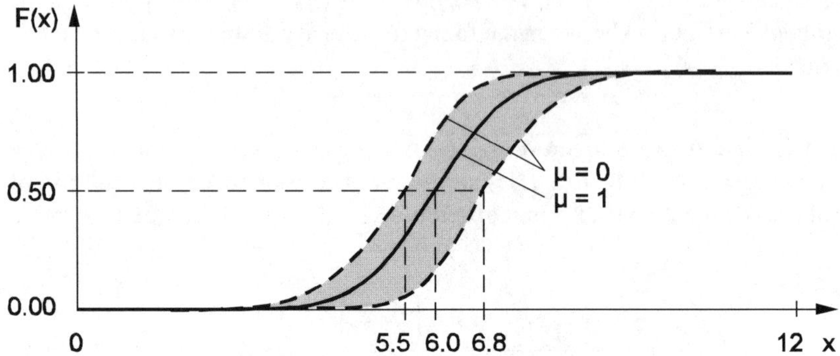

Fig. 4.2. Fuzzy probability distribution function $\tilde{F}(x)$ of a fuzzy normal distribution

Example 4.2. The type of probability distribution of the originals is now assumed to be an extreme value distribution of Ex-Max Type I (Gumbel). The following then holds for the fuzzy probability distribution function

$$\tilde{F}(x) = \exp(-\exp(-\tilde{a} \cdot (x-\tilde{b}))) . \tag{4.3}$$

From the fuzzy parameters $\tilde{m}_x$ and $\tilde{\sigma}_x$ (see Example 4.1) the fuzzy functional parameters

$$\tilde{a} = \frac{\pi}{\tilde{\sigma}_x \cdot \sqrt{6}} , \tag{4.4}$$

and

$$\tilde{b} = \tilde{m}_x - 0.45 \cdot \tilde{\sigma}_x \tag{4.5}$$

are computed. This results in interaction between $\tilde{a}$ and $\tilde{b}$. The interaction for the membership level $\alpha = 0$ (as an abbreviation for the limit $\alpha \to +0$) is shown in Fig. 4.3.

The corresponding fuzzy functions $\tilde{f}(x)$ und $\tilde{F}(x)$ of the extreme value distribution are shown in Figs. 4.4 and 4.5.

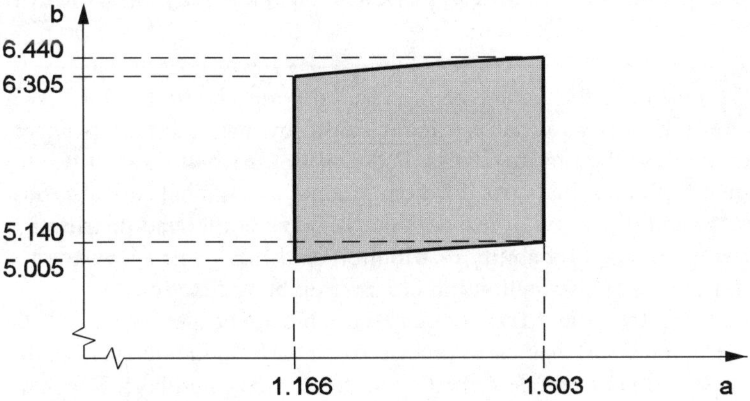

Fig. 4.3. Interaction between the fuzzy functional parameters $\tilde{a}$ and $\tilde{b}$

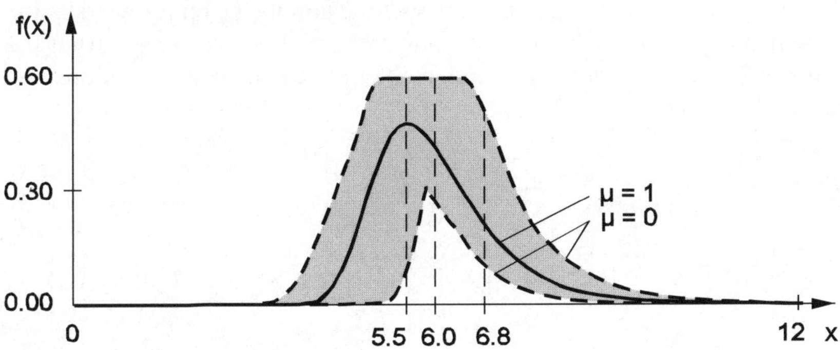

Fig. 4.4. Fuzzy probability density function $\tilde{f}(x)$ of an extreme value distribution of Ex-Max Type I (Gumbel)

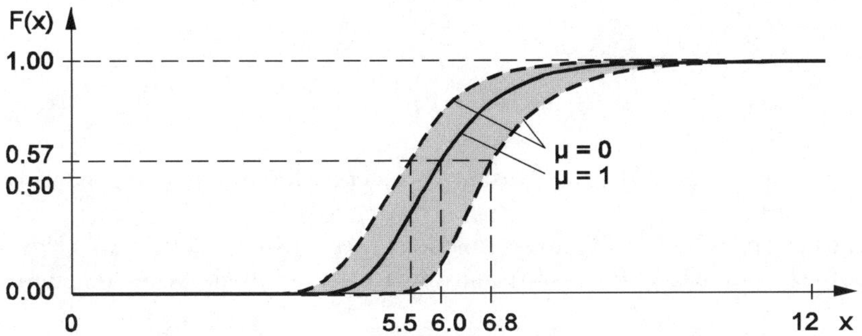

Fig. 4.5. Fuzzy probability distribution function $\tilde{F}(x)$ of an extreme value distribution of Ex-Max Type I (Gumbel)

4.1.2 Fuzzy Probability Distribution Function with a Fuzzy Distribution Type

Not all originals must have the same type of probability distribution. If the distribution type is not uniquely known, i.e., characterized by uncertainty, this uncertainty may then be described by fuzziness. Probability distributions of different types are assigned to the originals. As different gradual assignments are also possible, fuzzy compound distributions are defined. A fuzzy compound distribution defines the type of a fuzzy probability distribution as a fuzzy term constructed from different types of real distributions and fuzzy weighting functions.

In order to ensure that the fuzzy random variable to be described for the membership level $\mu = \alpha = 1.0$ possesses precisely one original the fuzzy variables present in the fuzzy term must be described by means of fuzzy numbers. The mapping model is the deterministic expression of the compound distribution as a function comprised of different distribution types.

The mapping model of a compound distribution may be developed with the aid of different mathematical formulations. It may, for example, be represented by the sum of n real distributions. The n summands are formed in each case from the product of the probability density function $f_i(x)$ of a real distribution and the weighting function $g_i(x)$

$$f(x) = \sum_{i=1}^{n} g_i(x) \cdot f_i(x)\,, \tag{4.6}$$

$$F(x) = \int_{t=-\infty}^{t=x} \left(\sum_{i=1}^{n} g_i(t) \cdot f_i(t) \right) dt = \int_{t=-\infty}^{t=x} f(t)\, dt\,. \tag{4.7}$$

Shape functions with free values are chosen for the weighting functions $g_i(x)$. A requirement in this respect is that the $g_i(x)$ are integrable and that the compound distribution fulfills the conditions

$$f(x) \geq 0; \; \forall x \in \mathbf{X}\,, \tag{4.8}$$

and

$$\int_{x=-\infty}^{x=+\infty} f(x)\, dx = 1\,. \tag{4.9}$$

For weighting functions that are constant in x a constant ratio of components is obtained for all x.

If the $g_i(x)$ are replaced by fuzzy functions $\tilde{g}_i(x)$ (whose functional values must be fuzzy numbers), Eqs. (4.6) and (4.7) also represent fuzzy functions (Sect. 2.1.11)

$$\tilde{f}(x) = \sum_{i=1}^{n} \tilde{g}_i(x) \cdot f_i(x)\,, \tag{4.10}$$

$$\tilde{F}(x) = \int_{t=-\infty}^{t=x} \left(\sum_{i=1}^{n} \tilde{g}_i(t) \cdot f_i(t) \right) dt = \int_{t=-\infty}^{t=x} \tilde{f}(t)\, dt \, . \tag{4.11}$$

An integration over $\tilde{f}(x)$ must thereby be performed as a normal integration for each original of the fuzzy random variable (Sect. 2.3.1). The conditions specified by Eqs. (4.8) and (4.9) must also be complied with for each original. These give rise to interaction between the fuzzy free values of the shape functions, which must be accounted for in the fuzzy compound distribution. In the general case, fuzzy compound distributions also possess uncertain parameters.

Example 4.3. A fuzzy compound distribution may, for example, be constructed from a normal distribution $f_1(x)$ with $m_{x\,1} = 6.8$ and $\sigma_{x\,1} = 1.1$, and a logarithmic normal distribution $f_2(x)$ with $m_{x\,2} = 5.5$, $\sigma_{x\,2} = 0.8$, and a minimum value $x_{0\,2} = 2$. A constant ration of components is adopted over x, and fuzzy numbers are introduced for the weighting functions. With $\tilde{g}_1(x) = \tilde{a}$ and $\tilde{g}_2(x) = \tilde{b}$ the expressions

$$\tilde{f}(x) = \tilde{a} \cdot f_1(x) + \tilde{b} \cdot f_2(x), \tag{4.12}$$

and

$$\tilde{F}(x) = \int_{t=-\infty}^{t=x} \tilde{a} \cdot f_1(t) + \tilde{b} \cdot f_2(t)\, dt = \tilde{a} \cdot F_1(x) + \tilde{b} \cdot F_2(x) \tag{4.13}$$

are obtained for the fuzzy compound distribution. According to Eq. (4.9) the following holds

$$\tilde{b} = 1 - \tilde{a} \, . \tag{4.14}$$

With the fuzzy free value $\tilde{a} = <0.3, 0.5, 0.7>$ the condition according to Eq. (4.8) is also complied with. For the fuzzy compound distribution the fuzzy expected value

$$\tilde{m}_x = \tilde{a} \cdot m_{x1} + (1-\tilde{a}) \cdot m_{x2} = <5.89, 6.15, 6.41> \tag{4.15}$$

is obtained. The membership function of the fuzzy standard deviation

$$\tilde{\sigma}_x = \sqrt{\tilde{a} \cdot \sigma_{x1}^2 + (1-\tilde{a}) \cdot \sigma_{x2}^2 + \tilde{a} \cdot (1-\tilde{a}) \cdot (m_{x1} - m_{x2})^2} \tag{4.16}$$

becomes nonlinear (Fig. 4.6). Owing to the interaction between $\tilde{m}_x$ and $\tilde{\sigma}_x$, precisely one σ_x belongs to each m_x value (Fig. 4.7).

The fuzzy probability density function $\tilde{f}(x)$ is shown in Fig. 4.8, while the fuzzy probability distribution function $\tilde{F}(x)$ is shown in Fig. 4.9.

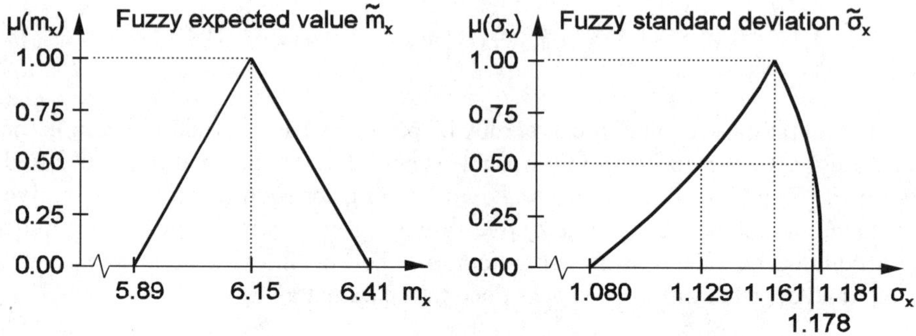

Fig. 4.6. Fuzzy expected value $\tilde{m}_x$ and fuzzy standard deviation $\tilde{\sigma}_x$ of the fuzzy compound distribution

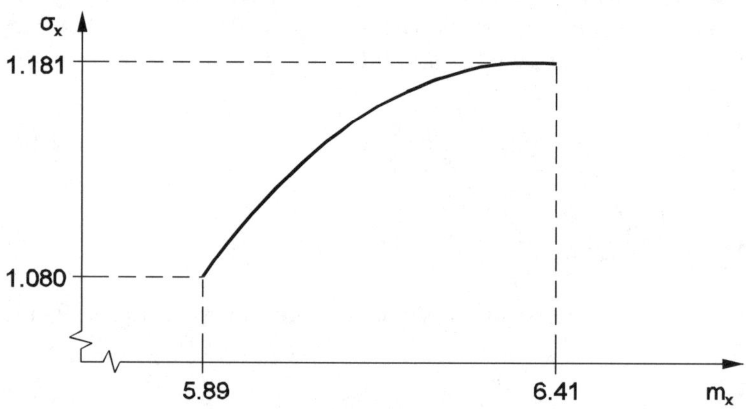

Fig. 4.7. Interaction between the fuzzy parameters $\tilde{m}_x$ and $\tilde{\sigma}_x$ of the fuzzy compound distribution

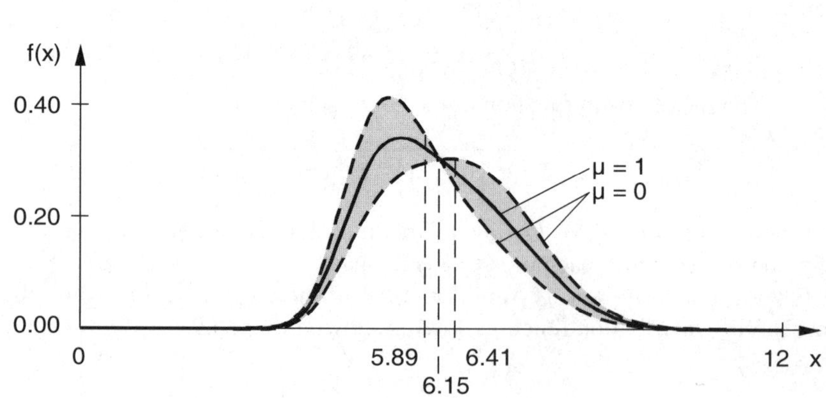

Fig. 4.8. Fuzzy probability density function $\tilde{f}(x)$ of the fuzzy compound distribution

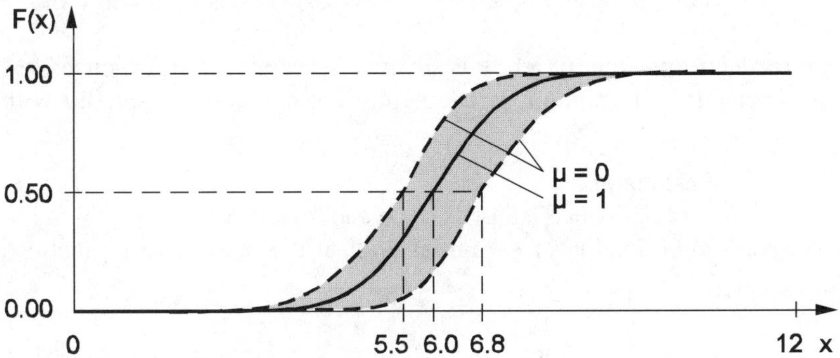

Fig. 4.9. Fuzzy probability distribution function $\tilde{F}(x)$ of the fuzzy compound distribution

4.2 Fuzzy Random Variables and Typical Data Situations

4.2.1 Small Sample Size

A concrete sample of small size is available. The sample elements possess uncertainty with the characteristic randomness. The available information on the sample is insufficient, however, to describe a real-valued random variable free of doubt. The type of the distribution function and the parameters cannot be uniquely determined; additional uncertainty exists. This uncertainty may be accounted for as informal uncertainty with the characteristic fuzziness. Statistical aids may be used, however, to fuzzify the informal uncertainty. Depending on the available information it is possible to formulate an uncertain parametric or nonparametric estimation problem. The fuzzy distribution parameters, the fuzzy type, and the fuzzy functional parameters of the sought fuzzy probability distribution (lumped together as $\tilde{p}_t(\tilde{X})$ in the following) are determined from the empirical fuzzy parameter values or from the empirical fuzzy probability distribution function of the concrete sample.

If, for example, the type of distribution is known with sufficient certainty, this implies an uncertain, parametric estimation problem. The sample functions applied in statistical methods yield more or less acceptable estimation values for the parameters of a distribution. In order to take account of the uncertainty of the estimator, confidence intervals may be determined for the estimator in question. The probabilistic propositions for confidence intervals applied in statistical methods only serve as additional information for the fuzzification of the $\tilde{p}_t(\tilde{X})$ in the present case. The only information obtained concerns the particular region in which a parameter value "may possibly lie". On the basis of the foregoing a fuzzification proposition for $\tilde{p}_t(\tilde{X})$ is derived, which serves as an initial draft of the membership function $\mu(p_t(X))$.

In order to determine the $\tilde{p}_t(\tilde{X})$ it is always necessary to draw on expert knowledge. Subjective information is taken into consideration especially with regard to

- The choice of the estimator
- The construction of confidence intervals (type and level), and
- The subsequent modification of the initial draft of the membership functions $\mu(p_t(X))$

Example 4.4. The cylinder compressive strength f_c of a C20/25 concrete is modeled as a fuzzy random variable. The (here assumed) measured values are listed in Table 4.1. Due to the small sample size fuzziness arises when specifying the parameters by statistical estimation.

Table 4.1. Realizations of the cylinder compressive strength f_c of concrete C 20/25

Number i of realization	Compressive strength $x_i = f_{ci}$ [N/mm²]	Number i of realization	Compressive strength $x_i = f_{ci}$ [N/mm²]
1	28.3	11	26.8
2	31.5	12	35.3
3	35.2	13	26.3
4	29.8	14	23.1
5	27.6	15	20.2
6	30.7	16	29.2
7	25.2	17	25.7
8	34.6	18	34.2
9	28.9	19	24.8
10	19.2	20	22.8

In an initial investigation a normal distribution is assumed and the parameters m_X and σ_X are determined as fuzzy values $\tilde{m}_x$ and $\tilde{\sigma}_x$. For this purpose interval estimations are applied. From the 20 measured values of the compressive strength the central confidence intervals for the confidence levels 0.50, 0.75, 0.90, and 0.99 are determined. Dependencies between the parameters are not taken into account. Additionally, common point estimations are used to specify crisp values for the expected value (as the mean value of the sample) and the standard deviation (based on the sample variance). The results (Table 4.2) are then taken as a basis for fuzzifying the parameters. Membership values are assigned to the estimation results by subjective assessment, i.e., the confidence intervals are interpreted as being α-level sets of the fuzzy values $\tilde{m}_x$ and $\tilde{\sigma}_x$ to be specified (Table 4.2). The mean values of the fuzzy numbers are taken from the point estimations. The result of this fuzzification process is shown in Fig. 4.10. Due to the fact that dependencies between the parameters in the interval estimations are neglected, interaction between $\tilde{m}_x$ and $\tilde{\sigma}_x$ is not obtained.

Table 4.2. Statistical estimation of the fuzzy parameters $\tilde{m}_x$ and $\tilde{\sigma}_x$

	Confidence level	Estimation, confidence interval m_X	σ_X	α-level
Point estimation	—	27.97	4.75	1.00
Interval estimation	0.50	[27.24, 28.70]	[4.35, 5.43]	0.75
	0.75	[26.71, 29.23]	[4.05, 5.92]	0.50
	0.90	[26.13, 29.81]	[3.77, 6.52]	0.25
	0.99	[24.93, 31.01]	[3.34, 7.92]	0.00

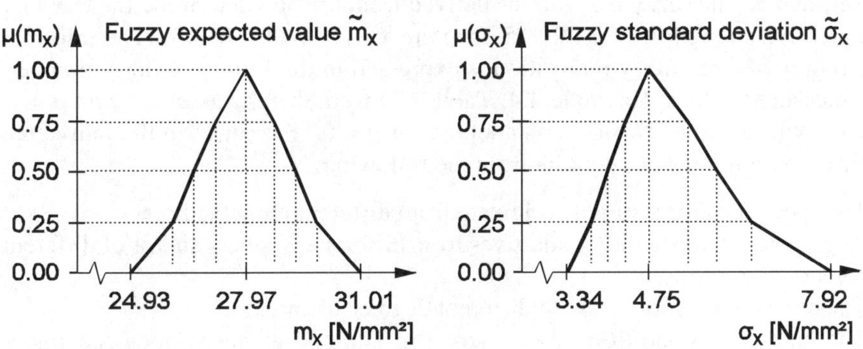

Fig. 4.10. Fuzzy expected value $\tilde{m}_x$ and fuzzy standard deviation $\tilde{\sigma}_x$

4.2.2 Samples with Unknown, Nonconstant Reproduction Conditions

For a sufficiently large sample size and constant reproduction conditions a sample exclusively possesses uncertainty with the characteristic randomness. If the reproduction conditions are not constant, however, the uncertainty characteristic alters accordingly. The additional uncertainty may be accounted for as fuzziness.

In the following treatment a distinction is made between known and unknown, nonconstant reproduction conditions. In the case of unknown, nonconstant reproduction conditions the observed realizations possess informal uncertainty. If this uncertainty in the realizations is described using fuzzy numbers, it is possible to construct a fuzzy random variable from the data material.

On the basis of the fuzzy realizations of the concrete sample the corresponding fuzzy distribution parameters, the fuzzy type, as well as the fuzzy functional parameters of the fuzzy probability distribution (lumped together in the following as $\tilde{p}_t(\tilde{X})$) must be estimated and substituted in the fuzzy probability distribution function as fuzzy bunch parameters. Algorithms of mathematical statistics are applied as the mapping model for this purpose. Each fuzzy realization is an input parameter of the mapping model.

Due to the fact that all parameters of a fuzzy probability distribution are computed from the same uncertain data material these are coupled via the individual originals of the fuzzy random variable, i.e., interaction exists between the estimated fuzzy variables $\tilde{p}_t(\tilde{X})$ (including the fuzzy type of the distribution). Each permissible point in the space of the $\tilde{p}_t(\tilde{X})$ describes one original X_j of the fuzzy random variable $\tilde{X}$, and hence one probability distribution function $F_j(x)$ (trajectory) from the bunch of functions $\tilde{F}(x)$.

Example 4.5. The fuzzy realizations derived from tests to determine the cylinder compressive strength f_c of a C20/25 concrete are given (Table 4.3). The informal uncertainty of each fuzzy realization is expressed in the form of a fuzzy number. The measured values (Example 4.4, Table 4.1) were thereby assessed with $\mu = 1$, and a deviation of ±2 N/mm² was adopted for $\mu = 0$. Fuzziness in the individual realizations could result inter alia from the following:

- The specimens for testing f_c originate from different manufacturers.
- Aggregates, cement and/or additives from different suppliers and/or of different composition are used.
- The concrete is produced with different degrees of care.
- The hardening conditions for concrete in situ and at the various specimen-storage locations differ due to different environmental factors, e.g., temperature and humidity.
- Various measuring devices of the same or different type are used for determining f_c, whereby each device has a different error of measurement.
- Measurements are performed by personnel with individual, i.e., different, degrees of conscientiousness.

Table 4.3. Fuzzy realizations of the cylinder compressive strength f_c of concrete C 20/25

Number i of fuzzy realization	Fuzzy compressive strength $\tilde{x}_i = \tilde{f}_{ci}$ [N/mm²]	Number i of fuzzy realization	Fuzzy compressive strength $\tilde{x}_i = \tilde{f}_{ci}$ [N/mm²]
1	<26.3, 28.3, 30.3>	11	<24.8, 26.8, 28.8>
2	<29.5, 31.5, 33.5>	12	<33.3, 35.3, 37.3>
3	<33.2, 35.2, 37.2>	13	<24.3, 26.3, 28.3>
4	<27.8, 29.8, 31.8>	14	<21.1, 23.1, 25.1>
5	<25.6, 27.6, 29.6>	15	<18.2, 20.2, 22.2>
6	<28.7, 30.7, 32.7>	16	<27.2, 29.2, 31.2>
7	<23.2, 25.2, 27.2>	17	<23.7, 25.7, 27.7>
8	<32.6, 34.6, 36.6>	18	<32.2, 34.2, 36.2>
9	<26.9, 28.9, 30.9>	19	<22.8, 24.8, 26.8>
10	<17.2, 19.2, 21.2>	20	<20.8, 22.8, 24.8>

In order to compute the empirical parameters common statistics (sample functions) are applied as the mapping model. With the fuzzy variables $\tilde{x}_i$ for the x_i the sample mean

$$\bar{x} = \frac{1}{n} \sum_{i=1}^{n} x_i \tag{4.17}$$

of the n realizations x_i becomes the fuzzy sample mean

$$\tilde{\bar{x}} = \frac{1}{n} \sum_{i=1}^{n} \tilde{x}_i \,. \tag{4.18}$$

For the linear mapping model given by Eq. (4.17) the evaluation for $\mu = 1$ and $\mu = 0$ is sufficient. The fuzzy triangular number $\tilde{\bar{x}}$ = <25.97, 27.97, 29.97> N/mm² is obtained for the fuzzy sample mean, see Fig. 4.11. From the mapping model for the sample variance

$$s_x^2 = \frac{1}{n-1}\left[\sum_{i=1}^{n} x_i^2 - \frac{1}{n}\left(\sum_{i=1}^{n} x_i\right)^2\right] \tag{4.19}$$

the $\tilde{x}_i$ may be used to compute the fuzzy standard deviation of the sample

$$\tilde{s}_x = \sqrt{\frac{1}{n-1}\left[\sum_{i=1}^{n} \tilde{x}_i^2 - \frac{1}{n}\left(\sum_{i=1}^{n} \tilde{x}_i\right)^2\right]} \,. \tag{4.20}$$

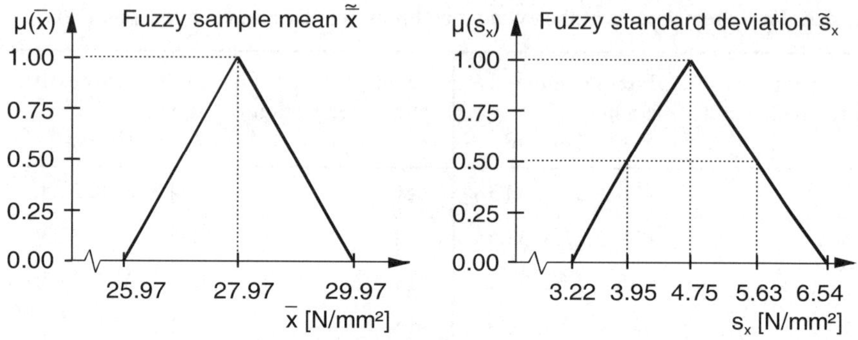

Fig. 4.11. Fuzzy mean $\tilde{\bar{x}}$ and fuzzy standard deviation $\tilde{s}_x$ of the sample given in Table 4.3

An α-level optimization (Sect. 5.2) must be applied to evaluate Eq. (4.20), as the mapping model given by Eq. (4.19) is nonmonotonic. The weakly nonlinear membership function $\mu(s_x)$ is shown in Fig. 4.11.

The fuzzy realizations $\tilde{x}_i$ enter Eq. (4.18) as well as Eq.(4.20). These establish a relationship between the fuzzy sample mean and the fuzzy standard deviation of the sample; interaction exists between the fuzzy variables $\tilde{\bar{x}}$ and $\tilde{s}_x$. Due to the fact that the interaction relationship includes all $\tilde{x}_i$ the analytical or numerical evaluation of the latter is problematic for a large number of samples. Considerable computational effort is required even for the 20 fuzzy realizations listed in Table 4.3. A numerical approximation solution was determined with the aid of systematic and random-oriented simulations. The result for the membership level $\alpha = 0$ is shown in Fig. 4.12. For the purpose of comparison the corresponding surface without interaction is also shown in the figure.

The effect of the number of fuzzy realizations on the interaction relationship becomes apparent when only the first seven sample elements in Table 4.3 are considered (Fig. 4.13). Compared with Fig. 4.12 the shape and position of the fuzzy set $\{\tilde{\bar{x}}, \tilde{s}_x\}$ have altered. As a consequence of the same support widths of the fuzzy realizations $\tilde{x}_i$ the minimum and maximum sample means are in each case coupled with the same standard deviation of the sample. This property is lost in the general case. The fact that the fuzzy realizations themselves may also be interactive may even lead to disjoint sets for the empirical parameters.

As already shown for $\tilde{\bar{x}}$ and $\tilde{s}_x$, interaction may also exist between all empirical parameters and the distribution type. Due to the numerical complications involved in the evaluation, the "exact" description of the interaction relationship may be replaced by an approximation.

For example, by investigating the passage to the limit of an infinite number of sample elements, approximation functions may be found that include the permissible combinations of parameter values. Where necessary, a representation in curvilinear coordinates may be derived in such a way that the interaction relationship is described by the coordinate transformation, hence leading to "new", noninteractive coordinates.

A simplified estimation of the interaction is also possible by determining representative points in the space of the empirical parameters on the basis of targeted investigations combined with knowledge based on experience. In the simplest case the interaction between the empirical parameters is neglected. Although this means that impermissible parameter combinations are included and thus enter subsequent computations, the "exact" solution is completely contained in this approximation due to the fact that the envelope curve of the permissible parameter combinations is taken into consideration (see Fig. 4.12).

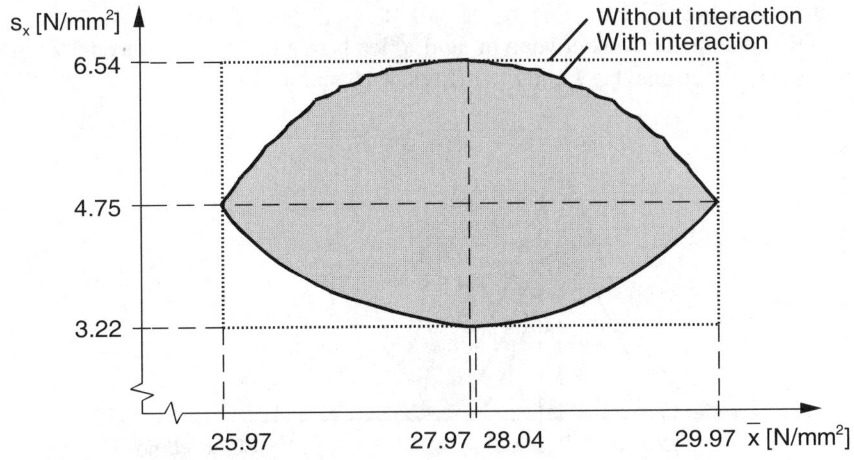

Fig. 4.12. Numerical approximation of the interaction between the fuzzy sample mean $\tilde{\bar{x}}$ and the fuzzy standard deviation $\tilde{s}_x$ for the 20 fuzzy realizations listed in Table 4.3

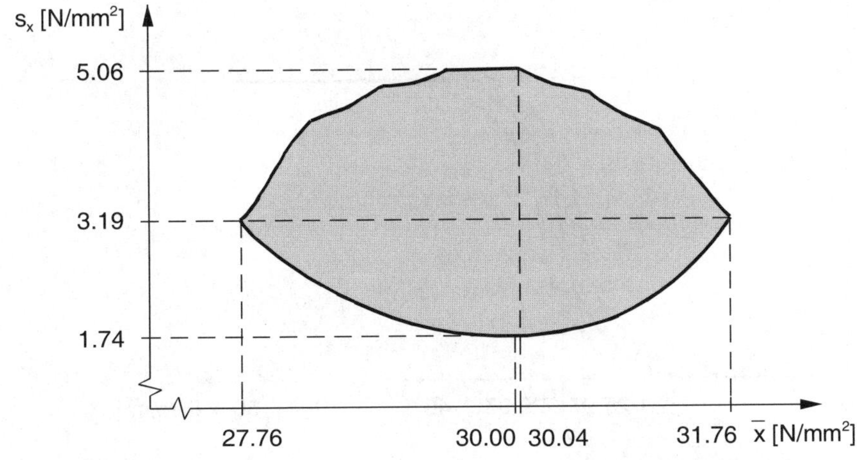

Fig. 4.13. Numerical approximation of the interaction between $\tilde{\bar{x}}$ and $\tilde{s}_x$ for the first seven fuzzy realizations listed in Table 4.3

For the fuzzy realizations listed in Table 4.3 it is still necessary to determine the fuzzy probability distribution function. If it is assumed that the fuzzy random variable is normally distributed, $\tilde{f}(x)$ and $\tilde{F}(x)$ are then defined by Eqs. (4.1) and (4.2). Only the fuzzy parameters $\tilde{\bar{x}}$ and $\tilde{s}_x$, already determined, are required to evaluate these equations. The fuzzy sample mean $\tilde{\bar{x}}$ is substituted as the fuzzy expected value $\tilde{m}_x$, while the fuzzy standard deviation $\tilde{s}_x$ of the sample is substituted as the fuzzy standard deviation $\tilde{\sigma}_x$ of the fuzzy probability distribution. The fuzzy probability density function $\tilde{f}(x)$ and the fuzzy probability distribution function $\tilde{F}(x)$ are shown in Figs. 4.14 and 4.15, respectively, under consideration of the interaction between $\tilde{m}_x$ and $\tilde{\sigma}_x$.

Neglecting the interaction between $\tilde{m}_x$ and $\tilde{\sigma}_x$ leads to envelope curves enclosing the exact fuzzy functions $\tilde{f}(x)$ and $\tilde{F}(x)$ (Figs. 4.16 and 4.17).

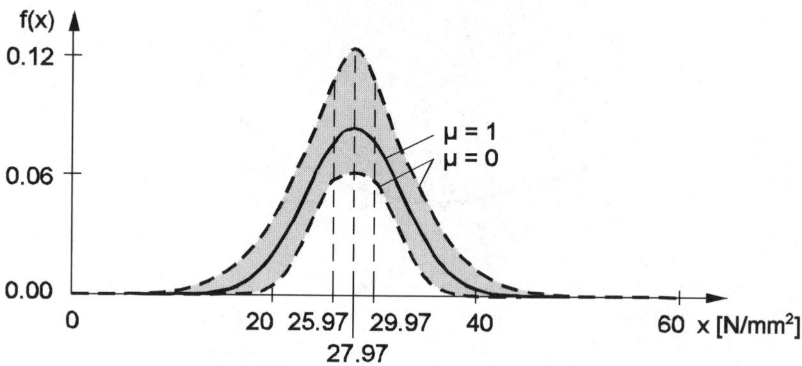

Fig. 4.14. Fuzzy probability density function $\tilde{f}(x)$ under consideration of the interaction between $\tilde{m}_x$ and $\tilde{\sigma}_x$

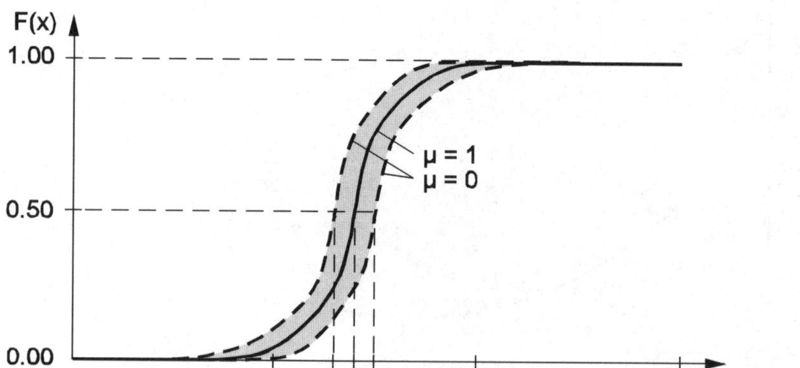

Fig. 4.15. Fuzzy probability distribution function $\tilde{F}(x)$ under consideration of the interaction between $\tilde{m}_x$ and $\tilde{\sigma}_x$

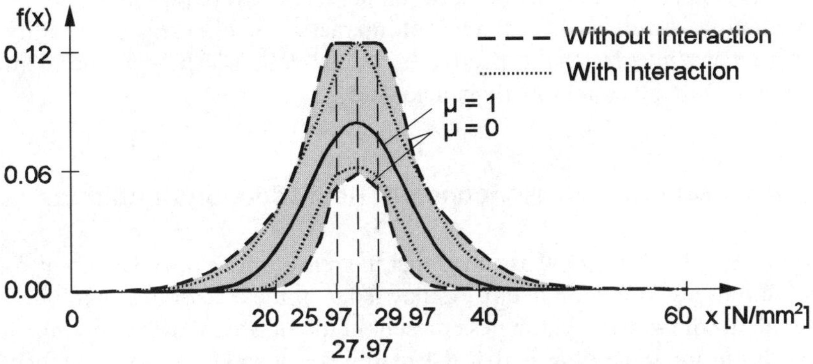

Fig. 4.16. Fuzzy probability density function $\tilde{f}(x)$ with and without consideration of the interaction between $\tilde{m}_x$ and $\tilde{\sigma}_x$

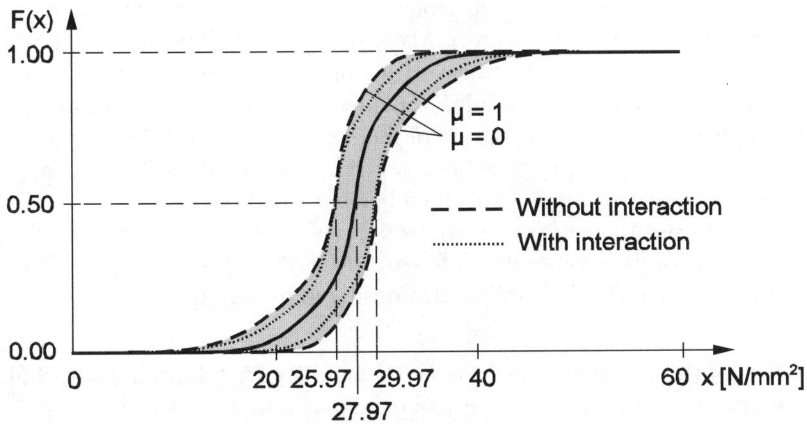

Fig. 4.17. Fuzzy probability distribution function $\tilde{F}(x)$ with and without consideration of the interaction between $\tilde{m}_x$ and $\tilde{\sigma}_x$

In this example the behavior of the fuzzy probability density function $\tilde{f}(x)$ is primarily influenced by the interaction between the fuzzy expected value $\tilde{m}_x$ and the fuzzy standard deviation $\tilde{\sigma}_x$ (Fig. 4.16). Especially in the boundary regions (tails) of $\tilde{f}(x)$, which yield significant contributions in the computation of the failure probability, differences between the solutions with and without consideration of interaction are evident. The interaction relationship shown in Fig. 4.12 excludes the simultaneous occurrence of extrema of the expected value and standard deviation. By taking interaction into account the functional values constituting the upper bound of $\tilde{f}(x)$ in the boundary regions become smaller.

Propositions concerning interaction between the empirical parameters and the type of distribution as well as the influence of interaction on the fuzzy probability density function are linked to the respective fuzzy realizations. These are problem specific and may only be conditionally generalized.

4.2.3 Samples with Known, Nonconstant Reproduction Conditions

In contrast to Sect. 4.2.2 it is now assumed that the reasons for nonconstant reproduction conditions are known in detail. A knowledge of these reasons is exploited to separate the fuzziness and randomness present in the statistical data material.

A precondition for separation is that the reason for nonconstant reproduction conditions may be characterized by attributes. Observed realizations with the same attributes are lumped together in a single *group*. These groups are subsets of the universe. Each group of realizations with the same attributes is treated as a sample and evaluated using statistical methods. The statistical evaluation yields empirical parameter values for each group. For all groups the set $\underline{S}$ of statistical propositions is obtained. Each element of $\underline{S}$ is assigned to a subset of the universe. The set $\underline{S}$ is uncertain, and characterizes the fuzziness of the universe. The fuzzy set $\tilde{\underline{S}}$ thus describes the set of real random variables contained in the observed realizations. These real random variables are originals of the sought fuzzy random variable. By means of the fuzzy set $\tilde{\underline{S}}$ membership values are assigned to the empirical parameter values of the originals. The membership functions of the empirical parameters may be constructed using histograms.

An alternative possibility is the direct fuzzification of the probability distribution function curve. Both possibilities are demonstrated by way of example.

Determination of the Parameters of Fuzzy Probability Distributions with the Aid of Histograms. With regard to the parameters of a fuzzy probability distribution to be fuzzified it is presumed that the groups and their corresponding empirical parameters are known. The parameter values constitute a sample for which a histogram is constructed. The parameter to be fuzzified is plotted along the abscissa, which is subdivided in advance into subsets. In the normal manner the number of sample elements, i.e., the number of empirical parameter values per subset, is plotted on the ordinate. The evaluation of the histogram for constructing the membership function is carried out according to Sect. 3.1; variants are possible by selecting different subset widths.

Example 4.6. Specimens of concrete C 20/25 from different concrete plants are available for tests to determine their cylinder compressive strength f_c. The specimens are labeled, and the concrete plant and work team are registered. The reproduction conditions are thus nonconstant but known. Specimens with the same identification (same attributes) are each lumped together in a group; this results in twelve groups containing a different number of specimens (sample size). By this means, randomness and fuzziness are separated. The statistical evaluation of the measured cylinder compressive strength f_c yields empirical parameters for each group; in the present case, the sample mean $\overline{x}$ and the sample variance s_x^2, which may be determined from Eqs. (4.17) and (4.19). The results of the statistical evaluation are listed in Table 4.4.

Table 4.4. Sample mean $\overline{x}$ and standard deviation s_x of the cylinder compressive strength f_c of concrete C 20/25 for twelve groups of specimens (twelve samples)

Label of group	Sample size	Sample mean $\overline{x}$ [N/mm²]	Standard deviation s_x [N/mm²]	Label of group	Sample size	Sample mean $\overline{x}$ [N/mm²]	Standard deviation s_x [N/mm²]
1	54	27.3	5.3	7	55	26.4	5.0
2	48	26.6	4.9	8	47	30.1	4.6
3	42	29.2	4.2	9	64	28.3	5.9
4	38	31.4	3.8	10	53	27.9	3.8
5	44	28.3	5.6	11	75	29.6	6.3
6	48	29.4	3.2	12	52	27.8	4.7

The values listed in Table 4.4 are used to construct histograms for the sample mean $\overline{x}$ and the standard deviation s_x of the samples (see Fig. 4.18). The chosen subset widths are 1.0 N/mm² for $\overline{x}$ and 0.75 N/mm² for s_x. Each of the empirical parameters is modeled using fuzzy triangular numbers. The method of least squares (Sect. 3.1) is applied to determine the linear membership functions. The derived fuzzification suggestions are shown in Fig. 4.18.

Due to the fact that the values $\overline{x}$ and s_x for each group originate from the same sample, interaction exists between the fuzzy variables $\tilde{\overline{x}}$ and $\tilde{s}_x$. Analogous to the analysis of stochastic dependencies between random variables, the *interaction relationship* may be determined by evaluating the value pairs ($\overline{x}$, s_x) obtained. The realizations of $\tilde{\overline{x}}$ and $\tilde{s}_x$ are plotted in an appropriate coordinate system, and the interaction relationship is determined for different membership levels. The value pairs listed in Table 4.4 are presented in Fig. 4.19 for the membership level $\alpha = 0$, and the Cartesian product of $\tilde{\overline{x}}$ and $\tilde{s}_x$ is illustrated in plan view. A suggestion for the interaction relationship is also shown.

The empirical fuzzy parameters $\tilde{\overline{x}}$ and $\tilde{s}_x$ are adopted as the fuzzy distribution parameters $\tilde{m}_x$ and $\tilde{\sigma}_x$ of the fuzzy probability distribution. Assuming a normal distribution for the type of probability distribution of the originals, $\tilde{m}_x$ and $\tilde{\sigma}_x$ are directly substituted as fuzzy functional parameters in the fuzzy probability density function and the fuzzy probability distribution function in accordance with Eqs. (4.1) and (4.2) (Sect. 4.1.1).

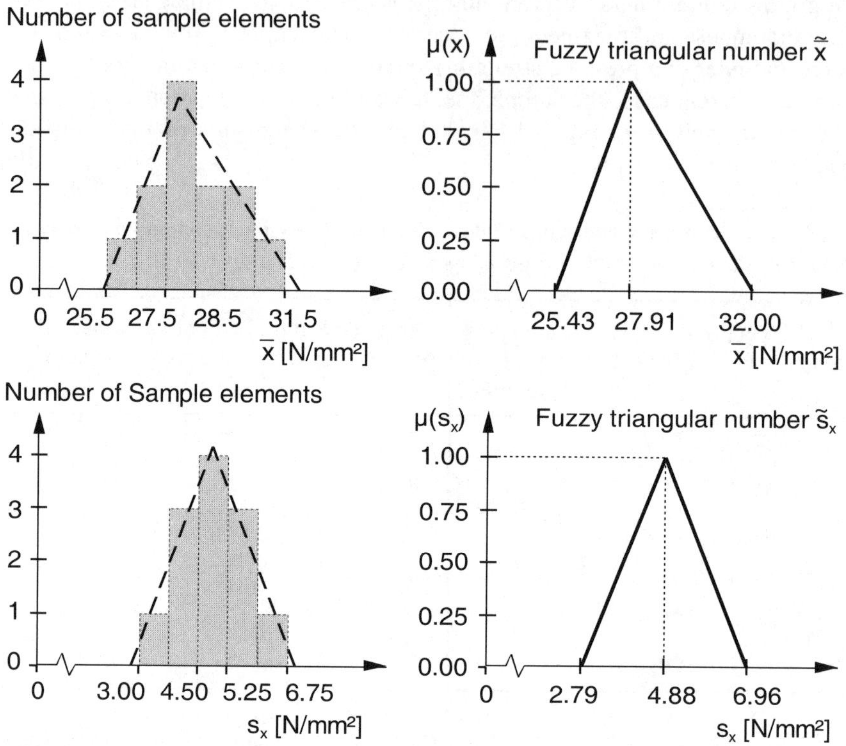

Fig. 4.18. Histograms and fuzzification of the sample mean $\bar{x}$ and the standard deviation s_x assigned to the groups (samples) of the cylinder compressive strength f_c of concrete C 20/25

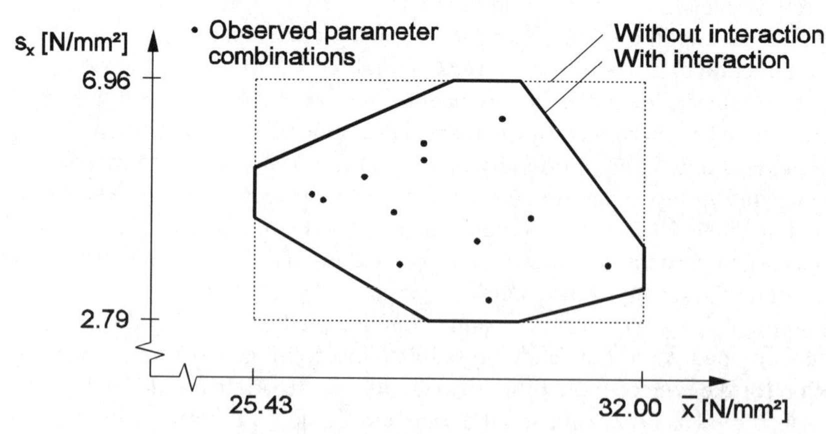

Fig. 4.19. Estimation of the interaction between $\tilde{\bar{x}}$ and $\tilde{s}_x$

Both the fuzzy parameters $\tilde{m}_x$ and $\tilde{\sigma}_x$ as well as their interaction relationship are comparable with the results given in Sect. 4.2.2; the curves of the fuzzy functions $\tilde{f}(x)$ and $\tilde{F}(x)$ are thus analogous to those shown in Figs. 4.14 to 4.17.

If it is also intended to model the uncertain distribution type, this may be achieved, e.g., using a compound distribution with a constant ratio of components between a normal distribution and a logarithmic normal distribution. An empirical distribution function is constructed for each group of observed realizations. On the basis of the latter the estimated values required for the parameters of the probability distribution as well as the ratio of components are determined and finally fuzzified.

Direct Fuzzification of the Probability Distribution Function Curve. The direct fuzzification of the functional curve is demonstrated by considering the problem of Example 4.6.

Example 4.6. The starting point is again the separation of randomness and fuzziness by constructing groups of observed realizations. A histogram of the corresponding realizations is constructed for each group. The construction of the histograms, i.e., subset widths and subset positioning on the abscissa, must be the same for all groups. The subsets are defined as half-closed intervals $[x_l, x_r)$ on the real number line. The number of observed realizations in the subsets is generally different for the individual groups. The histograms for the first two groups listed in Table 4.4 are shown in Fig. 4.20.

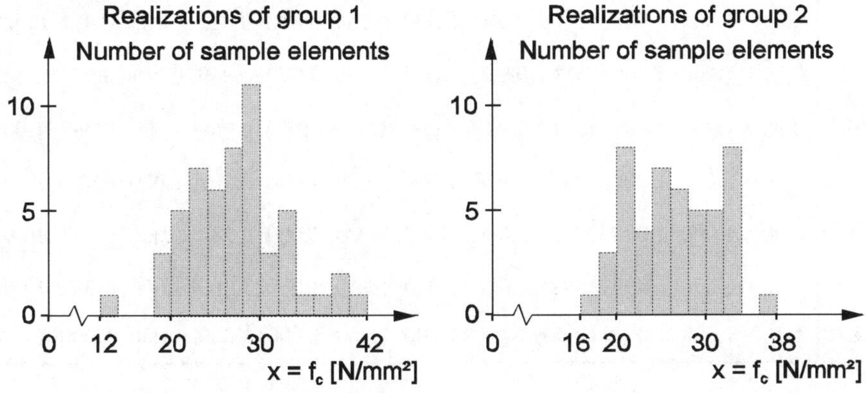

Fig. 4.20. Histograms for the realizations of groups 1 and 2 listed in Table 4.4

Table 4.5. Functional values of the empirical distribution functions $F_i^e(x = f_c)$ for all groups i of specimens with $x = f_c$ [N/mm²]

x	Group i											
	1	2	3	4	5	6	7	8	9	10	11	12
12	0.000	0.000	0.000	0.000	0.000	0.000	0.000	0.000	0.000	0.000	0.000	0.000
14	0.019	0.000	0.000	0.000	0.000	0.000	0.000	0.000	0.016	0.000	0.000	0.000
16	0.019	0.000	0.000	0.000	0.000	0.000	0.000	0.000	0.016	0.000	0.000	0.000
18	0.019	0.021	0.000	0.000	0.000	0.000	0.036	0.000	0.016	0.000	0.013	0.000
20	0.074	0.083	0.000	0.000	0.023	0.000	0.091	0.000	0.063	0.000	0.067	0.000
22	0.167	0.250	0.000	0.000	0.114	0.000	0.273	0.043	0.172	0.094	0.120	0.096
24	0.296	0.333	0.071	0.026	0.295	0.042	0.327	0.064	0.266	0.170	0.227	0.231
26	0.407	0.479	0.262	0.105	0.409	0.167	0.436	0.191	0.328	0.283	0.320	0.346
28	0.556	0.604	0.476	0.184	0.523	0.354	0.655	0.340	0.422	0.509	0.400	0.577
30	0.759	0.708	0.595	0.395	0.705	0.563	0.764	0.532	0.625	0.717	0.507	0.731
32	0.815	0.813	0.786	0.632	0.750	0.833	0.855	0.702	0.766	0.887	0.653	0.788
34	0.907	0.979	0.810	0.711	0.795	0.917	0.927	0.766	0.844	0.962	0.733	0.904
36	0.926	1.000	0.905	0.816	0.909	0.979	0.964	0.830	0.922	0.981	0.827	0.923
38	0.944	1.000	1.000	1.000	0.932	1.000	1.000	0.979	0.969	0.981	0.933	0.962
40	0.981	1.000	1.000	1.000	0.977	1.000	1.000	0.979	0.969	1.000	0.947	1.000
42	1.000	1.000	1.000	1.000	0.977	1.000	1.000	1.000	0.969	1.000	0.973	1.000
44	1.000	1.000	1.000	1.000	1.000	1.000	1.000	1.000	0.984	1.000	0.987	1.000
46	1.000	1.000	1.000	1.000	1.000	1.000	1.000	1.000	1.000	1.000	0.987	1.000
48	1.000	1.000	1.000	1.000	1.000	1.000	1.000	1.000	1.000	1.000	1.000	1.000

For each group the empirical probability distribution function

$$F_i^e(x) = \frac{n_{i,k}(x)}{n_i} \tag{4.21}$$

is developed from the corresponding histogram. In the above, i denotes the group number, n_i the number of all elements (realizations) in group i, and $n_{i,k}(x)$ the number of those elements k (in group i), whose values x_k are smaller than x. The values x of the observed realizations are determined by the left-hand subset boundaries (i.e., by the x_l of the half-closed intervals $[x_l, x_r)$) in the histograms; these mark discrete positions on the abscissa. The evaluation of all groups yields a bunch of discrete empirical distribution functions. The functional values $F_i^e(x = f_c)$ are listed in Table 4.5.

The functional values $F^e(x)$ are directly fuzzified. At each (discrete) position x a histogram is constructed using the functional values of the empirical distribution functions. The abscissa is subdivided into suitable subsets in the interval [0, 1]; the number of functional values assigned to each subset is plotted on the ordinate. With the aid of the elementary fuzzification methods outlined in Sect. 3.1 fuzzy numbers are generated from the histograms. In this generation process the properties of the probability measure must be observed (Sect. 2.2.2). In the present case fuzzy triangular numbers and fuzzy numbers with a polygonal membership function are chosen. The fuzzification process for three selected $x = f_c$ is shown in Fig. 4.21. The fuzzification results for all $x = f_c$ are listed in Table 4.6. The interval bounds of the support as well as the mean value are indicated for each fuzzy probability $\tilde{F}^e(x)$.

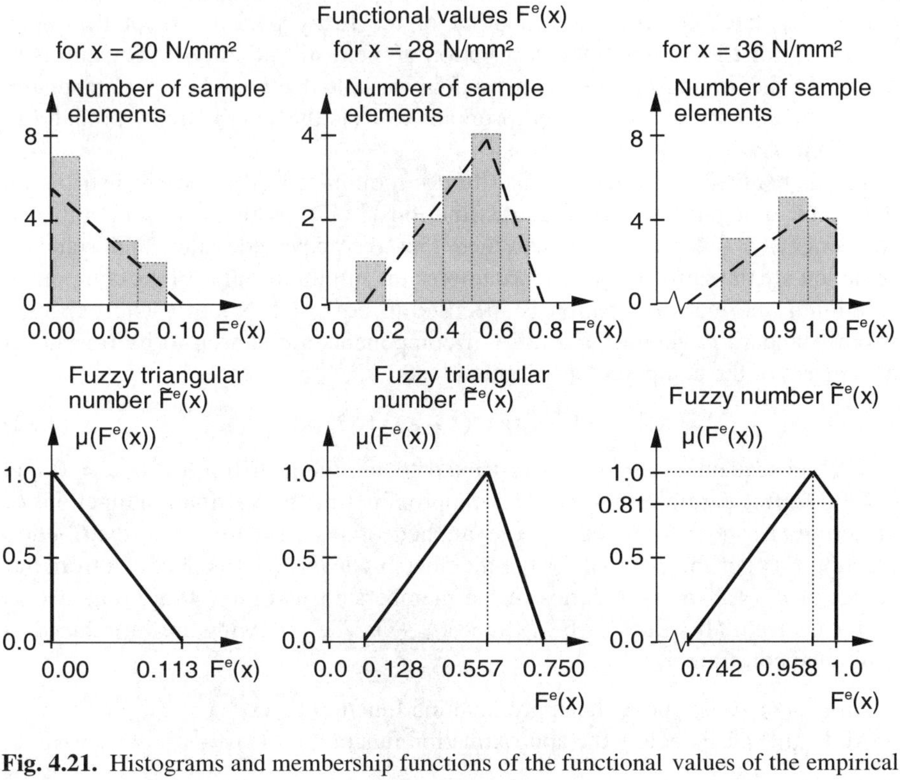

Fig. 4.21. Histograms and membership functions of the functional values of the empirical distribution function $F^e(x)$ for $x = f_c = 20$ N/mm², $x = f_c = 28$ N/mm², and $x = f_c = 36$ N/mm²

Table 4.6. Support bounds $F^e_{0\,l}(x)$ and $F^e_{0\,r}(x)$, and mean values $F^e_1(x)$ of the fuzzy probability $\tilde{F}^e(x)$ for all $x = f_c$ [N/mm²] given in Table 4.5

x	$F^e_{0\,l}(x)$	$F^e_1(x)$	$F^e_{0\,r}(x)$	x	$F^e_{0\,l}(x)$	$F^e_1(x)$	$F^e_{0\,r}(x)$	x	$F^e_{0\,l}(x)$	$F^e_1(x)$	$F^e_{0\,r}(x)$
12	0.000	0.000	0.000	26	0.025	0.358	0.492	40	0.949	1.000	1.000
14	0.000	0.000	0.018	28	0.128	0.557	0.750	42	0.966	1.000	1.000
16	0.000	0.000	0.018	30	0.331	0.763	0.825	44	0.983	1.000	1.000
18	0.000	0.000	0.035	32	0.603	0.799	0.975	46	0.984	1.000	1.000
20	0.000	0.000	0.113	34	0.652	0.925	1.000	48	1.000	1.000	1.000
22	0.000	0.000	0.369	36	0.742	0.958	1.000	–	–	–	–
24	0.000	0.283	0.417	38	0.913	1.000	1.000	–	–	–	–

The fuzzified probabilities $\tilde{F}^e(x)$ for discrete $x = f_c$ are functional values of the sought fuzzy probability distribution function $\tilde{F}(x)$. Different membership levels α are considered for the determination of the fuzzy parameters of the fuzzy probability distribution and the description of the distribution type. The aim is to determine the originals of the fuzzy random variable that bound each membership level. The entirety of all included originals reflects the sought fuzzy probability distribution.

In this example a compound distribution comprised of a normal distribution (ND) and a logarithmic normal distribution (LND) with a constant ratio of components is adopted. It is assumed that the expected value and standard deviation are the same for both distributions; the minimum value of the component logarithmic normal distribution is specified to be $x_0 = 5$ N/mm². The expected value, standard deviation, and ratio of components are chosen to be free fuzzy parameters of the compound distribution

$$\tilde{F}(x) = \tilde{a} \cdot \tilde{F}^{NV}(x) + (1 - \tilde{a}) \cdot \tilde{F}^{LNV}(x)\,. \tag{4.22}$$

The subsequent evaluation is restricted to the membership levels $\alpha = 0$ and $\alpha = 1$. The free parameters required for approximating the distribution functions of the originals are determined by the method of least squares. The distribution function $F_1(x)$ of the original for the membership level $\alpha = 1$ is obtained from the values of $F^e_1(x)$. The boundaries of the membership level $\alpha = 0$ are obtained in each case from all values of $F^e_{0\,l}(x)$ and $F^e_{0\,r}(x)$. The following are introduced as additional constraints:

- All $F^e_{0\,l}(x) > 0$ lie above the approximation function $F_{0\,l}(x)$
- All $F^e_{0\,r}(x) < 1$ lie below the approximation function $F_{0\,r}(x)$

The following values are obtained for the free distribution parameters and the functional parameter a of the implemented distribution function:

- Approximation of $F^e_1(x)$: $m_x = 27.66$ N/mm², $\sigma_x = 4.34$ N/mm², a = 0.00
- Approximation of $F^e_{0\,l}(x)$: $m_x = 34.29$ N/mm², $\sigma_x = 4.81$ N/mm², a = 0.00
- Approximation of $F^e_{0\,r}(x)$: $m_x = 23.30$ N/mm², $\sigma_x = 4.44$ N/mm², a = 1.00

The computed distribution functions of the originals are shown in Fig. 4.22 together with the adopted functional values of the fuzzified empirical distribution function from Table 4.6.

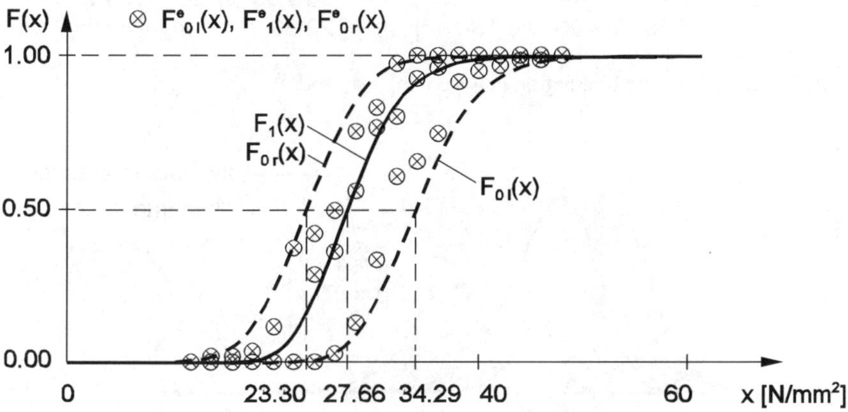

Fig. 4.22. Functional values of the empirical probability distribution functions $F^e_1(x)$, $F^e_{0\,l}(x)$, and $F^e_{0\,r}(x)$ as well as the approximation functions $F_1(x)$, $F_{0\,l}(x)$, and $F_{0\,r}(x)$

The fuzzy distribution parameters and the fuzzy functional parameter $\tilde{a}$ of the sought fuzzy probability distribution according to Eq. (4.22) may be expressed as fuzzy triangular numbers (confined to $\alpha = 0$ and $\alpha = 1$):

- $\tilde{m}_x$ = < 23.30, 27.66, 34.29 > N/mm²
- $\tilde{\sigma}_x$ = < 4.34, 4.34, 4.81 > N/mm², and
- $\tilde{a}$ = < 0.00, 0.00, 1.00 >

The interaction relationships for $\tilde{m}_x$, $\tilde{\sigma}_x$ and $\tilde{a}$ may be determined numerically (Sect. 4.2.2), or may be approximately estimated on the basis of the available information. A possible estimation of the interaction is shown in Fig. 4.23.

In the example, the interaction between $\tilde{m}_x$, $\tilde{\sigma}_x$ and $\tilde{a}$ has only a very slight effect, and may thus be neglected. The fuzzy probability density functions and the fuzzy probability distribution functions are compared in Figs. 4.24 and 4.25, with and without consideration of interaction. The approximation functions $F_{0\,l}(x)$ and $F_{0\,r}(x)$ as well as the corresponding probability density functions $f_{0\,l}(x)$ and $f_{0\,r}(x)$ are also shown in the figures. Variants of the fuzzy probability distribution function may be obtained by choosing different subset widths in the histogram.

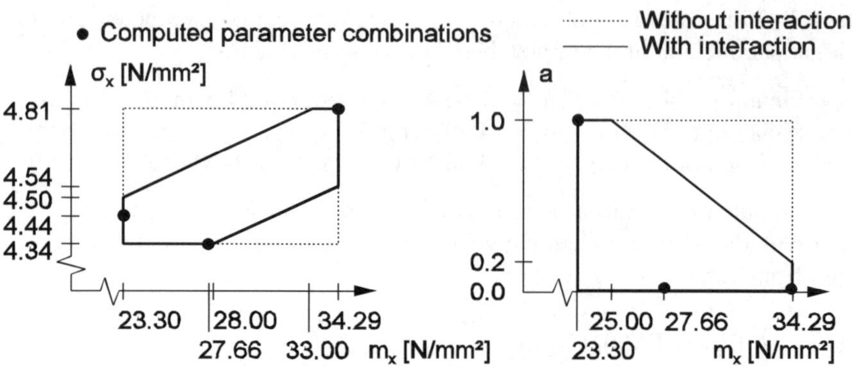

Fig. 4.23. Estimation of the interaction between $\tilde{m}_x$, $\tilde{\sigma}_x$ and $\tilde{a}$

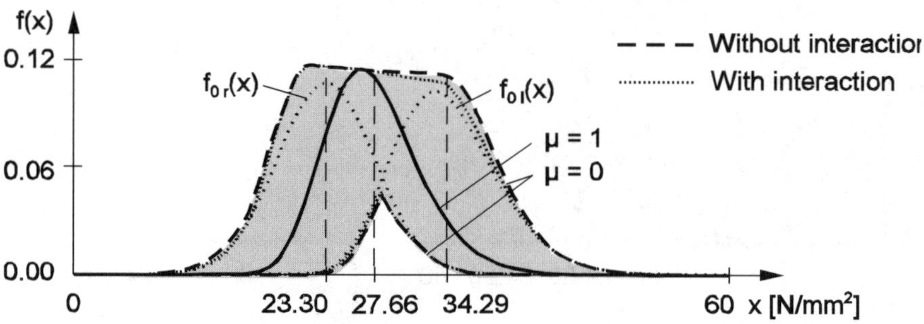

Fig. 4.24. Fuzzy probability density function $\tilde{f}(x)$ with and without consideration of interaction between $\tilde{m}_x$, $\tilde{\sigma}_x$ and $\tilde{a}$; probability density functions $f_{0\,l}(x)$ and $f_{0\,r}(x)$ belonging to $F_{0\,l}(x)$ and $F_{0\,r}(x)$

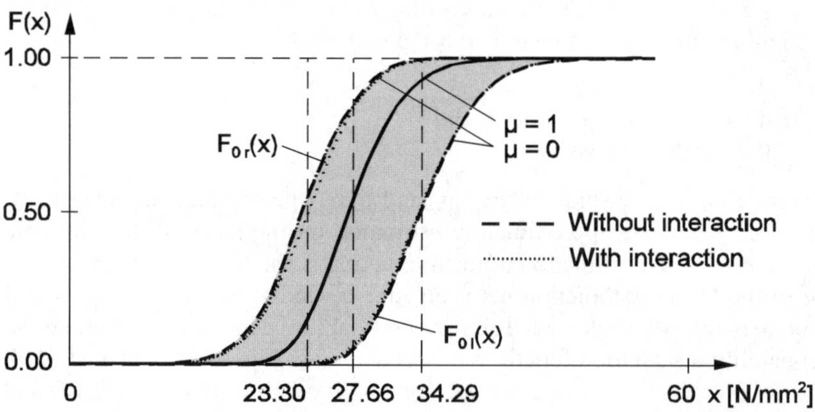

Fig. 4.25. Fuzzy probability distribution function $\tilde{F}(x)$ with and without consideration of interaction between $\tilde{m}_x$, $\tilde{\sigma}_x$ and $\tilde{a}$; probability distribution functions $F_{0\,l}(x)$ and $F_{0\,r}(x)$

5 Fuzzy and Fuzzy Stochastic Structural Analysis

5.1 Uncertain Structural Analysis and Deterministic Fundamental Solution

In deterministic structural analysis crisp input vectors $\underline{x} \subseteq \underline{\mathbf{X}}$ representing load, geometry, and material parameters are known. With the aid of the particular computational model M the structural responses (result vectors) $\underline{z} \in \underline{\mathbf{Z}}$ like, e.g., stresses, internal forces, or displacements are to be determined. This problem may be described by the mapping

$$\underline{x} \rightarrow \underline{z}, \tag{5.1}$$

which may also be denoted in the form $\underline{z} = f(\underline{x})$. The computational model, which characterizes the crisp dependency between the crisp vectors $\underline{x}$ and $\underline{z}$ is referred to as the *deterministic fundamental solution*. It represents the *mapping model* f = M indicated by the arrow in Eq. (5.1). Deterministic fundamental solutions that are often applied are, e.g., finite element models, bar models based on differential formulations, and boundary element models. In structural analysis, as a rule, the mapping model f represents a complex algorithm, which makes the computation of $\underline{z}$ numerically extensive.

If the input vectors of structural analysis are characterized by uncertainty and if the mapping model possesses model uncertainty, the mapping according to Eq. (5.1) then becomes an uncertain mapping of uncertain input vectors onto uncertain result vectors. Therefore, the nature of the uncertainty may be described with the types *stochastic*, *informal*, or *lexical uncertainty*, from which the uncertainty characteristics *fuzziness*, *randomness*, and *fuzzy randomness* may be identified.

By applying different uncertainty models to the mapping according to Eq. (5.1) or to parts of this mapping, respectively, several variants of *uncertain structural analysis* may be developed. The particular form of uncertain structural analysis depends on the chosen uncertainty description.

Besides the established probabilistic models the following two formal uncertainty descriptions presented in Chap. 2 are available:

Uncertainty Description with Fuzziness.

- Fuzzy vector $\underset{\sim}{\tilde{x}}$
- Fuzzy function $\tilde{x}(\underline{t}) = x(\underline{\tilde{s}},\underline{t})$
- Fuzzy field $\tilde{x}(\underline{\theta}) = x(\underline{\tilde{s}},\underline{\theta})$
- Fuzzy process $\tilde{x}(\underline{\tau}) = x(\underline{\tilde{s}},\underline{\tau})$

Uncertainty Description with Fuzzy Randomness.

- Fuzzy random vector $\underline{\tilde{X}}$
- Fuzzy random function $\underline{\tilde{X}}(\underline{t}) = \underline{X}(\underline{\tilde{s}},\underline{t})$
- Fuzzy random field $\underline{\tilde{X}}(\underline{\theta}) = \underline{X}(\underline{\tilde{s}},\underline{\theta})$
- Fuzzy random process $\underline{\tilde{X}}(\underline{\tau}) = \underline{X}(\underline{\tilde{s}},\underline{\tau})$

If all uncertainty is exclusively considered as fuzziness, *fuzzy structural analysis* may formally be developed from Eq. (5.1)

$$\underline{\tilde{x}} \;\tilde{\rightarrow}\; \underline{\tilde{z}}. \tag{5.2}$$

The input vectors of structural analysis are then fuzzy vectors $\underline{\tilde{x}} \subseteq \underline{\mathbf{X}}$. On the basis of fundamental operations with fuzzy sets like, e.g., the extension principle (Sect. 2.1.7), fuzzy result vectors $\underline{\tilde{z}} \subseteq \underline{\mathbf{Z}}$ may be determined. These represent fuzzy structural responses. In general, the mapping itself is also characterized by fuzziness, i.e., it describes an uncertain dependency between elements of $\underline{\mathbf{X}}$ and $\underline{\mathbf{Z}}$. This uncertainty may be understood as model uncertainty described by the fuzzy model parameter vector $\underline{\tilde{m}} = (\tilde{m}_1, \tilde{m}_2, ..., \tilde{m}_p)$. The mapping model is then represented by an uncertain computational model or deterministic fundamental solution including fuzzy model parameters. As stated in Sect. 5.2, the extension principle is barely applicable to numerically sophisticated structural analysis; instead of that the α-level optimization is introduced. This generally applicable, numerical solution technique may be extended to fuzzy functions

$$\underline{\tilde{x}}(\underline{t}) \;\tilde{\rightarrow}\; \underline{\tilde{z}}(\underline{t}). \tag{5.3}$$

On this basis the *Fuzzy Finite Element Method (FFEM)* is elaborated as a special, but very significant form of fuzzy structural analysis.

If uncertainty with the characteristic fuzzy randomness is to be taken into account, from Eq. (5.1) the formal notation

$$\underline{\tilde{X}} \;\tilde{\rightarrow}\; \underline{\tilde{Z}} \tag{5.4}$$

for *fuzzy stochastic structural analysis* is obtained. Uncertain input vectors are then fuzzy random vectors $\underline{\tilde{X}}$ with their fuzzy probability distribution functions $\tilde{F}(\underline{x})$. In fuzzy stochastic structural analysis the representation of the $\underline{\tilde{X}}$ by their originals $\underline{X}_j$ is used (Sect. 2.3.1). In a discretization process originals $\underline{X}_j$ and their assigned probability distribution functions $F_j(\underline{x})$ are chosen. For each original $\underline{X}_j$

representing a real-valued random vector the fuzzy stochastic structural analysis reduces to a common stochastic structural analysis. This may be carried out, e.g., with the aid of a Monte Carlo simulation in combination with a deterministic fundamental solution. If required the deterministic fundamental solution may include fuzzy model parameters as model uncertainty. Each common stochastic structural analysis leads to one real probability distribution functions $F(\underline{z}_j)$ for the associated original $\underline{Z}_j$ of the fuzzy random result vector $\tilde{\underline{Z}}$. Thereby the membership value of the originals $\underline{X}_j$ remains and is assigned to the result originals, $\mu(\underline{Z}_j) = \mu(\underline{X}_j)$. Combining the results $\underline{Z}_j$ and their membership values $\mu(\underline{Z}_j)$ to the fuzzy set $\tilde{\underline{Z}}$ yields the fuzzy probability distribution function $\tilde{F}(\underline{z})$ of the fuzzy random result vector, which represents a fuzzy random structural response. This approach may be extended to fuzzy random functions

$$\tilde{\underline{X}}(\underline{t}) \tilde{\rightarrow} \tilde{\underline{Z}}(\underline{t}) . \tag{5.5}$$

If the deterministic fundamental solution makes use of a finite element model, the general fuzzy stochastic structural analysis is then formulated in terms of the *Fuzzy Stochastic Finite Element Method (FSFEM)*.

5.2 Fuzzy Structural Analysis

5.2.1 Fuzzy Structural Analysis with the Aid of the Extension Principle

The fuzzy results $\tilde{\underline{z}}$ of the fuzzy structural analysis are fuzzy vectors. They may be computed from the n fuzzy input variables $\tilde{x}_i$ and the p fuzzy model parameters $\tilde{m}_r$ by means of the extension principle in combination with the Cartesian product between fuzzy sets (Sects. 2.1.6 and 2.1.7). The mapping model

$$\underline{z} = (z_1, \ldots, z_j, \ldots, z_m) = f(x_1, \ldots, x_i, \ldots, x_n) , \tag{5.6}$$

which is represented here by the analysis algorithm, transforms all points $\underline{x}$ from the space of the fuzzy input variables $\tilde{x}_i$ (x-space) into the space of the fuzzy result variables $\tilde{z}_j$ (z-space). The max-min operator determines the membership values $\mu(\underline{z})$ for the result points.

For fuzzy structural analysis the mapping model $f(\underline{x})$ is represented by the computational model M. The fuzzy model parameters $\tilde{m}_r$ are introduced into the model M, which, owing to

$$\tilde{f} = \tilde{M}(\tilde{m}_1, \ldots, \tilde{m}_r, \ldots, \tilde{m}_p) \tag{5.7}$$

becomes the uncertain mapping model $\tilde{f}$. The fuzzy model parameters $\tilde{m}_r$ are also treated according to the rules of fuzzy set theory. The uncertain mapping model $\tilde{f}$ is determined from Eq. (5.7) as a mapping of the fuzzy model parameters $\tilde{m}_r$ onto $\tilde{f}$. This mapping is treated mathematically in the same way as the mapping of the fuzzy input variables $\tilde{x}_i$ onto the fuzzy result variables $\tilde{z}_j$. For this reason only the processing of the fuzzy input variables $\tilde{x}_i$ using the crisp mapping model f is described in the following.

Applying the extension principle, the fuzzy input variables $\tilde{x}_i$ in the x-space form the fuzzy input set $\underline{\tilde{X}}$ and are mapped onto the fuzzy result set $\underline{\tilde{Z}}$ in the z-space. The fuzzy result variables $\tilde{z}_j$ are contained in $\underline{\tilde{Z}}$.

The extension principle is hardly practicable in the case of complex mapping model, as its application requires discretization of the support of the fuzzy input set $\underline{\tilde{X}}$ (if this represents an at least partially continuous fuzzy set) – e.g., using a point mesh. This leads to numerical problems.

Problem 1. The number of combinations of elements from the fuzzy input variables (= the number of points in the point mesh when discretizing the support of the fuzzy input set $\underline{\tilde{X}}$) is quasi-infinite (Sect. 2.1.7).

Problem 2. The limited numerical accuracy may lead to difficulties when evaluating the value pairs ($\underline{z}$, $\mu(\underline{z})$) to constitute the fuzzy result $\tilde{\underline{z}}$. With the aid of the relation *equal to* the decision is to be made if the elements $\underline{z}_1$ and $\underline{z}_2$ of the fuzzy result vector $\tilde{\underline{z}}$ are considered as equal to each other or not. Only if $\underline{z}_1 = \underline{z}_2$ holds can the max operator of the extension principle be applied. The limited numerical accuracy thus yields a jagged membership function of the fuzzy result. This effect may be partially compensated by introducing a tolerance limit for the comparison, i.e., rounding of the values to be compared, or by "smoothing" the numerically determined membership function. A rounding that is too coarse leads to inaccurate fuzzy results, whereas tolerance limits that are too fine require the determination of a very large number of elements for the fuzzy result.

Problem 3. The governing combinations of elements from the fuzzy input variables $\tilde{z}_j$ for determining the membership values of the elements of the fuzzy result are, in general, not found exactly. The max-min operator is exclusively applied to that combinations of elements from the fuzzy input variables that are defined by the discretizing point mesh. This reduces the accuracy of the obtained membership values, which thus only represent lower bounds for the graph of the actual membership function.

Example 5.1. We consider the mapping of the continuous fuzzy numbers $\tilde{x}_1$ and $\tilde{x}_2$ from Fig. 5.1 onto the continuous fuzzy result variable $\tilde{z}$. The mapping model of the extension principle is given by the function

$$z = f(x_1, x_2) = 0.01 \cdot x_1^3 + 0.17 \cdot x_1^2 + 0.48 \cdot x_1 \\ + 0.01 \cdot x_2^3 + 0.13 \cdot x_2^2 + 0.03 \cdot x_2 + 0.65 \,. \tag{5.8}$$

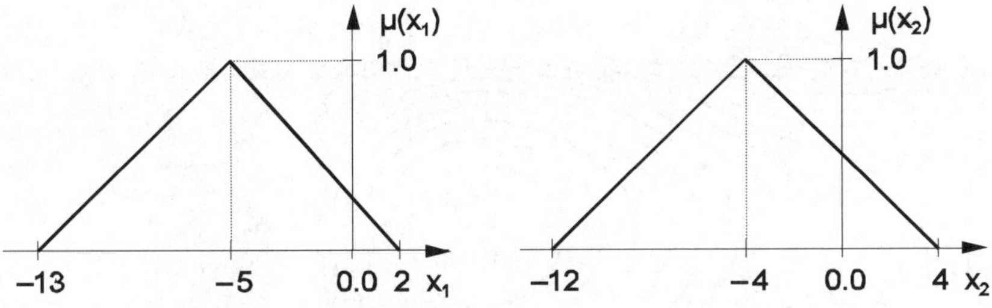

Fig. 5.1. Continuous fuzzy numbers $\tilde{x}_1$ and $\tilde{x}_2$

The function in Eq. (5.8) does not represent a one-to-one mapping, each functional value z may be computed from an infinite number of different combinations of arguments x_1 and x_2. The mapping model $z = f(x_1, x_2)$ is continuous, the fuzzy result variable $\tilde{z}$ is obtained as a continuous fuzzy number. The function according to Eq. (5.8) is shown in Fig. 5.2. It is plotted on that part of the domain of definition that is formed by the supports of $\tilde{x}_1$ and $\tilde{x}_2$. The function possesses one local minimum at $(x_1, x_2) = (-1.653, -0.117)$ with the functional value $z_{min} = 0.274$ and one local maximum at $(x_1, x_2) = (-9.681, -8.550)$ with the functional value $z_{max} = 5.859$. On the domain considered, these local extreme values simultaneously represent global extreme values. In Fig. 5.3 the level curves of the function are displayed and the locations of the extreme values are marked.

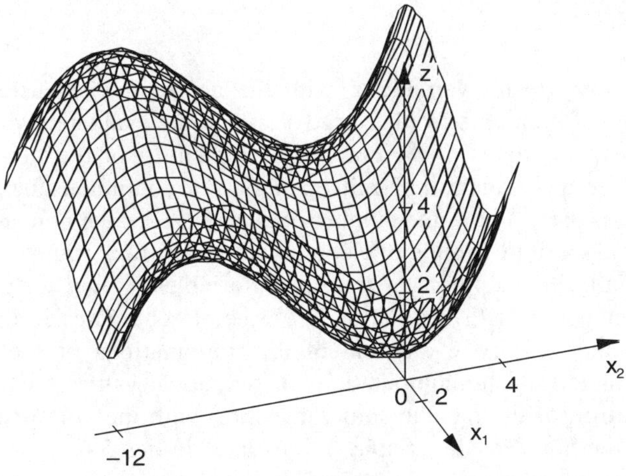

Fig. 5.2. Function $z = f(x_1, x_2)$ according to Eq. (5.8) on all elements of the fuzzy numbers $\tilde{x}_1$ and $\tilde{x}_2$

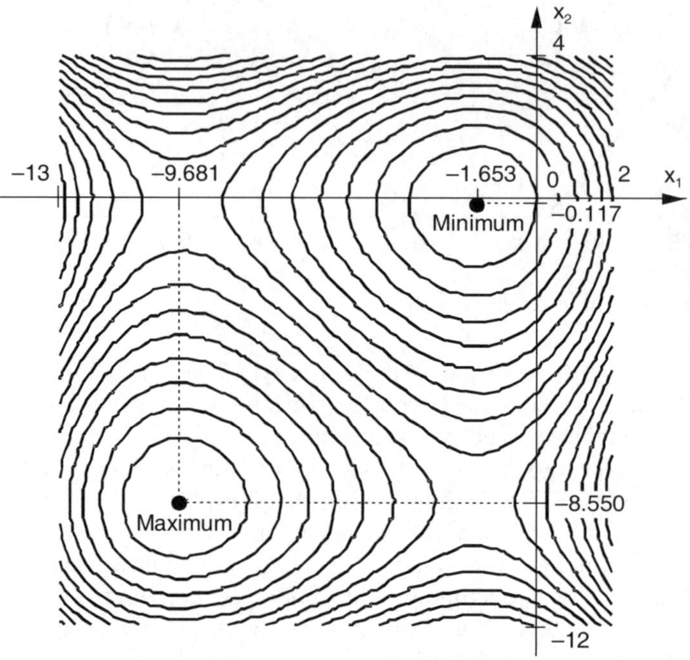

Fig. 5.3. Level curves of the function $z = f(x_1, x_2)$ according to Eq. (5.8) and locations of the extreme values

For determining the fuzzy result variable $\tilde{z}$ with the aid of the extension principle all combinations of elements from $\tilde{x}_1$ and $\tilde{x}_2$ are to be evaluated with Eq. (5.8) and the membership values of the computed elements z of the fuzzy result are to be calculated by applying the max-min operator. The continuous fuzzy input variables $\tilde{x}_1$ and $\tilde{x}_2$ possess an infinite number of elements. For the numerical treatment, however, only a finite sample of elements from both fuzzy numbers may be considered. Initially, the supports of the fuzzy variables $\tilde{x}_1$ and $\tilde{x}_2$ are subdivided by equidistant points with $\Delta x = 1.0$. These points characterize the various elements in $\tilde{x}_1$ and $\tilde{x}_2$. The evaluation of all combinations of these elements comprises 272 deterministic computations of functional values z with Eq. (5.8). The determination of the membership values $\mu(z)$ with the aid of the max-min operator yields the fuzzy result variable $\tilde{z}$ illustrated in Fig. 5.4.

If the distances between the elements of the fuzzy variables $\tilde{x}_1$ and $\tilde{x}_2$ are all chosen to be $\Delta x = 0.5$, a total of 1023 combinations are then to be evaluated. The associated membership function $\mu(z)$ is plotted in Fig. 5.5.

With an increasing number of combinations of elements from the fuzzy input variables the numerical solution converges to the exact result for $\tilde{z}$. In Fig. 5.6 both the numerical result and the exact graph of the membership function $\mu(z)$ are displayed. Additionally, an approximation solution for μ(z) is illustrated, which has been derived by smoothing the numerical result.

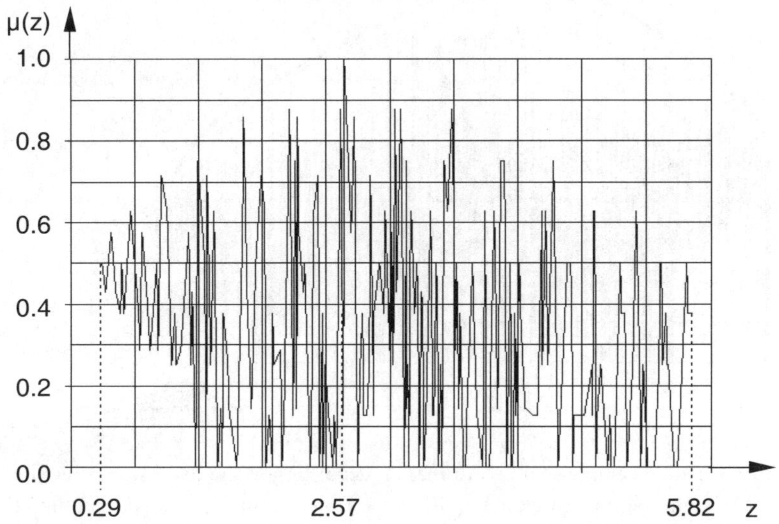

Fig. 5.4. Fuzzy result $\tilde{z}$: numerical determination according to the extension principle by evaluating 272 combinations of elements from the fuzzy input variables $\tilde{x}_1$ and $\tilde{x}_2$

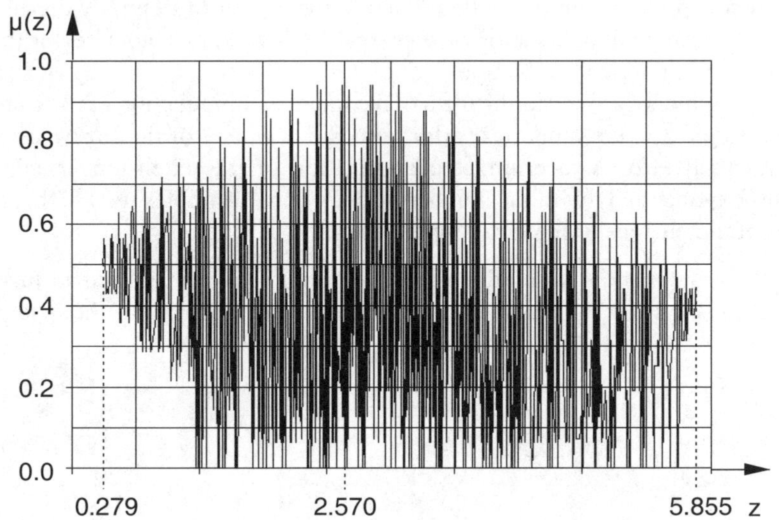

Fig. 5.5. Fuzzy result $\tilde{z}$: numerical determination according to the extension principle by evaluating 1023 combinations of elements from the fuzzy input variables $\tilde{x}_1$ and $\tilde{x}_2$

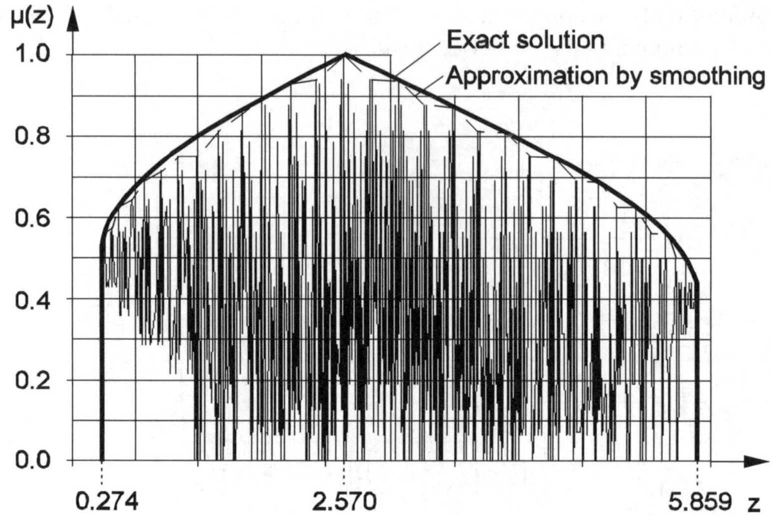

Fig. 5.6. Fuzzy-result $\tilde{z}$: numerical determination according to the extension principle by evaluating 1023 combinations of elements from the fuzzy input variables; approximation solution by smoothing the numerical result; exact solution

Summarizing, for the numerical treatment of the extension principle the following two basic propositions may be drafted:

- The accuracy of the membership function of a fuzzy result essentially depends on the number and on the location of elements from the fuzzy input vector that are chosen for numerical evaluation.
- The mapping model as well as the form of the membership function of the fuzzy input vector affects the numerical effort and the accuracy of the fuzzy result. The numerical effort concerning the application of the extension principle increases exponentially with the number of fuzzy input variables, i.e., with the number of dimensions of the fuzzy input vector.

Owing to the problems discussed the extension principle is not applied in fuzzy structural analysis. For this reason a capable numerical technique – the α-level optimization – is developed in the subsequent sections.

5.2.2 Fuzzy Structural Analysis with the Aid of α-level Optimization

5.2.2.1 Developing α-level Optimization from the Extension Principle

In order to develop a suitable method for processing fuzzy input vectors and fuzzy model parameters the concept of α-discretization is adopted. Procedures that exploit special properties of the mapping model or additional information concerning the mapping are suggested, e.g., in [19, 20, 45, 135, 200, 201]. In the following, a technique is derived that permits the use of mapping model without special properties [114, 117, 118, 127]. This is realized starting from the α-discretization of fuzzy sets (Sect. 2.1.9).

Replacing the Max-min Operator by an Optimization Problem. All fuzzy input variables are discretized using the same number of α-levels α_k, k = 1, ..., r. For each fuzzy input variable $\tilde{x}_i = \tilde{A}_i$ on the level α_k the α-level set A_{i,α_k} is then assigned to $\tilde{x}_i$, and all A_{i,α_k} form the crisp subspace $\underline{X}_{\alpha_k}$. With the aid of the mapping model $\underline{z} = f(x_1, \ldots, x_n)$ it is possible to compute elements of the α-level sets B_{j,α_k} of the fuzzy result variables $\tilde{z}_j = \tilde{B}_j$, j = 1, ..., m on the α-level α_k. The mapping of all elements of $\underline{X}_{\alpha_k}$ yields the crisp subspace $\underline{Z}_{\alpha_k}$ of the z-space.

Once the largest element $z_{j,\alpha_k r}$ and the smallest element $z_{j,\alpha_k l}$ of the α-level set B_{j,α_k} have been found, two points of the membership function $\mu(z_j) = \mu_{B_j}(z_j)$ are known (Fig. 5.10). In the case of convex fuzzy result variables the $\mu(z_j)$ are thus completely described. The determination of $z_{j,\alpha_k r}$ and $z_{j,\alpha_k l}$ replaces the max-min operator of the extension principle.

The procedure may be verified as follows:
Provided that

- The max-min operator is used in the extension principle, and
- The same α-level α_k is considered for all fuzzy input variables and fuzzy result variables

the following holds:

The application of the extension principle to the α-level sets A_{i,α_k} of the fuzzy input variables yields the α-level sets B_{j,α_k} of the fuzzy result variables (Sect. 2.1.7). With the mapping of the crisp input subspace $\underline{X}_{\alpha_k}$ into the result space the α-level sets B_{j,α_k} of the fuzzy result variables $\tilde{z}_j = \tilde{B}_j$ are completely specified. If the $\tilde{z}_j$ are convex fuzzy sets, the membership functions $\mu(z_j)$ are then completely described by the largest and smallest elements ($z_{j,\alpha_k r}$ and $z_{j,\alpha_k l}$) of the α-level sets B_{j,α_k}. The membership values of these elements are $\mu(z_{j,\alpha_k r}) = \mu(z_{j,\alpha_k l}) = \alpha_k$. The membership values $\mu(z_j)$ of all other elements z_j with $z_{j,\alpha_k l} < z_j < z_{j,\alpha_k r}$ are not needed – thus the application of the max-min operator is dispensable. The determination of $z_{j,\alpha_k r}$ and $z_{j,\alpha_k l}$ replaces the max-min operator. Exact mathematical conditions for the validity of this replacement are provided, e.g., in [47].

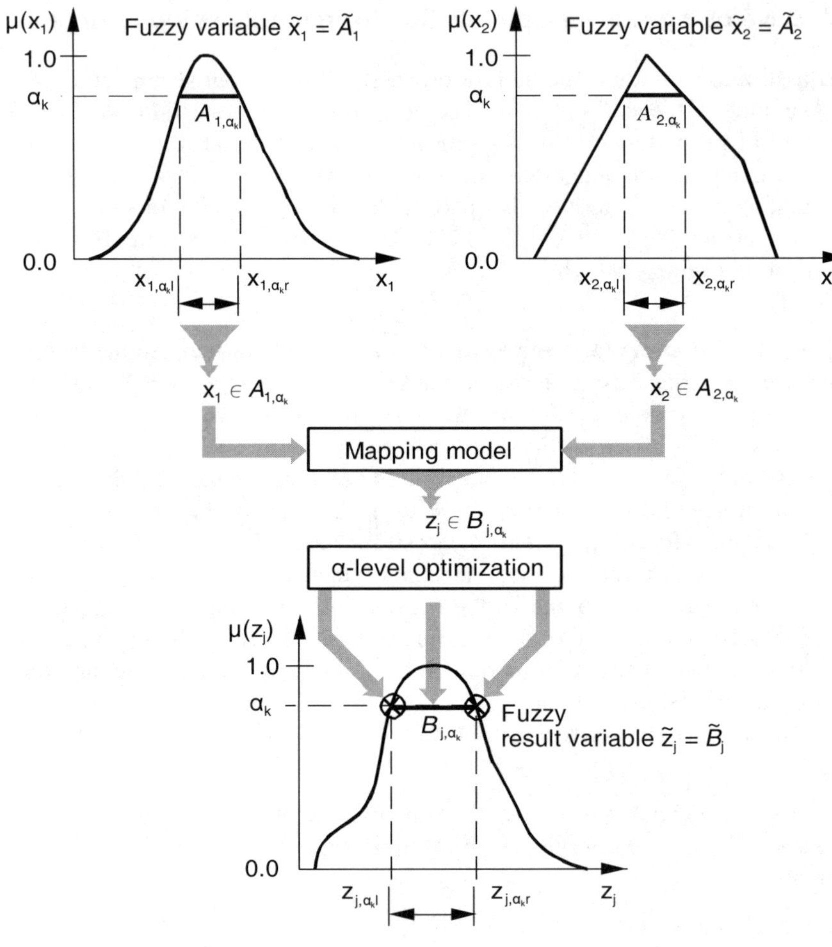

Fig. 5.10. Mapping of the fuzzy input variables $\tilde{x}_1$ and $\tilde{x}_2$ onto the fuzzy result variable $\tilde{z}_j$ by mapping of all α-level sets A_{i,α_k} onto the α-level sets B_{j,α_k}

The search for the smallest and largest elements is formulated as an optimization problem. The objective functions

$$z_j = f_j(x_1, \dots, x_n) \Rightarrow \text{Max} \mid (x_1, \dots, x_n) \in \underline{X}_{\alpha_k}, \tag{5.9}$$

$$z_j = f_j(x_1, \dots, x_n) \Rightarrow \text{Min} \mid (x_1, \dots, x_n) \in \underline{X}_{\alpha_k} \tag{5.10}$$

must be satisfied. The requirements $(x_1, \dots, x_n) \in \underline{X}_{\alpha_k}$ represent the constraints of the optimization problem.

The objective functions in Eqs. (5.9) and (5.10) are satisfied by the optimum points $\underline{x}_{opt}$. For each fuzzy result variable and each α-level α_k two optimum points have to be determined in the crisp subspace $\underline{X}_{\alpha_k}$, respectively. The optimization task according to Eqs. (5.9) and (5.10) for all α-levels α_k and all fuzzy result variables $\tilde{z}_j$ is referred to as *α-level optimization* [117, 118, 127]. In order to solve the α-level optimization problem special properties of the mapping model $\underline{z} = f(x_1, \ldots, x_n)$ may be used (Sect. 5.2.2.2).

If the mapping model has no special properties, the optimum points $\underline{x}_{opt}$ are located arbitrarily in $\underline{X}_{\alpha_k}$; otherwise the search for the $\underline{x}_{opt}$ may be limited to parts of $\underline{X}_{\alpha_k}$ - e.g., on its "boundary".

If

- Every crisp subspace $\underline{X}_{\alpha_k}$ is a connected set, and
- The mapping model is continuous and single valued

the fuzzy result variables $\tilde{z}_j$ are then convex fuzzy sets. If no interaction exists between the fuzzy input variables $\tilde{x}_i$, the first condition is satisfied when all $\tilde{A}_i = \tilde{x}_i$ are convex fuzzy sets. If the second condition is not complied with, α-level optimization yields exact envelope curves of the actual membership functions of the fuzzy result variables (Sect. 5.2.2.2).

Solution techniques for α-level optimization are described in Sect. 5.2.3.

The application of α-level optimization is demonstrated for the example from Sect. 5.2.1. With the aid of the mapping model according to Eq. (5.8) the continuous fuzzy input variables $\tilde{x}_1$ and $\tilde{x}_2$ illustrated in Fig. 5.1 are mapped onto the fuzzy result variable $\tilde{z}$. The fuzzy input variables are convex and the mapping model represents a continuous function, interaction does not exist. For computing the fuzzy result variable eleven α-levels that subdivide the interval (0, 1] equidistantly are chosen. For each α-level the largest and the smallest value for z are to be determined. In Table 5.1 the largest and smallest elements of the α-level sets of the fuzzy input variables as well as the values min z = $z_{\alpha_k l}$ and max z = $z_{\alpha_k r}$ are indicated. Moreover, the points from the crisp subspace $\underline{X}_{\alpha_k}$ that belong to the extreme values $z_{\alpha_k l}$ and $z_{\alpha_k r}$ on the respective α-levels are recorded. For example, introducing the points $x1_{opt,\alpha_k,z_l}$ and $x2_{opt,\alpha_k,z_l}$ in Eq. (5.8) yields min z = $z_{\alpha_k l}$ and max z = $z_{\alpha_k r}$. The optimum points lie partially inside the subspaces $\underline{X}_{\alpha_k}$ and partially on their "boundaries".

Table 5.1. Determination of the fuzzy result variable $\tilde{z}$ by α-level optimization; mapping model according to Eq. (5.8); $\tilde{x}_1$ and $\tilde{x}_2$ from Fig. 5.1

α_k	Boundary points of the α-level sets of the fuzzy input variables				Minimum value $z_{\alpha_k l}$ and assigned elements from $\tilde{x}_1$ and $\tilde{x}_2$			Maximum value $z_{\alpha_k r}$ and assigned elements from $\tilde{x}_1$ and $\tilde{x}_2$		
	$x_{1,\alpha_k l}$	$x_{1,\alpha_k r}$	$x_{2,\alpha_k l}$	$x_{2,\alpha_k r}$	x_{1_{opt},α_k,z_l}	x_{2_{opt},α_k,z_l}	$z_{\alpha_k l}$	x_{1_{opt},α_k,z_r}	x_{2_{opt},α_k,z_r}	$z_{\alpha_k r}$
0.0	-13.0	2.0	-12.0	4.0	-1.653	-0.117	0.274	-9.681	-8.550	5.859
0.1	-12.2	1.3	-11.2	3.2	-1.653	-0.117	0.274	-9.681	-8.550	5.859
0.2	-11.4	0.6	-10.4	2.4	-1.653	-0.117	0.274	-9.681	-8.550	5.859
0.3	-10.6	-0.1	-9.6	1.6	-1.653	-0.117	0.274	-9.681	-8.550	5.859
0.4	-9.8	-0.8	-8.8	0.8	-1.653	-0.117	0.274	-9.681	-8.550	5.859
0.5	-9.0	-1.5	-8.0	0.0	-1.653	-0.117	0.274	-9.000	-8.000	5.770
0.6	-8.2	-2.2	-7.2	-0.8	-2.200	-0.800	0.364	-8.200	-7.200	5.422
0.7	-7.4	-2.9	-6.4	-1.6	-2.900	-1.600	0.688	-7.400	-6.400	4.866
0.8	-6.6	-3.6	-5.6	-2.4	-3.600	-2.400	1.197	-6.600	-5.600	4.165
0.9	-5.8	-4.3	-4.8	-3.2	-4.300	-3.200	1.842	-5.800	-4.800	3.379
1.0	-5.0	-5.0	-4.0	-4.0	-5.000	-4.000	2.570	-5.000	-4.000	2.570

The values for z in the shaded columns and the predefined values for α_k lead to a polygonal shape for the membership function of the fuzzy result variable $\tilde{z}$. In Fig. 5.11 this result is compared to both the exact graph of the membership function μ(z) and the solution according to the extension principle (Sect. 5.2.1). The differences from the exact result may only be recognized at the stronger curved parts of the membership function, which are marked by dashed lines.

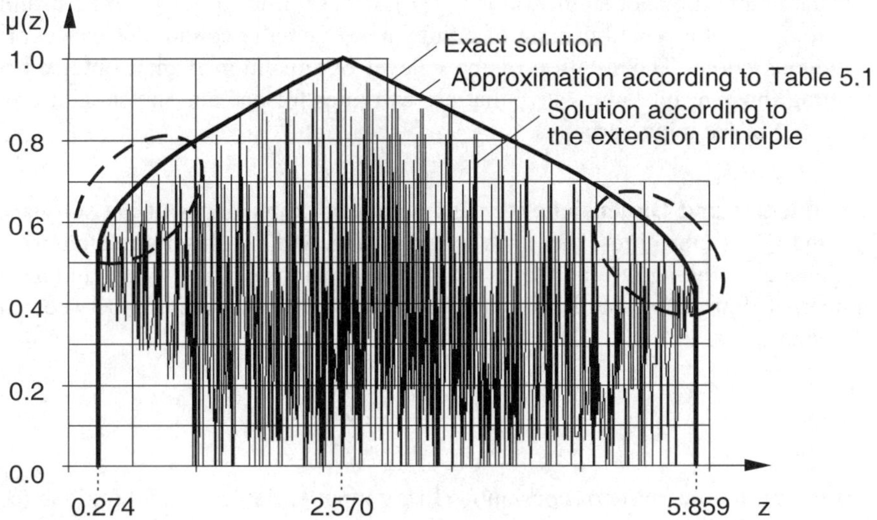

Fig. 5.11. Fuzzy result $\tilde{z}$: numerical determination according to the extension principle by evaluating 1023 combinations of elements from the fuzzy input variables; approximation solution by α-level optimization; exact solution

5.2.2.2 Properties of the Mapping Model

When solving the optimization problem (Eqs. (5.9) and (5.10)) special properties of the mapping model $f(x_1, ..., x_n)$ are of interest. Owing to the constraint $(x_1, ..., x_n) \in \underline{X}_{\alpha_k}$ it is necessary to investigate the properties of $f(x_1, ..., x_n)$ for mappings of the crisp subspace $\underline{X}_{\alpha_k}$ onto the crisp subspace $\underline{Z}_{\alpha_k}$, i.e., for mapping on the respective α-level α_k.

In α-level optimization the following properties of the mapping model are of interest:

- Uniqueness
- Biuniqueness
- Continuity
- Monotonicity
- Dimensionality of the input subspace $\underline{X}_{\alpha_k}$ and the result subspace $\underline{Z}_{\alpha_k}$

Uniqueness. In the case of single-valued mapping models Eqs. (5.6) and (2.36) yield precisely one result $(z_1, ..., z_m)$ for each element set $(x_1, ..., x_n)$. Due to the fact that the mapping models in fuzzy structural analysis are functions in a closed form for simple cases only, and numerical algorithms in the general case, uniqueness is not ensured in all cases. Mapping models that are not single-valued arise, e.g., in problems of kinetic stability or in analyses in the supercritical range.

If the mapping is not single-valued, $\underline{z} = (z_1, ..., z_m)$ in Eq. (5.6) for an n-tuple $(x_1, ..., x_n)$ is not a point but a set of points, a segmentally continuous curve, or a continuous curve. The solution of the α-level optimization is then obtained by selecting those result values z_{j,α_k} that are extremes for one and the same n-tuple $(x_1, ..., x_n)$ of the input variables.

Biuniqueness and Dimensionality. The number of dimensions of the subspaces $\underline{X}_{\alpha_k}$ and $\underline{Z}_{\alpha_k}$ determines the type of mapping and is closely related to its biuniqueness. Taking n to be the number of dimensions of the x-space and m the number of dimensions of the z-space, a distinction must be made between the following cases:

- $n > m$
- $n = m$
- $n < m$

If $n > m$, the mapping of continuous fuzzy input variables is not biunique (see also Sect. 2.1.7). Biuniqueness may occur in the case of discrete fuzzy input variables. The mapping of a crisp subspace $\underline{X}_{\alpha_k}$ onto a crisp subspace $\underline{Z}_{\alpha_k}$ for $n = 3$ and $m = 2$ is shown in Fig. 5.12. The points $\underline{x}_1$ and $\underline{x}_2$ of the x-space lead to the same result point $\underline{z}_1 = \underline{z}_2$ in the z-space. The mapping is irreversible.

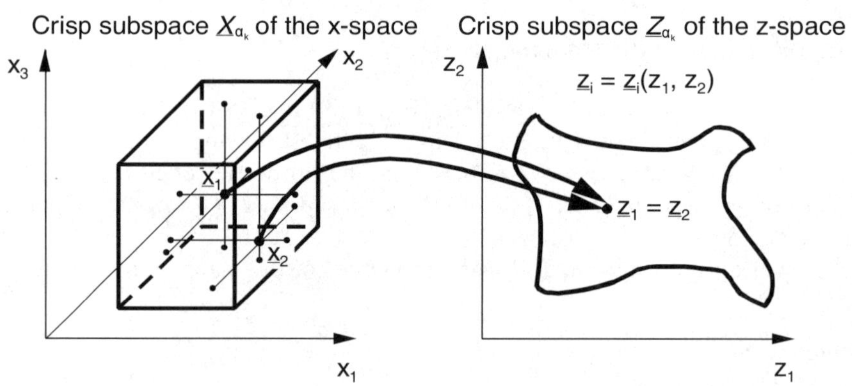

Fig. 5.12. Mapping of the three-dimensional crisp subspace $\underline{X}_{\alpha_k}$ onto the two-dimensional crisp subspace $\underline{Z}_{\alpha_k}$; nonbiunique mapping

A mapping in which $n = m$ may be biunique. A biunique and a nonbiunique mapping of a crisp subspace are shown in Fig. 5.13 for $n = 2$ and $m = 2$.

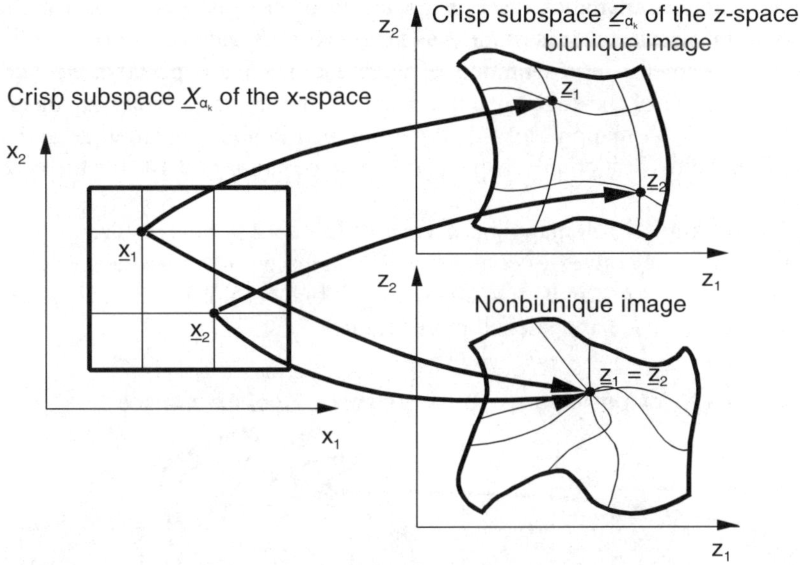

Fig. 5.13. Mapping of the two-dimensional crisp subspace $\underline{X}_{\alpha_k}$ onto the two-dimensional crisp subspace $\underline{Z}_{\alpha_k}$; biunique and nonbiunique mapping

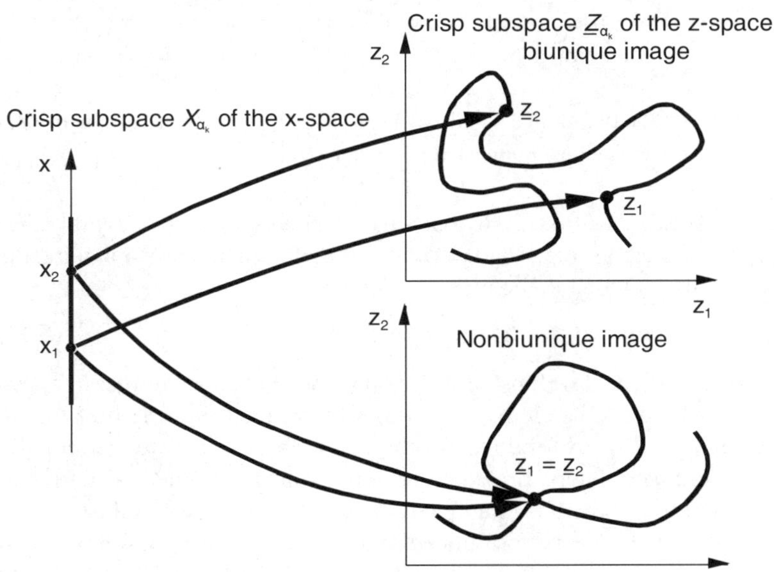

Fig. 5.14. Mapping of the one-dimensional crisp subspace X_{α_k} onto the two-dimensional crisp subspace $\underline{Z}_{\alpha_k}$; biunique and nonbiunique mapping

In the case of n < m the z-space is generally m-dimensional in extent. In differential terms, however, it is at most n-dimensional at each of its points. The z-space is the parameter representation of an n-dimensional hypersurface in the m-dimensional space of the fuzzy result variables. The input variables x_i constitute the parameters. A mapping of this type may be biunique. A biunique and a nonbiunique mapping of a crisp subspace are compared in Fig. 5.14 for the case n = 1 and m = 2.

Generally speaking, for all mappings according to Eq. (2.38) the result subspace must not necessarily be convex or connected. The mapping of a crisp subspace for this general case with n = 2 and m = 2 is shown in Fig. 5.15. The crisp subspaces of the x-space and z-space are hatched in this figure.

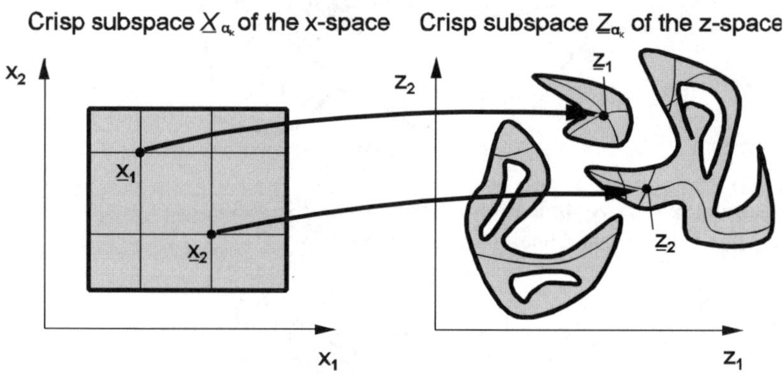

Fig. 5.15. Mapping of the two-dimensional crisp subspace $\underline{X}_{\alpha_k}$ onto the two-dimensional crisp subspace $\underline{Z}_{\alpha_k}$, which is neither connected nor convex

A mapping according to Eq. (2.36) may also depend on one or several *crisp parameters* $t_1, \ldots, t_p$, such as, e.g., load or time parameters. These crisp parameters are then additionally included in the mapping model

$$(z_1, \ldots, z_m) = f(x_1, \ldots, x_n, t_1, \ldots, t_p). \tag{5.11}$$

The result subspace then "develops" in the z-space with changes in these parameters through translation of its elements. Besides the influence of the fuzzy input variables, the shape and size of the result subspace in the z-space are additionally determined by these parameters. If a continuous mapping is assumed (also in relation to crisp parameters), which is biunique for each fixed parameter adjustment $\underline{t}_q$, the following proposition may be formulated for the case of continuous fuzzy input variables:

All points $\underline{z}$ that lie on the "boundary" of the crisp subspace $\underline{Z}_{\alpha_k}$ belonging to α_k for an arbitrary parameter vector $\underline{t}_q$ and a selected α-level α_k are also *boundary points* of $\underline{Z}_{\alpha_k}$ for all other possible parameter vectors $\underline{t}_r$.

The points $\underline{z}$ sought in the α-level optimization are always *boundary points* due to min z_{j,α_k} or max z_{j,α_k}. The corresponding optimum points $\underline{x}_{opt}$ in the space of the fuzzy input variables are also *boundary points* for the case n = m. In [19] this proposition is verified and specifically applied for the solution of systems of ordinary differential equations containing fuzzy input variables. A biunique and a nonbiunique mapping that "develop" in the z-space as a function of the crisp parameter t are shown in Fig. 5.16.

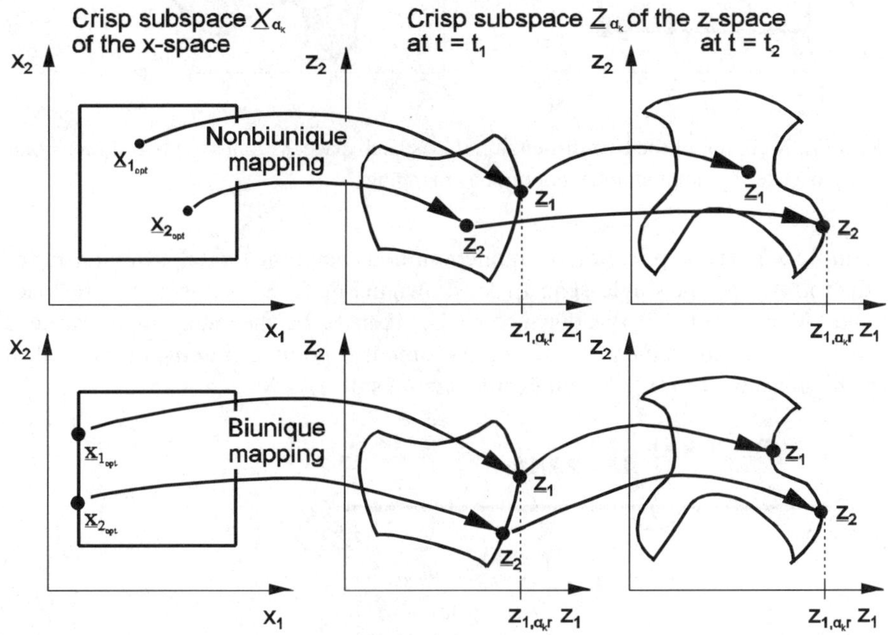

Fig. 5.16. Mapping of the two-dimensional crisp subspace $\underline{X}_{\alpha_k}$ onto the two-dimensional crisp subspace $\underline{Z}_{\alpha_k}$ and its development in the z-space as a function of the crisp parameter t for a biunique and nonbiunique, continuous mapping

Continuity. In the case of continuous fuzzy input variables with a continuous single-valued mapping model and a connected subspace $\underline{X}_{\alpha_k}$ the subspace $\underline{Z}_{\alpha_k}$ is connected, but may exhibit "holes" (Fig. 5.17). If the mapping is additionally biunique, "holes" are ruled out.

For a continuous, single-valued mapping of a connected subspace $\underline{X}_{\alpha_k}$ with continuous fuzzy input variables, convex fuzzy result variables are obtained. Convexity of the corresponding crisp subspace $\underline{Z}_{\alpha_k}$ is not guaranteed, however.

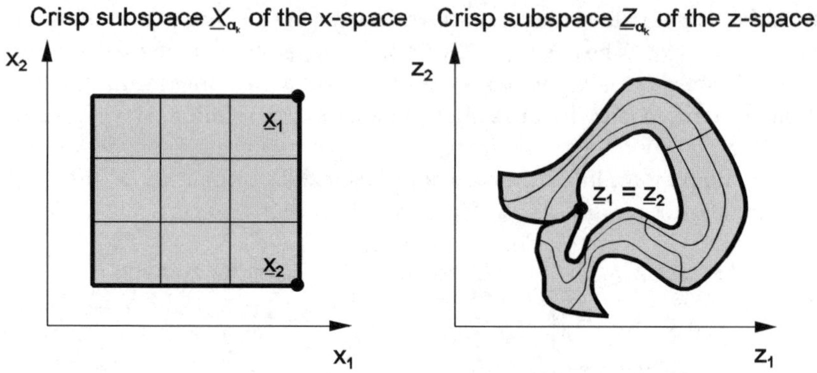

Fig. 5.17. Mapping of the two-dimensional crisp subspace $\underline{X}_{\alpha_k}$ onto the two-dimensional crisp subspace $\underline{Z}_{\alpha_k}$; nonbiunique, continuous mapping

Example 5.2. The application of a noncontinuous mapping model is demonstrated in the following. The single-span girder shown in Fig. 5.18 is subjected to the loads P and M at point x_p; the distance x_p is taken to be the fuzzy input variable $\tilde{x} = \tilde{x}_p$. The result value is the bending moment at point x_k (middle of the girder) which, on account of $\tilde{x}_p$, becomes the fuzzy result $\tilde{z} = \tilde{M}_k$.

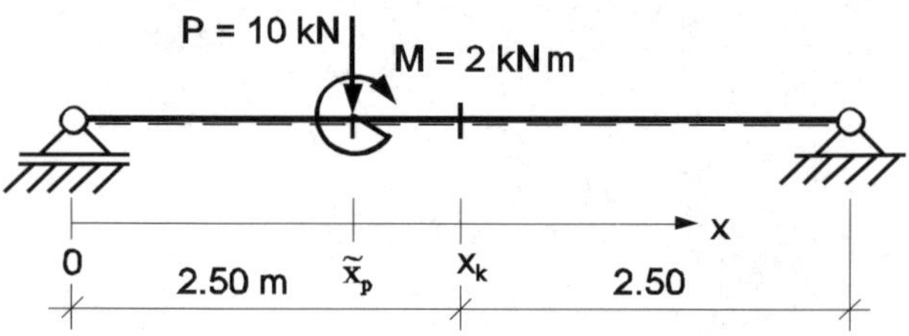

Fig. 5.18. Structural system, loading

The bending moment at point x_k is given by

$$M_k(x_p) = \begin{cases} 1 + 5 \cdot x_p & [\mathrm{kNm}] \mid x_p < x_k \\ 24 - 5 \cdot x_p & [\mathrm{kNm}] \mid x_p > x_k \end{cases} . \tag{5.12}$$

In this case, Eq. (5.12) is the mapping model and exhibits a discontinuity at the point $x_p = x_k$. The fuzzy input variable $\tilde{x}_p$ is prescribed as a fuzzy triangular number $\tilde{x}_p = <2.3, 2.4, 2.6>$ m (Fig. 5.19). The point of discontinuity at $x_p = 2.50$ m is an element of the fuzzy input set. For several α-levels the subspace $\underline{Z}_{\alpha_k}$ is nonconnected. The computation of the fuzzy result $\tilde{z} = \tilde{M}_k$ with the aid of the extension principle results in a disjointed fuzzy set; the solution of this problem by means of α-level optimization on the other hand yields a connected fuzzy result set (Fig. 5.19).

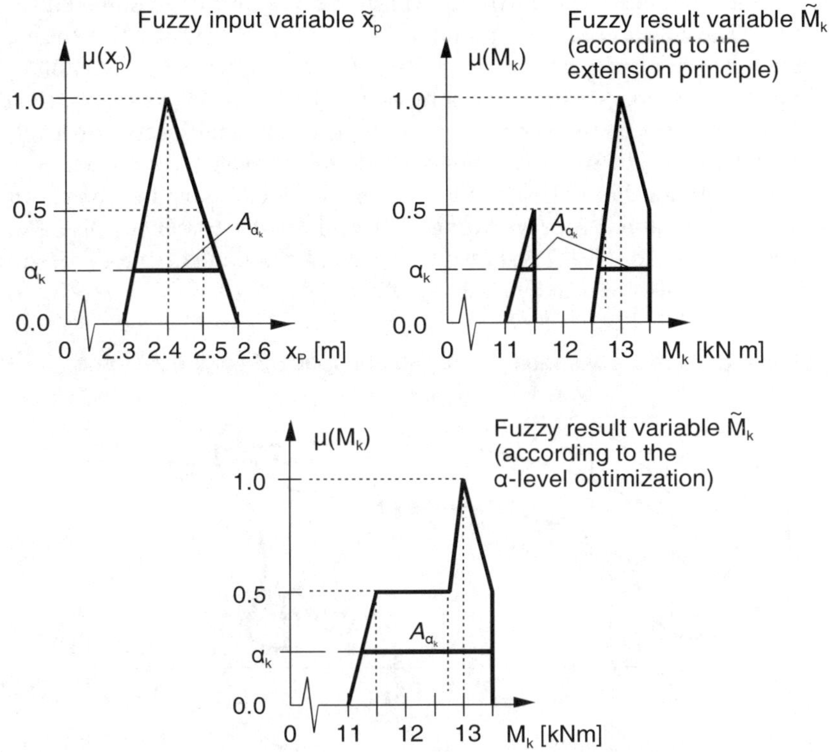

Fig. 5.19. Fuzzy input variable $\tilde{x} = \tilde{x}_p$ and fuzzy result variable $\tilde{z} = \tilde{M}_k$

Due to the fact that min z_{α_k} and max z_{α_k} are always determined in α-level optimization the fuzzy result variable is always convex. Its membership function represents the envelope curve of the membership function according to the extension principle.

In structural-engineering problems the possible errors resulting from the computation of fuzzy result variables by means of α-level optimization using discontinuous mapping model (determination of the *envelope curves* of the $\mu(z_j)$) are only of secondary importance. As a rule, only the extreme values of the result variables (e.g., static and kinematic variables, safety level, service life) are of interest here rather than possible intermediate values.

Monotonicity. The monotonicity properties of the mapping model determine the position of the optimum points in the space of the fuzzy input variables. A mapping in accordance with Eq. (2.36) is given. The mapping model is intended to describe a monotonic relationship between the input variable x_i and the result variable z_j. The fuzzy result $\tilde{z}_j$ is sought.

In the α-level optimization the extreme values for z_{j,α_k} must be computed for each α-level. The position of the optimum points $\underline{x}_{opt}$ that lead to the extreme values sought may be partly determined a priori by exploiting the given monotonicity property between x_i and z_j. The coordinates x_i corresponding to the $\underline{x}_{opt}$ then take on the values of the interval bounds $x_{i,\alpha_k l}$ or $x_{i,\alpha_k r}$ for the respective α-level. The optimum points $\underline{x}_{opt}$ lie on the boundary of the respective crisp subspace $\underline{X}_{\alpha_k}$. If a monotonic relationship exists between *all* input variables x_i and the result variable z_j, the $\underline{x}_{opt}$ required for computing $\tilde{z}_j$ then lie in the corners of $\underline{X}_{\alpha_k}$. The mapping according to Eq. (2.36) is shown for different monotonicity properties in Figs. 5.20 to 5.22 for n = 3 and m = 2.

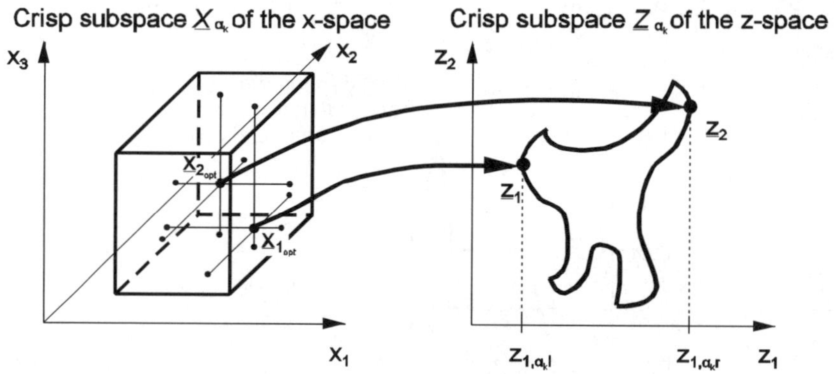

Fig. 5.20. Mapping of the crisp subspace $\underline{X}_{\alpha_k}$ (n = 3) onto the crisp subspace $\underline{Z}_{\alpha_k}$ (m = 2) for computing the fuzzy result variable $\tilde{z}_1$; no monotonicity properties of the mapping model

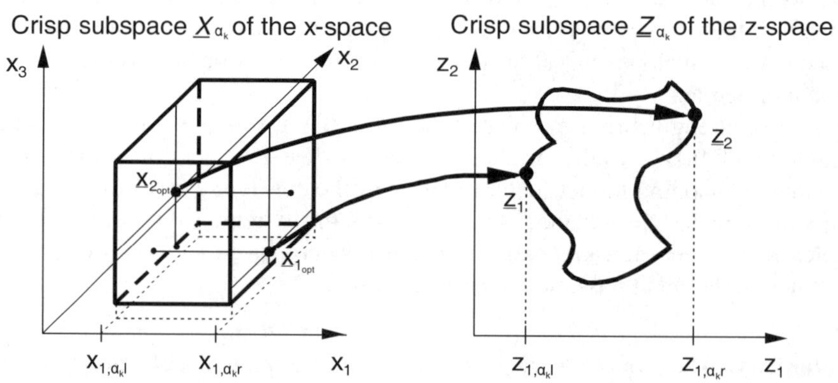

Fig. 5.21. Mapping of the crisp subspace $\underline{X}_{\alpha_k}$ (n = 3) onto the crisp subspace $\underline{Z}_{\alpha_k}$ (m = 2) for computing the fuzzy result variable $\tilde{z}_1$; monotonic relationship between x_1 and z_1

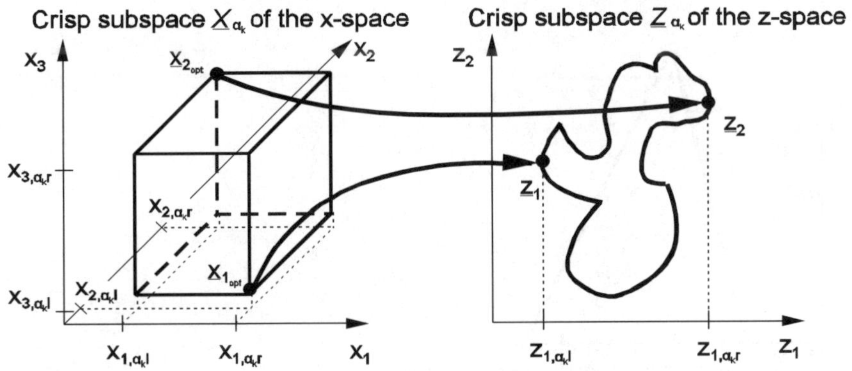

Fig. 5.22. Mapping of the crisp subspace $\underline{X}_{\alpha_k}$ (n = 3) onto the crisp subspace $\underline{Z}_{\alpha_k}$ (m = 2) for computing the fuzzy result variable $\tilde{z}_1$; monotonic relationship between x_1, x_2, x_3 and z_1

A knowledge of the monotonicity properties of the mapping model may be advantageously used to limit the search domain when searching for the optimum points, e.g., on the boundary, the edges, or the corners of the input subspace. This may significantly reduce the numerical effort required in the α-level optimization. Generally valid propositions concerning monotonicity cannot be made for mapping model in structural-engineering problems, as these may involve both monotonic and nonmonotonic mappings. It is only possible to take advantage of the monotonicity properties in special cases. In general, the optimum points $\underline{x}_{opt}$ may lie in the entire input subspace.

Example 5.3. For a single-span girder with a uniformly distributed line load (Fig. 5.23) the nonmonotonic mapping of a one-dimensional fuzzy input set onto a one-dimensional fuzzy result set is considered. The fuzzy input variable is the fuzzy position of the commencement of the line load $\tilde{x}_p$ (Fig. 5.24), whereas the fuzzy result variable is the fuzzy bending moment $\tilde{M}_k$ at x_k in the middle of the girder.

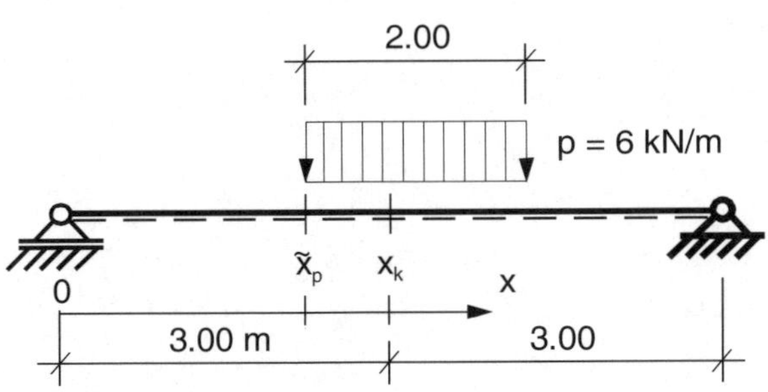

Fig. 5.23. Structural system, loading

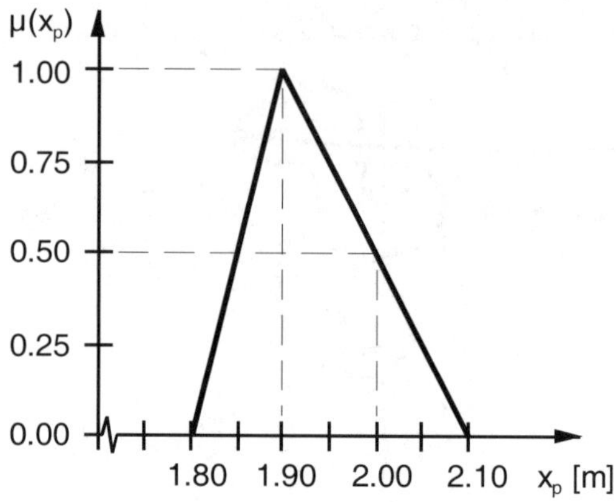

Fig. 5.24. Fuzzy input variable $\tilde{x}_p$ (fuzzy position of the commencement of the line load)

The mapping model

$$M_k = -3 \cdot x_p^2 + 12 \cdot x_p + 3 \text{ [kNm]} \tag{5.13}$$

exhibits a local maximum of M_k at $x_p = 2.00$ m. This value is an element of the input subspace for $\alpha_k \leq 0.50$. For α-level optimization it is concluded from the foregoing that the optimum points in the search for max M_k may lie in the interior of the input subspace. For the elements of the α-level sets with $\alpha_k \geq 0.50$ the mapping model is monotonic, and $x_p = 2.00$ m lies outside the crisp input subspaces for $\alpha_k > 0.50$. This means that the optimum points for $\alpha_k \geq 0.50$ lie in the corners of the input subspace, i.e., they coincide with the interval bounds of the α-level sets of $\tilde{x}_p$. The solution obtained from α-level optimization using five α-levels is illustrated in Fig. 5.25. The optimum points x_{opt} and the extreme result values of M_k for the investigated α-levels are indicated in the figure. For $\alpha_k < 0.50$ the optimum points for the maximum search lie within the respective crisp subspace for x_p. The computed interval bounds of the α-level sets for $\tilde{M}_k$ yield the curve of the membership function of the fuzzy result. Due to the fact that the mapping model is continuous, α-level optimization yields the same result as computed using the extension principle.

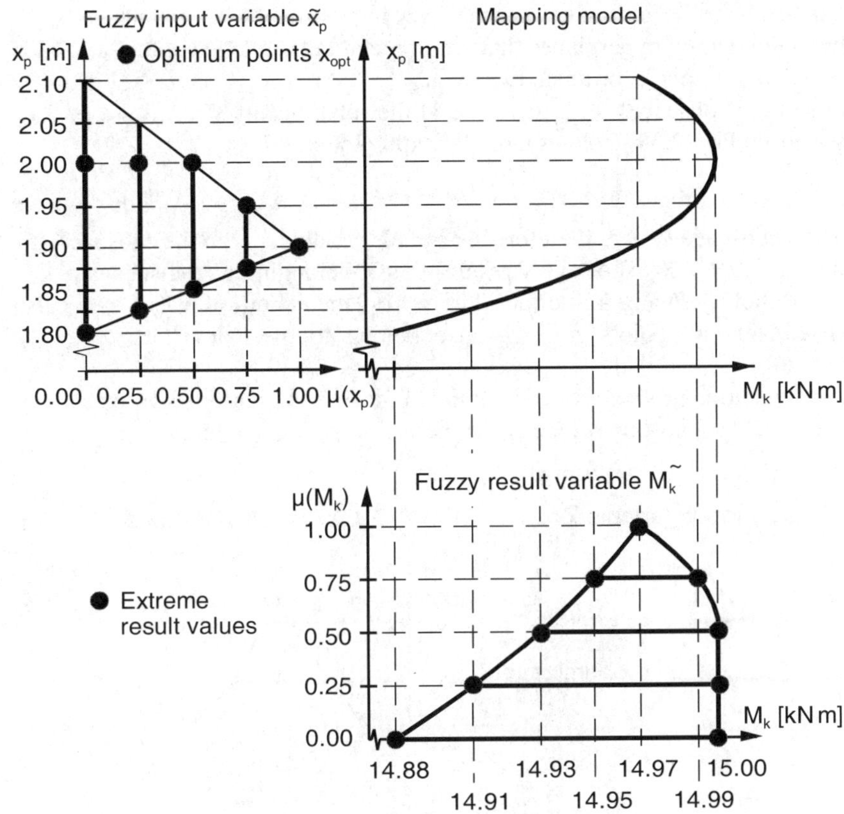

Fig. 5.25. Mapping of the fuzzy position of the commencement of the line load $\tilde{x}_p$ onto the fuzzy bending moment $\tilde{M}_k$ at x_k by means of α-level optimization using five α-levels; optimum points x_{opt} in the space of the input variables and extreme result values for the investigated α-levels

5.2.3 Solution Techniques for α-level Optimization

5.2.3.1 Specification of the Optimization Problem

The optimization problem according to Eqs. (5.10) and (5.11) is characterized by the following features:

1. The optimization variables are continuous.
2. The objective function is generally represented in the form of an algorithm, i.e., not in a closed form. It is only possible to formulate the objective function explicitly in simple cases.

3. The value range of the $x_1, ..., x_n$ is defined on each α-level by the crisp subspace $\underline{X}_{\alpha_k}$. If no interaction exists between the fuzzy input variables $\tilde{x}_i$, the subspace $\underline{X}_{\alpha_k}$ is then bounded by hyperplanes that are perpendicular to one another and each perpendicular to one coordinate axis x_i.
4. Between the subspaces $\underline{X}_{\alpha_k} \mid \alpha_k \in (0, 1]$ the relationship $\underline{X}_{\alpha_i} \subseteq \underline{X}_{\alpha_k} \mid \alpha_i \geq \alpha_k$ exists. From Eq. (2.38) follows that all result variables

$$\underline{z}_t = f(x_1, ..., x_n) \mid (x_1, ..., x_n) \in \underline{X}_{\alpha_i} \tag{5.14}$$

from the subspace $\underline{Z}_{\alpha_i}$ are therefore also result variables $\underline{z}_t$ in the subspace $\underline{Z}_{\alpha_k}$ with $(x_1, ..., x_n) \in \underline{X}_{\alpha_k}$. For every result point $\underline{z}_t$ belonging to the subspace $\underline{Z}_{\alpha_i}$, $\mu(\underline{z}_t) \geq \alpha_i$ holds. Owing to the fact that $\alpha_i \geq \alpha_k$, the condition $\mu(\underline{z}_t) \geq \alpha_k$ is also satisfied, i.e., all points $\underline{z}_t \in \underline{Z}_{\alpha_i}$ also belong to the α-level α_k, $\underline{z}_t \in \underline{Z}_{\alpha_k}$ (Fig. 5.26).
5. The optimization problem for each of the m fuzzy result variables $\tilde{z}_j$ on each of the r α-levels α_k must be solved twice, i.e., $(2 \cdot m \cdot r)$ times in total.

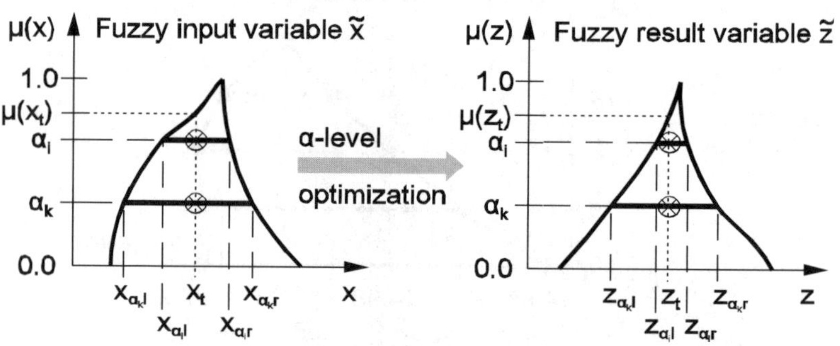

Fig. 5.26. Mapping of the fuzzy input variable $\tilde{x}$ onto the fuzzy result variable $\tilde{z}$ by α-level optimization with $z_t = f(x_t)$ as an element of the α-level sets assigned to α_i and α_k

Features 1 to 4 characterize an optimization problem without special properties that might be exploited numerically. The extent of the numerical computation is mainly determined by the objective function representing the mechanical model adopted (e.g., linear, nonlinear). The optimum points may lie within the subspace $\underline{X}_{\alpha_k}$ as well as on its "boundary" or in its "corners". As hardly any information is available a priori the solution technique should be independent of assumptions concerning the position of the optimum points. Owing to feature 5 repeated solution of the optimization problem is necessary.

An α-level optimization thus demands a robust optimization technique that is independent of the type and behavior of the objective function or constraints (restrictions) and is capable of reliably finding global optima.

Standard optimization methods are only partly suitable for this purpose. For this reason a compromise solution is developed by combining *evolution strategy*, *the gradient method* and the *Monte Carlo method*.

5.2.3.2 Monte Carlo Simulation and Mesh Search Techniques

The solution of optimization problems by a systematic or random, nondirected search in the variable space (space of the input variables) for the optimum points is very simple and robust. In the case of a systematic search a uniform mesh is laid over the entire domain of the variable space under consideration (input subspace $\underline{X}_{\alpha_k}$) and the objective function is evaluated at each of the mesh points. As a simple stochastic method, the Monte Carlo simulation makes use of the uniform distribution of the points to be evaluated in the space of the variables.

In both methods no requirements are placed on the objective functions or constraints; the behavior of the latter may be arbitrary. In general, the results are approximations; it is not possible to precisely determine the exact solution in the case of continuous optimization problems. The numerical effort required to solve the problem is considerable; this increases exponentially with the number of input variables for the same accuracy of the results. The efficiency of these methods may be increased by restricting the search domain to the neighborhood of the best solution obtained so far. Compared with other techniques, however, these methods are generally less efficient.

Owing to their robustness mesh search techniques are suitable for obtaining a reference solution for assessing the results of other methods. In the present case a multilayer mesh technique is chosen. By refining the mesh incrementally in restricted domains this technique generally yields reliable results. The use of a start-up mesh over the entire subspace $\underline{X}_{\alpha_k}$ and the computation of the corresponding functional values of the objective function are thereby equivalent to the incremental treatment of the mapping model by means of the extension principle (Sect. 5.2.1).

The use of systematic search techniques or simple random search techniques in α-level optimization offers no advantages regarding computational effort compared with the application of the extension principle; they are thus considered here to be unsuitable. There is virtually no difference between the computation of the complete membership function of the fuzzy result and the computation of selected parts of the membership function. In α-level optimization on the support level the results for the remaining levels are always simultaneously obtained provided an appropriate mesh is used. A systematic search does have advantages if the optimum points may be predicted to lie in the corners of the input subspace a priori, and a search is necessary for these points only.

5.2.3.3 Gradient method and evolution strategy

Directed search methods yield a solution of the optimization problem in successive steps with an incremental improvement in the solution. In order to reliably attain the optimization objective in as few steps as possible the most suitable direction of advancement must be specified in each step. In order to realize this in the gradient method the direction of the gradient of the objective function at the current point in the space of the input variables is followed. The efficiency of the method may be improved if a new computation of the gradient and hence the specification of a new step direction is first carried out when an improvement in the solution may no longer be obtained along the previous step direction. A combination with the diagonal-step technique results in a further improvement in efficiency. The results obtained by these methods are generally approximation solutions for the optimum values sought, the accuracy of which may be influenced by the use of control parameters.

The numerical effort (as characterized by the number of computations necessary to determine objective function values) required to solve an optimization problem by means of the gradient method is considerably less than that required by systematic and simple random search techniques. The reliability of the solution is lower; this depends on the type and behavior of the objective function and the constraints. It is not always possible to find the global optimum. Discontinuous or highly nonlinear objective functions (e.g., jumps or "narrow canyons") and constraints are considered to be particularly problematic in this respect. If the objective function together with its directional derivatives cannot be stated in a closed form, the computation of the gradient may be falsified due to numerical inaccuracies in the evaluation of the objective function.

The gradient method is only suitable to a limited degree for application in α-level optimization. In special problem areas, however, this is considered to be the optimum method of solution [19].

Evolution concepts are modeled on the "natural" solution of problems [154]. Oriented to biological evolution processes, an attempt is made to find an optimum strategy adapted to surrounding conditions. The search for optimum points is carried out by a process of mutation and selection. The "favorable" properties of the parent points are thereby retained, while the offspring points evolve from random changes in their neighborhood. The next step is first performed once an improvement in the optimization objective has been realized, and the offspring point then becomes the new parent point. A distinction is made between (1+1) and higher-order evolution methods.

The step direction in the optimization procedure is determined randomly; the often time-consuming computation of the gradients is no longer necessary. Although the individual steps do not yield an optimum improvement in the value of the objective function, an acceptable improvement is generally obtained. Compared with the gradient method the number of steps necessary to arrive at the solution may be somewhat higher; the principle of random mutation in each optimization step may slightly "decelerate" the method. The numerical effort required by both methods is comparable in overall terms; according to [92] this increases almost linearly with the number of input variables. Due to the simple testing of the offspring points using the criterion *better than*, the evolution method is less sensitive with respect to numerical inaccuracies in the computation of the functional values of the objective function. With regard to the determination of the global optima and the accuracy of the results the same propositions apply as in the gradient method. Improvements in efficiency may be achieved, e.g., by adaptive control of the step increments.

In their classical form, evolution strategies are only conditionally suitable for application in α-level optimization; they are, however, considered to be more advantageous than the gradient method.

5.2.3.4 Modified Evolution Strategy

The Monte Carlo simulation, the mesh search technique, the gradient method, and the (classical) evolution strategy are only conditionally suitable for solving the optimization problem characterized in Sect. 5.2.3.1. By combining these methods an attempt is made to develop a more suitable optimization procedure.

The combination of directed and nondirected search techniques is advantageous compared with purely directed search methods regarding the determination of global optima; a hybrid method is less sensitive in relation to less "well-behaved" objective functions. By making use of existing information on the behavior of the objective function the number of "unnecessary" computations of objective function values, which do not lead to an improvement in the results, is reduced. If the amount of available information is insufficient, random-oriented methods are applied. The developed method of solution, as described in the following, is illustrated graphically in Fig. 5.27.

The optimization problem is described by the input variables $x_i \in \tilde{x}_i$, $i = 1, ..., n$, the result variable $z_j \in \tilde{z}_j$, and the algorithm for computing the objective function value (e.g., statical or dynamic analysis). The constraints $(x_1, ..., x_n) \in \underline{X}_{\alpha_k}$ are obtained from the interaction relationships between the fuzzy input variables $\tilde{x}_i$; these constrain the domain of definition of the objective function $z_j = f_j(x_1, ..., x_n)$. If no interaction exists between the convex fuzzy input variables $\tilde{x}_i$, the permissible domain is obtained as an n-dimensional hypercuboid (subspace) in the x-space. The permissible domain $\underline{X}_{\alpha_k}$ for the α-level $\alpha_k = \alpha_1 = 0$ (support subspace) is shown as a rectangle in Fig. 5.27.

The basis for the compromise solution is a (1+1) evolution strategy, which is matched to the optimization problem by modification. The starting point for the optimization is specified with the aid of uniformly distributed random numbers $U \in [0, 1)$ (or their realizations u_i) in the subspace $\underline{X}_{\alpha_k}$ (permissible domain for α_k); this serves as the first parent point $\underline{x}^{[0]}$ (represented as O in Fig. 5.27)

$$x_i^{[0]} = x_{i,\alpha_k l} + u_i \cdot \left(x_{i,\alpha_k r} - x_{i,\alpha_k l}\right);\; i = 1, ..., n. \tag{5.15}$$

A mutation of the properties of the parent point is then simulated by generating an offspring point in its neighborhood according to the random principle. Thereby a departure from the classical evolution strategy procedure is made in detail. For the offspring points $\underline{x}^{[q+1]}$ a maximum and a minimum distance

$$max\_d_i = \max | x_i^{[q+1]} - x_i^{[q]} | ; \ i = 1, ..., n, \tag{5.16}$$

$$min\_d_i = \min | x_i^{[q+1]} - x_i^{[q]} | ; \ i = 1, ..., n \tag{5.17}$$

from the parent point $\underline{x}^{[q]}$ are specified directionally, i.e., for each coordinate x_i. The subspace defined by max_d_i and min_d_i assigned to the parent point $\underline{x}^{[q]}$ (local search domain) is an n-dimensional hypercuboid. The specification of max_d_i and min_d_i by compressing the support subspace is advantageous

$$max\_d_i = c_1 \cdot (x_{i,\alpha_1 = 0r} - x_{i,\alpha_1 = 0l}) \ | \ 0 < c_1 < 1; \ i = 1, ..., n, \tag{5.18}$$

$$min\_d_i = c_2 \cdot max\_d_i \ | \ 0 < c_2 < 1; \ i = 1, ..., n. \tag{5.19}$$

If no interaction exists between the fuzzy input variables $\tilde{x}_i$, a similarity relationship exists between the local search domain and the support subspace $\underline{X}_{\alpha_1 = 0}$ (permissible domain for $\alpha_1 = 0$). In the case of interaction, the local search domain may be oriented to the form of $\underline{X}_{\alpha_1 = 0}$ using the constraints for the $\tilde{x}_i$. The distance bounds max_d_i and min_d_i are independent of the α-level α_k. For the representative run in Fig. 5.27 the parent points $\underline{x}^{[q]}$ are indicated by ⊕ and the bounds of the assigned local search domains are plotted as dotted (for min_d_i) and solid rectangles (for max_d_i).

The first offspring point $\underline{x}^{[q+1]}$ of $\underline{x}^{[q]}$ is generated within this local search domain (between the dotted and solid lines in Fig. 5.27) by means of a uniform distribution of $U \in [0, 1)$

$$x_i^{[q+1]} = x_i^{[q]} + 2 \cdot (u_i - 0.5) \cdot max\_d_i \ ; \ i = 1, ..., n, \tag{5.20}$$

with the requirement

$$min\_d_i \le d_i = | x_i^{[q+1]} - x_i^{[q]} | \ ; \ 1 \le i \le n \tag{5.21}$$

for at least one coordinate x_i. Based on the parent point $\underline{x}^{[q]}$ one offspring point $\underline{x}^{[q+1]}$ is first determined. A test is carried out to check whether its objective function value $z_j^{[q+1]} = f_j(x_1^{[q+1]}, ..., x_n^{[q+1]})$ is an improvement compared with $z_j^{[q]} = f_j(x_1^{[q]}, ..., x_n^{[q]})$ of the parent point. If an improvement is obtained, the search proceeds along the randomly selected direction until no further improvement in the objective function value occurs for $r = 1, 2, 3, ...$

$$\underline{x}^{[q+r+1]} = 2 \cdot \underline{x}^{[q+r]} - \underline{x}^{[q+r-1]} \ \left| \ \begin{array}{ll} z_j^{[q+r]} < z_j^{[q+r-1]} & \text{for minimum search} \\ z_j^{[q+r]} > z_j^{[q+r-1]} & \text{for maximum search} \end{array} \right. . \tag{5.22}$$

The step increment thereby remains constant. The last improved point $\underline{x}^{[q+r+1]}$ then becomes the current parent point. These step sequences are indicated by the points ⊕ in Fig. 5.27.

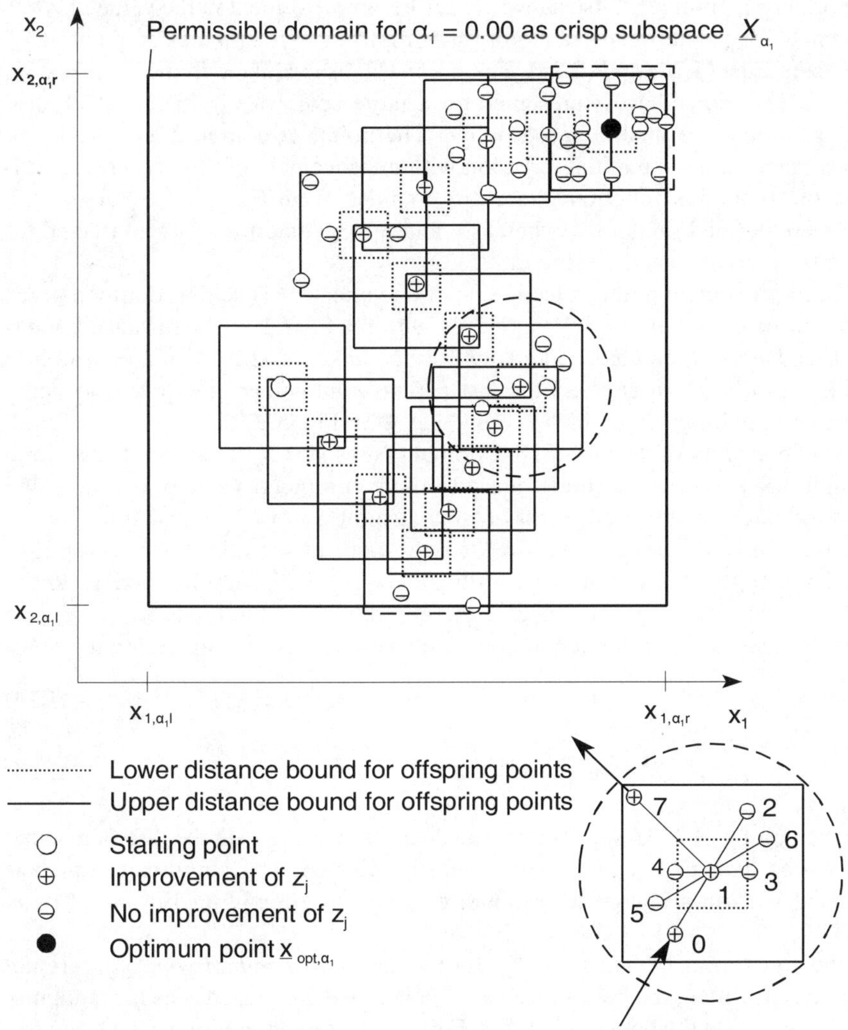

Fig. 5.27. Modified evolution strategy, representative run for $\alpha_k = \alpha_1 = 0.00$

If the offspring point $\underline{x}^{[q+1]}$ determined randomly from Eqs. (5.20) and (5.21) does not lead to an improvement in the objective function value $\underline{z}_j^{[q+1]}$ compared with $\underline{z}_j^{[q]}$ (denoted by ⊖ in Fig. 5.27), the next offspring point $\underline{x}^{[q+2]}$ is positioned at the same distance as $\underline{x}^{[q+1]}$ from the parent point $\underline{x}^{[q]}$ but in the opposite direction

$$\underline{x}^{[q+2]} = 2 \cdot \underline{x}^{[q]} - \underline{x}^{[q+1]} \left| \begin{array}{ll} z_j^{[q+1]} \geq z_j^{[q]} & \text{for minimum search} \\ z_j^{[q+1]} \leq z_j^{[q]} & \text{for maximum search} \end{array} \right. . \tag{5.23}$$

If the offspring point $\underline{x}^{[q+2]}$ also shows no further improvement in the value of $z_j^{[q+2]}$, the point $\underline{x}^{[q+3]}$ is determined randomly starting from $\underline{x}^{[q]}$ (as in the computation of $\underline{x}^{[q+1]}$) using Eqs. (5.20) and (5.21). The point $\underline{x}^{[q+3]}$ is evaluated in the same way as for $\underline{x}^{[q+1]}$. This procedure is presented on a large scale in Fig. 5.27, which also shows the sequence of the placed points. The points 0, 1, and 2 are a result of advancements along the initial direction with improvement of the functional values of the objective function; these are obtained from Eq. (5.22). Points 3, 5, and 7 were defined randomly, whereas 4 and 6 were determined in a targeted direction from points 3 and 5 using Eq. (5.23).

If no improvement in the objective function value $z_j^{[q]}$ is achieved after a given number n_P of point pairs ($\underline{x}^{[q+2r+1]}$, $\underline{x}^{[q+2r+2]}$) with r = 0, 1, 2,..., the distance bounds min_d_i and max_d_i are reduced. After further unsuccessful steps the search is considered to be at an end and the last parent point is then interpreted as being the optimum point $\underline{x}_{opt,\alpha_k}$, which is plotted as ● in Fig. 5.27.

On the boundary of the permissible domain (subspace $\underline{X}_{\alpha_k}$) the search algorithm distinguishes between randomly placed points computed from Eqs. (5.20) and (5.21) and directed specified points obtained from Eqs. (5.22) or (5.23).

If the *randomly placed point* $\underline{x}^{[q+1]}$ does not satisfy all of the constraints $(x_1, ..., x_n) \in \underline{X}_{\alpha_k}$, the respective coordinates $x_i^{[q+1]}$, which do not adhere to the constraints, are corrected. In this case the coordinates of the boundary of $\underline{X}_{\alpha_k}$ are chosen for these $x_i^{[q+1]}$. If no interaction exists between the $\tilde{x}_i$, the following holds

$$x_{i(new)}^{[q+1]} = x_{i,\alpha_k l} \mid x_i^{[q+1]} < x_{i,\alpha_k l};\ i = 1,...,n\,, \tag{5.24}$$

$$x_{i(new)}^{[q+1]} = x_{i,\alpha_k r} \mid x_i^{[q+1]} > x_{i,\alpha_k r};\ i = 1,...,n\,. \tag{5.25}$$

If the point $\underline{x}_{(new)}^{[q+1]}$ found in this way does not comply with the distance bound min_d_i according to Eq. (5.21), this point is rejected. The coordinate search is continued with the random determination of $\underline{x}^{[q+1]}$ by applying Eqs. (5.20) and (5.21).

If the constraints $(x_1, ..., x_n) \in \underline{X}_{\alpha_k}$ for the *directed specified point* $\underline{x}^{[q+r]}$ are not completely fulfilled, the coordinates $x_i^{[q+r]}$ concerned are placed – as for randomly placed points – on the boundary of $\underline{X}_{\alpha_k}$ (Fig. 5.28). If no interaction exists between the $\tilde{x}_i$, Eqs. (5.24) and (5.25) may be applied.

The distance bound min_d_i is then checked using Eq. (5.21). For this purpose that coordinate $x_i^{[q+r]}$ is determined that falls short of the minimum distance min_d_i by the least relative distance

$$d_{rel,min} = \max_{i=1,...,n} \left(\frac{d_i}{min\_d_i} \right). \tag{5.26}$$

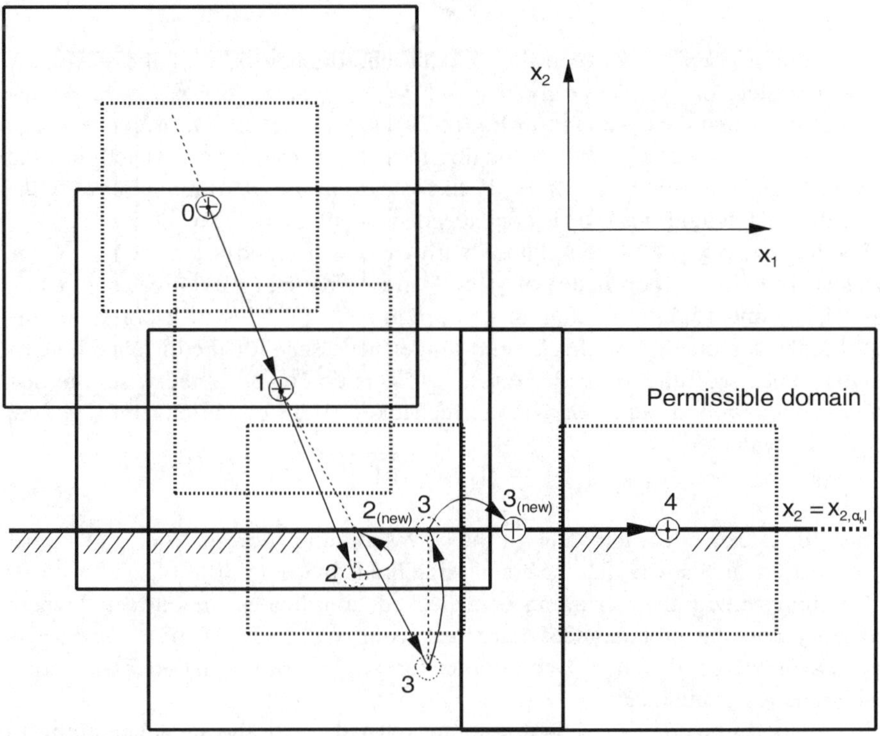

Fig. 5.28. Modified evolution strategy, behavior of the search algorithm along the boundary of the permissible domain $\underline{X}_{\alpha_k}$

If $d_{rel,min} \geq 1.0$, at least one coordinate $x_i^{[q+r]}$ does not fall short of the distance bound *min*_d$_i$, i.e., Eq. (5.21) is fulfilled; the point $\underline{x}_{(new)}{}^{[q+r]}$ has been found. For $d_{rel,min} < 1.0$ the requirement of minimum distance is not complied with and, as shown in Fig. 5.28, all d_i values from Eq. (5.21) are proportionally increased by the factor

$$c_3 = \frac{1}{2} \cdot \left(\frac{1}{d_{rel,\,min}} + \frac{1}{d_{rel,\,max}} \right) , \tag{5.27}$$

in which

$$d_{rel,\,max} = \max_{i=1,\ldots,n} \left(\frac{d_i}{max\_d_i} \right) \tag{5.28}$$

is the minimum relative deficit below the maximum distance *max*_d$_i$. In accordance with Eq. (5.17) $d_{rel,max}$ corresponds to the same coordinate $x_i^{[q+r]}$ as $d_{rel,min}$. The point $\underline{x}_{(new)}{}^{[q+r]}$ is determined from

$$\underline{x}^{[q+r]}_{(new)} = \underline{x}^{[q+r-1]} + c_3 \cdot \left(\underline{x}^{[q+r]} - \underline{x}^{[q+r-1]}\right). \tag{5.29}$$

If no interaction exists between the $\tilde{x}_i$, the coordinates $x_i^{[q+r]}$ that are already boundary values of $\underline{X}_{\alpha_k}$, i.e., for which $x_i^{[q+r]} = x_{i,\alpha_k l}$ or $x_i^{[q+r]} = x_{i,\alpha_k r}$, are retained and no longer considered in Eq. (5.29), see Fig. 5.28. On account of Eq. (5.27) the coordinate $x_{i\,(new)}^{[q+r]}$ belonging to $d_{rel,min}$ is thus placed exactly on the midpoint between min_d_i and max_d_i. In the case of *no interaction between the* $\tilde{x}_i$ the distance bound max_d_i is not exceeded for all $i = 1, ..., n$.

For the point $\underline{x}_{(new)}^{[q+r]}$ compliance with the constraints $(x_1, ..., x_n) \in \underline{X}_{\alpha_k}$ is checked; individual coordinates of $\underline{x}_{(new)}^{[q+r]}$ are corrected as required, e.g., using Eqs. (5.24) and (5.25). If, after this, the point $\underline{x}_{(new)}^{[q+r]}$ lies in a corner of the permissible domain $\underline{X}_{\alpha_k}$, a check is no longer necessary for the distance bounds and the objective function value for $\underline{x}_{(new)}^{[q+r]}$ is computed. Otherwise, the distance bounds min_d_i and max_d_i are rechecked. If all conditions are satisfied, $\underline{x}_{(new)}^{[q+r]}$ is evaluated

$$max\_d_i \geq d_i = \left| x_i^{[q+r]} - x_i^{[q+r-1]} \right| . \tag{5.30}$$

If max_d_i is exceeded, the value of the corresponding coordinate $x_{i\,(new)}^{[q+r]}$ must be reduced in such a way that the equals sign holds in Eq. (5.30).

Compliance with the constraints is rechecked; coordinates are corrected where necessary and the procedure is continued using Eq. (5.26). This procedure is repeated as required until all conditions for $\underline{x}_{(new)}^{[q+r]}$ are satisfied or a corner position in $\underline{X}_{\alpha_k}$ is attained.

If the condition for min_d_i is not complied with, only the steps according to Eqs. (5.26) to (5.29) are included in the sequence to be repeated.

When searching for optimum points $\underline{x}_{opt}$, use may be made of the inclusion properties of the subspaces $\underline{X}_{\alpha_k}$ and $\underline{X}_{\alpha_i}$ for $\alpha_k \leq \alpha_i$ according to Eq. (2.45). If it is necessary to check the point $\underline{x}^{[q]}$ belonging to α-level $\alpha_k \leq \alpha_i$ for optimality and this point has already been evaluated for $\alpha = \alpha_i$, the existing functional value $z_j^{[q]}$ for $\alpha = \alpha_i$ is then used. A recomputation of $z_j^{[q]}$ is no longer necessary; see feature 4 of the optimization problem.

The use of existing points $\underline{x}^{[q]}$ with a known objective function value $z_j^{[q]}$ leads to an improved efficiency of the method. An endeavor is thus made to include existing points in the "neighborhood" of a newly located point in the search for an extreme value. The neighborhood is defined with the aid of a direction-dependent distance Δd_i referred to the distance bound max_d_i

$$\Delta d_i = c_4 \cdot max\_d_i \;\; | \;\; 0 < c_4 < 1;\; i = 1, ..., n. \tag{5.31}$$

For each newly placed point $\underline{x}^{[q]}$ a test is carried out to check whether this point lies in the neighborhood of a known point $\underline{x}^{[p]}$ with the known objective function value $z_j^{[p]}$. If

$$\Delta d_i \geq \left| x_i^{[q]} - x_i^{[p]} \right| ;\; 1, ..., n \tag{5.32}$$

then

$$\underline{x}^{[q]} = \underline{x}^{[p]} \tag{5.33}$$

is specified and the optimum search is continued.

All points that have already been evaluated in *one* extreme value search – i.e., optimization for *one* α-level, *one* fuzzy result variable, and *one* optimization objective (minimum or maximum) – are designated by $\underline{x}^{[\bar{p}]}$. These points $\underline{x}^{[\bar{p}]}$ are unable to produce improved results within *the same* extreme value search. It is further assumed that an improvement in the results is also not possible using points in the neighborhood of the points $\underline{x}^{[\bar{p}]}$. For the randomly or directed specified points $\underline{x}^{[q]}$ determined in *this* extreme value search a test is carried out using Eq. (5.32) to check whether a point $\underline{x}^{[\bar{p}]}$ is located in their neighborhood. If this is the case, $\underline{x}^{[q]}$ is not evaluated; the optimization search is continued with a new point $\underline{x}^{[q]}$.

The described procedure combines elements of different optimization methods: evolution strategy, the gradient method, and the Monte Carlo method. The random specification of offspring points within the distance bounds *min*_d_i and *max*_d_i corresponds to the simple Monte Carlo method. In the event that no interaction exists between the $\tilde{x}_i$ and that the local search domains bounded by *min*_d_i and *max*_d_i represent n-dimensional hypercuboids, the most probable, randomly selected search direction runs parallel to one of the spatial diagonals of the permissible domain. In the case of uniform distribution of the randomly selected points $\underline{x}^{[q+r]}$ within the local search domain assigned to the point $\underline{x}^{[q]}$, the functional value of the probability density function is

$$\begin{aligned} p_0 = f(\underline{x}^{[q+r]}) &= \frac{1}{\prod_{i=1}^{n}(2 \cdot max\_d_i) - \prod_{i=1}^{n}(2 \cdot min\_d_i)} \\ &= \frac{1}{2^n \cdot (1 - c_2^{\,n}) \cdot \prod_{i=1}^{n}(max\_d_i)} . \end{aligned} \tag{5.34}$$

If the angles between the coordinate axes x_i and the vector $(\underline{x}^{[q+r]} - \underline{x}^{[q]})$ are designated by φ_i, the functional value of the probability density function for the randomly selected search direction as a function of φ_i may be stated as $f(\underline{\varphi}^{[q+r]})$. The ray determined by the direction of the vector $(\underline{x}^{[q+r]} - \underline{x}^{[q]})$ originating in $\underline{x}^{[q]}$ intersects the planes $x_i = x_i^{[q]} \pm min\_d_i$ and $x_i = x_i^{[q]} \pm max\_d_i$. The line segment Δl between the two intersection points for *min*_d_i and *max*_d_i (for the same i) located at the shortest distance from point $\underline{x}^{[q]}$ runs within the local search domain (Fig. 5.29).

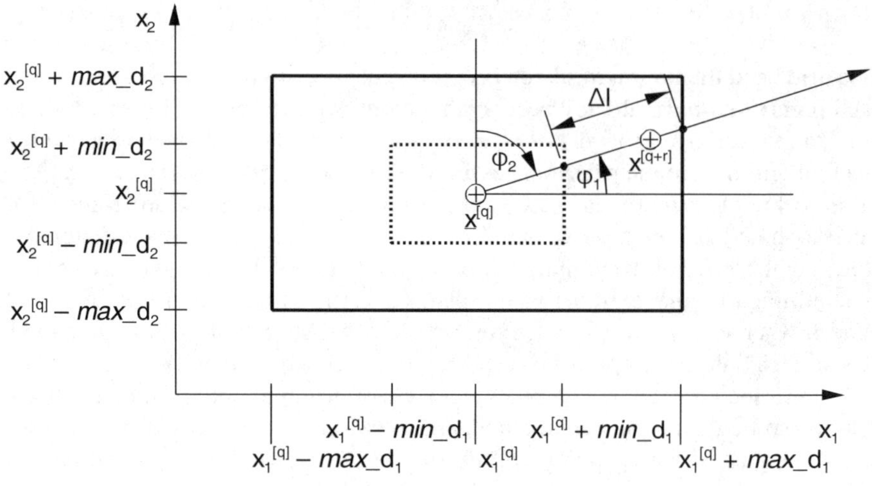

Fig. 5.29. Random determination of the search direction

The product of the length Δl and the constant p_0 yields the functional value of the probability density function

$$f(\underline{\varphi}^{[q+r]}) = \frac{p_0 \cdot (max\_d_i - min\_d_i)}{|\cos\varphi_i|} = \frac{p_0 \cdot (1 - c_2) \cdot max\_d_i}{|\cos\varphi_i|} . \tag{5.35}$$

The values of min_d_i, max_d_i, and $\cos\varphi_i$ must be referred to the coordinates x_i for which $x_i = x_i^{[q]} \pm min\_d_i$ and $x_i = x_i^{[q]} \pm max\_d_i$ form the limit points for Δl. The denominator in Eq. (5.35) is the absolute value of the direction cosine between x_i and $(\underline{x}^{[q+r]} - \underline{x}^{[q]})$. For $\cos\varphi_i = 1$, i.e., in the direction $\pm x_i$, the function $f(\underline{\varphi}^{[q+r]})$ possesses local minima; $f(\underline{\varphi}^{[q+r]})$ increases with decreasing $\cos\varphi_i$. For angles $\underline{\varphi}^{[q+r]}$ for which the reference coordinate x_i changes (edges or corners of the local search domain) breakpoints appear in the function $f(\underline{\varphi}^{[q+r]})$; local maxima of the functional value occur here. The global maximum of $f(\underline{\varphi}^{[q+r]})$ lies in the direction of a corner of the local search domain.

The search for optimum points is thus, by higher probability, guided towards the corners of the permissible domain (i.e., towards the corners of the subspace $\underline{X}_{\alpha_k}$), where the $\underline{x}_{opt}$ are located in the case of monotonic mappings. The direction of the largest extent of $\underline{X}_{\alpha_k}$ is preferred compared with other directions. The larger the factor c_2 the less pronounced are these characteristics.

The progressive search in a fixed direction for improvement of the solution corresponds to an improved gradient method. The random determination of this direction avoids the numerical computations necessary to determine the gradient.

The flexibility of the algorithm is realized by means of variable control parameters. These must be specified for the particular problem concerned. The following control parameters were introduced:

1. Maximum number n_P of randomly specified offspring points (starting from the same parent point) without improvement of the result.
2. Number of refinement stages n_f for the distance bounds min_d_i and max_d_i.
3. Maximum directional step increment max_d_i relating to the support of the fuzzy input variable $\tilde{x}_i$; factor c_1 in Eq. (5.18).
4. Minimum directional step increment min_d_i relating to max_d_i according to point 3; factor c_2 in Eq. (5.19).
5. Maximum directional distance Δd_i for the reuse of existing points instead of the newly placed ones, referred to the maximum step increment according to point 3; factor c_4 in Eq. (5.31).
6. Reduction of the maximum step increment for each refinement stage of the distance bounds, referred to the respective current maximum step increment according to point 3; factor c_5 in

$$max\_d_{i\,(new)} = c_5 \cdot max\_d_i . \tag{5.36}$$

7. Termination limits for the relative improvement of the result in the last n_z steps, referred to the maximum difference in the functional values of the objective function within one extreme value search; factor c_6 in

$$min\_\Delta z_j = c_6 \cdot \max \left| z_j^{[q]} - z_j^{[p]} \right| ; \; q \neq p , \tag{5.37}$$

termination occurs when

$$min\_\Delta z_j > \max_{r=1,\dots,n_z} \left| z_j^{[q]} - z_j^{[q-r]} \right| \left| \begin{array}{ll} z_j^{[q-r]} < z_j^{[q]} & \text{for maximum search} \\ z_j^{[q-r]} > z_j^{[q]} & \text{for minimum search} \end{array} \right. . \tag{5.38}$$

If the parameter according to point 1 is active, a jump occurs into the next refinement step for the distance bounds. If the number of refinement steps according to point 2 is also exhausted, the optimization is terminated and the optimum point is taken to be localized. Alternatively, termination occurs when the criterion according to point 7 becomes active.

Due to the direct or indirect orientation of all distance dimensions at the support of the respective fuzzy input variables $\tilde{x}_i$ (independent of the α-level) a decrease in the defined distances with increasing α is avoided. The computational effort and the accuracy of the results are thus the same for all α_k, i.e., all points defining the membership function of a fuzzy result variable are determined to the same accuracy. For large α_k the permissible domain $\underline{X}_{\alpha_k}$ may be very small compared with the support subspace $\underline{X}_{\alpha_1=0}$ and hence also very small compared with the local search domain. For an extreme-value search with coarse accuracy requirements this may mean that (starting from the first parent point $\underline{x}^{[0]}$) all additional points in the permissible domain $\underline{X}_{\alpha_k}$ do not satisfy the requirement for the minimum distance min_d_i according to Eq. (5.21). The optimization is then (corresponding to the selected control parameters) continued with the next refinement step or terminated. An orientation of the distance dimensions to the respective permissible domain $\underline{X}_{\alpha_k}$ would lead to an increase in the search accuracy and thus to an increase in the computational effort with increasing α_k.

The determination of the variable control parameters for the optimization depends on the particular problem concerned. Conducting test runs with different objective functions positive experiences have been gained with the following parameter combination, which may be suggested, e.g., as default if no information regarding the behavior of the objective function is available:

- Maximum number of randomly specified offspring points according to point 1: $n_P = 2^n$ with n as the number of fuzzy input variables (dimensionality of the input subspace)
- The number of refinement stages according to point 2: $n_f = 1$
- Maximum step increment according to point 3: 60% of the support, $c_1 = 0.60$
- Minimum step increment according to point 4: 40% of the maximum step increment according to point 3, $c_2 = 0.40$
- Maximum distance for the reuse of existing points according to point 5: 20% of the maximum step increment according to point 3, $c_4 = 0.20$
- Reduction of the maximum step increment according to point 6: 50% of the maximum step increment according to point 3, $c_5 = 0.50$
- Termination limit for the relative improvement of the result according to point 7: not used in the test runs

A reduction of the maximum number n_P of randomly specified offspring points may lead to the effect that the global optima are not found. The number of necessary computations for determining objective function values merely decreases slightly.

The number of refinement stages n_f, in contrast to the latter, possesses an essential influence on the numerical effort. Each refinement stage requires approximately the same number of computations of objective function values. The accuracy of the results increases with the number of refinement stages n_f, however, the greater n_f becomes the less the improvement of the solution.

A distance bound max_d_i that has been defined too small (too small values c_1) may cause the missing of global optima and leads to an increase of necessary evaluations of the objective function.

Accordingly, a distance bound min_d_i that is determined too small (too small values c_2) leads to the same effect. If min_d_i is defined too large (too large values c_2) and this yields a local search domain that is constrained too rigorously, the global optima may also be missed.

A too strong reduction of max_d_i per refinement stage (too small values c_5) may lead to an increased numerical effort, as the number of necessary computations of objective function values rises. The probability for detecting global optima decreases. If the reduction is specified too poor (too large values c_5), the computational effort is also expected to increase.

An appropriate determination of the maximum distance Δd_i for the reuse of existing points is advantageous, in particular in analyses with a fine α-discretization. A large distance (large values c_4) reduces the number of necessary evaluations of the objective function, but, due to the more extensive usage of existing points, simultaneously leads to a repeated picking up of local extreme values. The probability for detecting global optima abates. With a decrease of Δd_i (smaller values c_4) the number of computations of objective function values in total increases.

If the termination limit for the maximal relative improvement of the result in the last n_z steps is chosen too small (too small values c_6), this termination criterion then does not become active. In this case the optimization is continued until the other termination stipulations according to the parameters n_P and n_f are met. The computational effort rises without yielding a moderate improvement in the result. For a number of steps n_z that is defined too large the number of evaluations of the objective function also increases. On the other hand, too large values c_6 as well as too small numbers n_z, however, lead to a precipitate termination of the optimization with unsatisfactory results.

The identification of global optima is not always guaranteed even using well-matched parameters. A *postcomputation* is thus carried out in order to raise the probability for making hits. After the completion of all optimizations for the selected α-levels all of the stored results z_j of the points $\underline{x}$ considered are rechecked for optimality, whereby the respective constraints $(x_1, \ldots, x_n) \in \underline{X}_{\alpha_k}$ are observed. When improving the results the algorithm is restarted at the "best" points found. At the end of these additional computations the postcomputation is repeated. When no significant improvement is obtained, the α-level optimization is considered to be completed.

The principle of the postcomputation is illustrated in Fig. 5.30 for the α-level $\alpha_k = 0.40$. All points $\underline{x}$ are plotted for which the objective function values z_j are known from previous optimizations; the point ● has not yet been evaluated. The optimum points $\underline{x}_{opt}$ identified in the optimizations on the five α-levels considered (here: only for one optimization objective in each case) are denoted by ▲. All points $\underline{x}$ to be included in the recheck procedure for $\alpha_k = 0.40$ are represented by O; these lie in the marked (solid lines) permissible domain (subspace $\underline{X}_{\alpha_k}$). A comparison of the results z_j obtained at points O with the current optimum obtained so far (at ▲) yields an improved result at point ⊕. Repeated optimization for $\alpha_k = 0.40$ with the identified point ⊕ results in the new optimum point $\underline{x}_{opt}$ indicated by ●. The final postcomputation leads to no improvement in the result.

The combination of the modified evolution strategy with a postcomputation for each fuzzy result variable on all selected α-levels leads to a qualitatively improved solution of the α-level optimization. The probability of finding global optima increases considerably. Based on the inclusion relationship between the subspaces $\underline{X}_{\alpha_i}$ and $\underline{X}_{\alpha_k}$ for different α_i and α_k according to Eq. (2.45) it may be concluded that the magnitudes of the computed optima $z_{j,\alpha_k l}$ and $z_{j,\alpha_k r}$ behave monotonically in relation to the associated α-values α_k. In view of the latter the convexity of the fuzzy result variables $\tilde{z}_j$ is guaranteed.

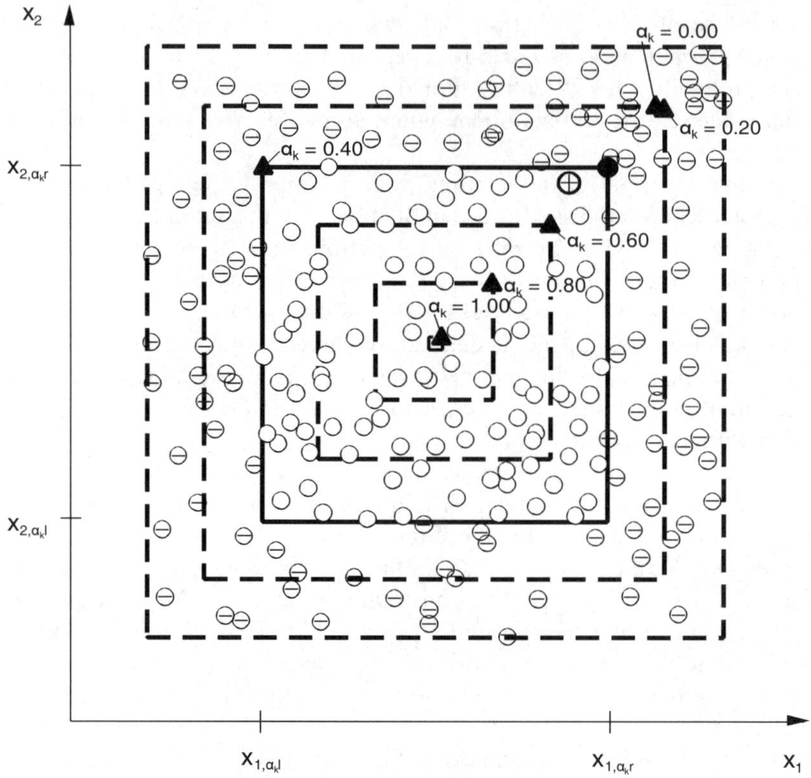

— — — Permissible domains for α_k = 0.00, 0.20, 0.60, 0.80 and 1.00

——— Permissible domain for α_k = 0.40

▲ Optimum points $\underline{x}_{opt}$ from α-level optimization

⊖ Nonpermissible points for α_k = 0.40

○ Permissible points for α_k = 0.40, points for rechecking $\underline{x}_{opt}$ for α_k = 0.40

⊕ "Best" point from the post-computation, new starting point

● New optimum point for α_k = 0.40

Fig. 5.30. Postcomputation for $\alpha_k = 0.40$

The new modified evolution strategy developed in combination with post-computations is considered advantageously compared with "pure" optimization methods. Test computations for solving selected, representative problems and applications of the new technique to the examples discussed in this book, however, only permit derivation of limited qualitative propositions concerning efficiency and reliability of the method. A general comparison with other optimization techniques is not possible. The huge number of available optimization algorithms and possible objective function types possessing considerably different characteristics do not permit this comparison. On the basis of the results from extensive comparative tests of optimization strategies presented in [92] and [176] it may be inferred that the modified evolution strategy in combination with post-computations, in general, with high probability leads to the global optima and requires acceptable numerical effort.

Proposals for improving the efficiency and reliability of the developed optimization technique are:

- Adaption of the shape of the local search domain to the shape of the current subspace $\underline{X}_{\alpha_k}$
- Implementation of an adaptive and success-dependent control of the step increment
- Application of a higher-order evolution method with parallelization of the descending successions
- Usage of a step-by-step optimization based on decomposition of the optimization model (dimension reduction), especially for problems with a large number of parameters (fuzzy input variables)
- Generation of several randomly specified offspring points descending from the same parent point and subsequent selection of the "best" descendant for determining the search direction
- Coupling of the modified evolution strategy with a diagonal step method
- Enhancement of the method by adding an adaptively learning algorithm for efficiently determining starting points and search directions

Consideration of Interaction Relationships. If interaction relationships $I_{i,\alpha}$ are a priori known, the n-dimensional subspace $\underline{X}_{\alpha_k}$ is then constrained by the according interaction surfaces. For the two-dimensional case this is illustrated in Fig. 5.31. When prescribing $I_{i,\alpha}$ the constraints of the optimization problem alter. The modified evolution strategy is to be enhanced accordingly.

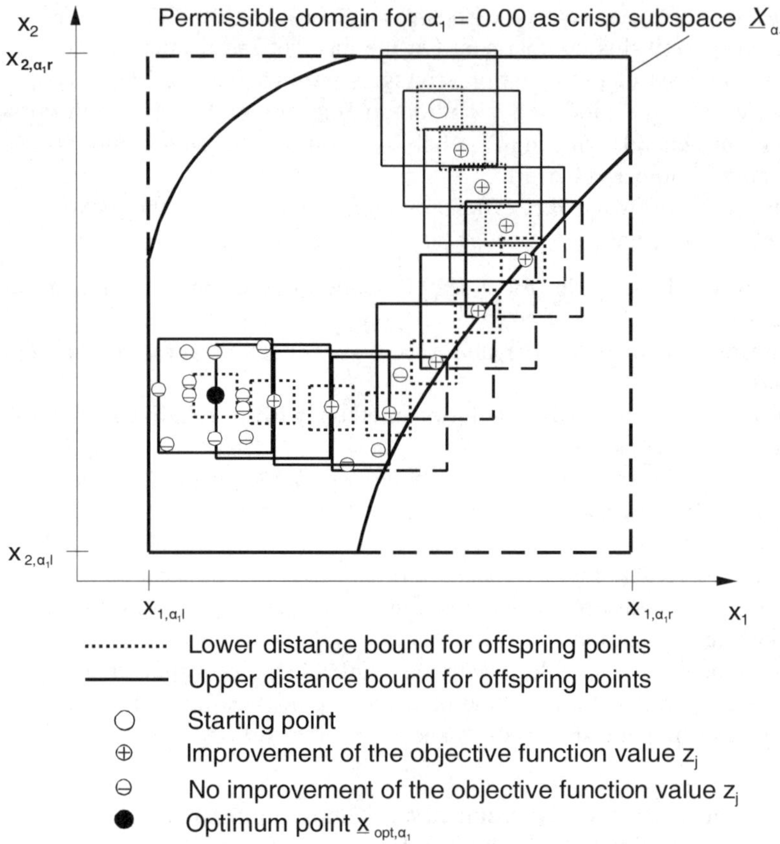

Fig. 5.31. α-level optimization considering interaction relationships

Numerical Solution of the Fuzzy Structural Analysis. The quality of the fuzzy results, i.e., of the prognosis of fuzzy structural response depends on the realistic description of all uncertain input and model parameters, and primarily on the quality of the deterministic fundamental solution for the structural analysis. The latter has a decisive influence on the prognosis regarding fuzzy structural behavior. For this reason a realistic numerical computational model must be applied for the deterministic fundamental solution. This must be capable of considering all essential geometrical and physical nonlinearities. Deterministic algorithms that meet these requirements are generally elaborated on an incremental-iterative basis, like, e.g., the computational model for the geometrically and physically nonlinear analysis of plane bar structures after [116] and [133]. The increments may also represent time steps for a dynamic analysis.

The algorithms of the α-level optimization are therefore numerically designed for the coupling with an arbitrary nonlinear computational model that is based on an incremental solution technique and provides restart options at each increment.

The general procedure of the fuzzy structural analysis possessing these features is schematically summarized in Figs. 5.32 and 5.33.

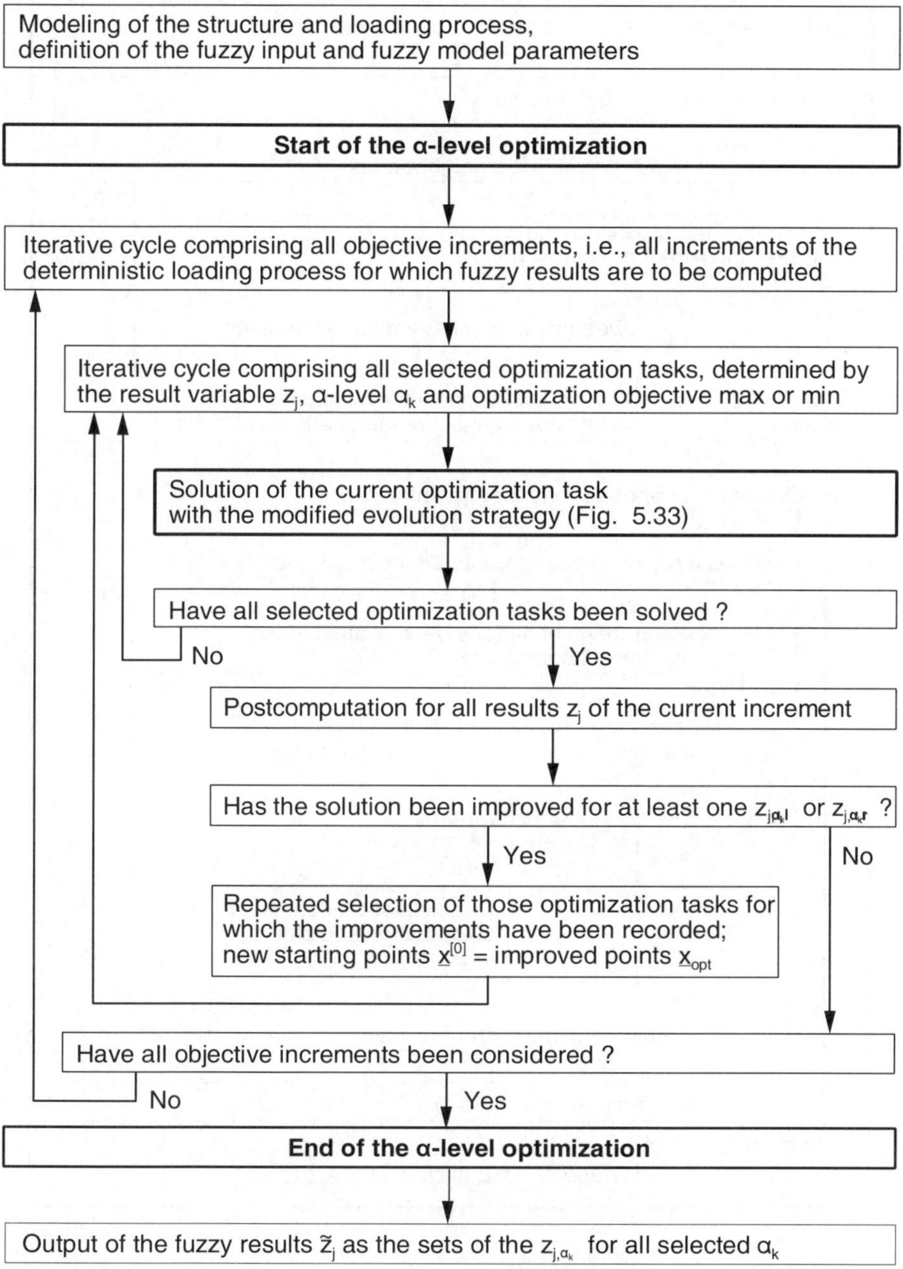

Fig. 5.32. Flowchart – numerical algorithm of the fuzzy structural analysis based on α-level optimization

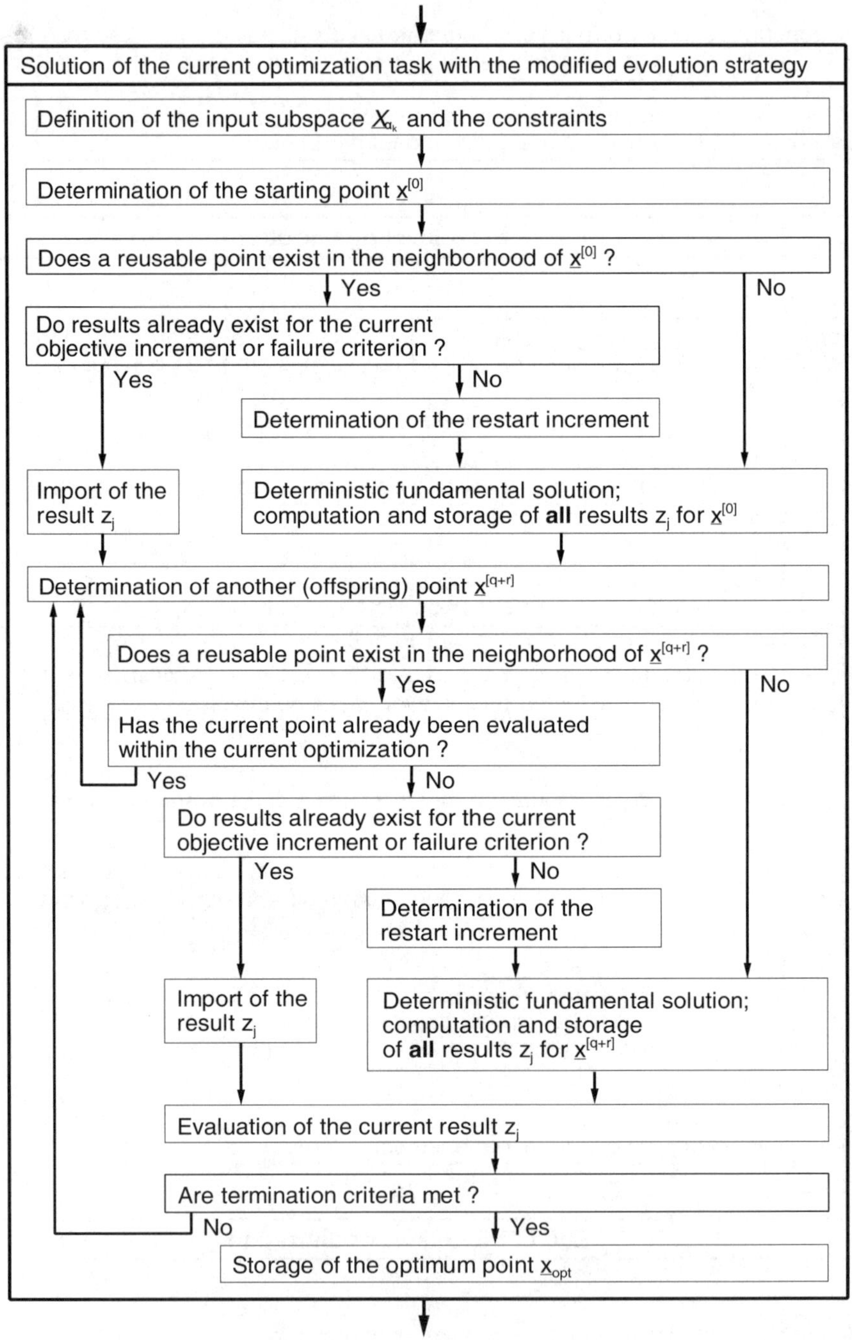

Fig. 5.33. Flowchart – modified evolution strategy for the fuzzy structural analysis with α-level optimization

5.2.4 Fuzzy Finite Element Method (FFEM)

Fuzzy Finite Elements. If in Eq. (5.7) a finite element model is chosen as mapping model, the fuzzy structural analysis then changes to the special form named the Fuzzy Finite Element Method. Considering problems that are independent of time uncertainty is thereby described with the aid of fuzzy vectors $\underline{\tilde{x}}$ and fuzzy fields $\underline{\tilde{x}}(\underline{\theta})$, when taking into account time-dependent problems fuzzy functions $\underline{\tilde{x}}(\underline{t})$ and fuzzy processes $\underline{\tilde{x}}(\underline{\tau})$ are additionally applied. The well-known finite element algorithms may be enhanced by introducing this uncertainty. Fuzziness may be accounted for in both the principle of virtual displacements and the particular variational principle adopted.

Fuzzy extension is demonstrated exemplarily for structures in $\mathbb{R}^1$ and $\mathbb{R}^2$ as well as the principle of virtual displacements.

For position vectors $\underline{\theta} = (\theta_1, \theta_2, \theta_3)$ it is presumed that the coordinates θ_1 and θ_2 do not exhibit fuzziness and the coordinate θ_3 representing the thickness of the structure may possess fuzziness.

A displacement field $\underline{\tilde{v}}(\underline{\theta})$ containing fuzziness is chosen

$$\begin{aligned} \underline{\tilde{v}}(\underline{\theta}) &= \underline{M}(\underline{\theta}) \cdot \underline{\tilde{p}}, \\ \underline{\tilde{v}}(e) &= \underline{A} \cdot \underline{\tilde{p}}, \\ \underline{\tilde{p}} &= \underline{A}^{-1} \cdot \underline{\tilde{v}}(e), \\ \underline{\tilde{v}}(\underline{\theta}) &= \underline{M}(\underline{\theta}) \cdot \underline{A}^{-1} \cdot \underline{\tilde{v}}(e) = \underline{N}(\underline{\theta}) \cdot \underline{\tilde{v}}(e), \end{aligned} \tag{5.39}$$

which is defined depending on the reduced crisp position vector $\underline{\theta} = (\theta_1, \theta_2)$.

With the restriction of a linear relationship between generalized strains and displacements the following is obtained

$$\underline{\tilde{\varepsilon}}(\underline{\theta}) = \underline{B} \cdot \underline{\tilde{p}} = \underline{B} \cdot \underline{A}^{-1} \cdot \underline{\tilde{v}}(e) = \underline{H} \cdot \underline{\tilde{v}}(e). \tag{5.40}$$

Moreover, a linear material law is assumed

$$\underline{\tilde{\sigma}}(\underline{\theta}) = \underline{\tilde{E}}(\underline{\theta}) \cdot \underline{\tilde{\varepsilon}}(\underline{\theta}) = \underline{\tilde{E}}(\underline{\theta}) \cdot \underline{H} \cdot \underline{\tilde{v}}(e). \tag{5.41}$$

For virtual displacements and generalized strains

$$\begin{aligned} \delta\underline{\tilde{v}}(\underline{\theta}) &= \underline{M}(\underline{\theta}) \cdot \delta\underline{\tilde{p}} = \underline{N}(\underline{\theta}) \cdot \delta\underline{\tilde{v}}(e) \\ \delta\underline{\tilde{\varepsilon}}(\underline{\theta}) &= \underline{H} \cdot \delta\underline{\tilde{v}}(e) \end{aligned} \tag{5.42}$$

is chosen. The virtual internal fuzzy work follows from

$$\begin{aligned} \delta\tilde{A}_i &= \int_{\tilde{V}} \delta\underline{\tilde{\varepsilon}}^T(\underline{\theta}) \cdot \underline{\tilde{\sigma}}(\underline{\theta}) \, d\tilde{V} \\ \delta\tilde{A}_i &= \delta\underline{\tilde{v}}^T(e) \cdot \int_{\tilde{V}} \underline{H}^T \cdot \underline{\tilde{E}}(\underline{\theta}) \cdot \underline{H} \, d\tilde{V} \cdot \underline{\tilde{v}}^T(e), \end{aligned} \tag{5.43}$$

with the fuzzy element stiffness matrix

$$\underline{\tilde{K}}(e) = \int_{\tilde{V}} \underline{H}^T \cdot \underline{\tilde{E}}(\underline{\theta}) \cdot \underline{H} \, d\tilde{V} . \tag{5.44}$$

For the virtual external fuzzy work the following holds under consideration of time-dependent fuzzy surface forces as well as mass, inertial, and damping forces possessing fuzziness:

$$\begin{aligned}
\delta\tilde{A}_a &= \delta\underline{\tilde{v}}^T(e) \cdot \underline{\tilde{F}}(e, \tau) && \text{virtual work of nodal forces} \\
&+ \int_{\tilde{V}} \delta\underline{\tilde{v}}^T(\underline{\theta}) \cdot \underline{\tilde{p}}_M(\underline{\theta}) \, d\tilde{V} && \text{virtual work of mass forces} \\
&+ \int_{O} \delta\underline{\tilde{v}}^T(\underline{\theta}) \cdot \underline{\tilde{\tilde{p}}}(\underline{\theta}, \tau) \, dA_O && \text{virtual work of time-dependent surface forces} \\
&- \int_{\tilde{V}} \delta\underline{\tilde{v}}^T(\underline{\theta}) \cdot \tilde{\rho}(\underline{\theta}) \cdot \underline{\ddot{\tilde{v}}}(\underline{\theta}, \tau) \, d\tilde{V} && \text{virtual work of inertial forces} \\
&- \int_{\tilde{V}} \delta\underline{\tilde{v}}^T(\underline{\theta}) \cdot \underline{\tilde{d}}(\underline{\theta}) \cdot \underline{\dot{\tilde{v}}}(\underline{\theta}, \tau) \, d\tilde{V} && \text{virtual work of damping forces.}
\end{aligned} \tag{5.45}$$

In $\delta\tilde{A}_a$ the time τ is introduced as a deterministic value but, in general, it may also be characterized by fuzziness.

Using the familiar abbreviations regarding the element e, the following is obtained from $\delta\tilde{A}_i - \delta\tilde{A}_a = 0$

$$\underline{\tilde{F}}(e, \tau) = \underline{\tilde{\tilde{F}}}(e, \tau) + \underline{\tilde{K}}(e) \cdot \underline{\tilde{v}}(e, \tau) + \underline{\tilde{M}}(e) \cdot \underline{\ddot{\tilde{v}}}(e, \tau) + \underline{\tilde{D}}(e) \cdot \underline{\dot{\tilde{v}}}(e, \tau) . \tag{5.46}$$

For the structure, the system of ordinary fuzzy differential equations of second order

$$\underline{\tilde{M}} \cdot \underline{\ddot{\tilde{v}}} + \underline{\tilde{D}} \cdot \underline{\dot{\tilde{v}}} + \underline{\tilde{K}} \cdot \underline{\tilde{v}} = \underline{\tilde{F}} \tag{5.47}$$

is obtained, which, for the static case, reduces to

$$\underline{\tilde{K}} \cdot \underline{\tilde{v}} = \underline{\tilde{F}} . \tag{5.48}$$

The system of fuzzy differential equations in Eq. (5.47) may be extended for its application in geometrically and physically nonlinear analysis.

Solution Technique for Time-independent Problems. For treating the time-independent problem according to Eq. (5.48) fuzziness is accounted for by fuzzy vectors and fuzzy functions. All n_p uncertain geometry, material, and load parameters are described by convex fuzzy variables

$$\begin{aligned}
\tilde{p}_1 &= \left(\left(x_1, \mu_{p_1}(x_1)\right) \mid x_1 \in \mathbf{X}\right) \\
\tilde{p}_2 &= \left(\left(x_2, \mu_{p_2}(x_2)\right) \mid x_2 \in \mathbf{X}\right) \\
&\vdots \\
\tilde{p}_{n_p} &= \left(\left(x_{n_p}, \mu_{p_{n_p}}(x_{n_p})\right) \mid x_{n_p} \in \mathbf{X}\right).
\end{aligned} \tag{5.49}$$

For preparing the problem for α-level optimization all fuzzy fields are formulated according to the bunch parameter representation. In compliance with Eq. (2.73) a fuzzy field may be described by

$$\tilde{x}(\underline{\theta}) = x(\underline{\tilde{s}}, \underline{\theta}) = \left\{\left(X_\alpha(\underline{\theta}), \mu(X_\alpha(\underline{\theta}))\right) \mid \mu(X_\alpha(\underline{\theta})) = \alpha \,\forall\, \alpha \in (0,1]\right\}, \tag{5.50}$$

in which $X_\alpha(\underline{\theta})$ denotes the α-function sets

$$X_\alpha(\underline{\theta}) = \left\{x(\underline{s}, \underline{\theta}) \;\; \forall\, \underline{s} \mid \underline{s} \in \underline{S}_\alpha \,;\;\; \alpha \in (0,1]\right\}, \tag{5.51}$$

with the bounding functions

$$x_{\alpha l}(\underline{\theta}) = \left\{x_{\theta\alpha l} \;\; \forall\, \underline{\theta} \mid \underline{\theta} \in \underline{\boldsymbol{\Theta}}\right\}, \tag{5.52}$$

and

$$x_{\alpha r}(\underline{\theta}) = \left\{x_{\theta\alpha r} \;\; \forall\, \underline{\theta} \mid \underline{\theta} \in \underline{\boldsymbol{\Theta}}\right\} \tag{5.53}$$

determined from

$$x_{\theta\alpha l} = \min\left[x_\theta(\underline{s}) \mid \underline{s} \in \underline{S}_\alpha\right], \tag{5.54}$$

$$x_{\theta\alpha r} = \max\left[x_\theta(\underline{s}) \mid \underline{s} \in \underline{S}_\alpha\right], \tag{5.55}$$

at all specified points $\underline{\theta}$.

If full interaction exists between the fuzzy functional values at all points $\underline{\theta} \in \underline{\boldsymbol{\Theta}}$ of the fuzzy field $x(\underline{\tilde{s}}, \underline{\theta})$, a point discretization is then not necessary. For such fuzzy fields the following holds:

$$\begin{aligned}
\tilde{x}_1(\underline{\theta}) &= x_1(\underline{\tilde{s}}_1, \underline{\theta}) = x_1((\tilde{s}_{1,1}, \ldots, \tilde{s}_{1,n_1}), \underline{\theta}) \\
\tilde{x}_2(\underline{\theta}) &= x_2(\underline{\tilde{s}}_2, \underline{\theta}) = x_2((\tilde{s}_{2,1}, \ldots, \tilde{s}_{2,n_2}), \underline{\theta}) \\
&\vdots \\
\tilde{x}_v(\underline{\theta}) &= x_v(\underline{\tilde{s}}_v, \underline{\theta}) = x_v((\tilde{s}_{v,1}, \ldots, \tilde{s}_{v,n_v}), \underline{\theta}).
\end{aligned} \tag{5.56}$$

In the v fuzzy fields a total of

$$n_{f_1} = n_1 + n_2 + \ldots + n_v \tag{5.57}$$

bunch parameters appear.

In the case that no interaction or an a priori prescribed interaction defined by an interaction relationship $I_{i,\alpha}$ (Sect. 2.1.11.2) exists between the functional values at the discrete points $\underline{\theta}_1, ..., \underline{\theta}_j, ..., \underline{\theta}_{k_p} \in \underline{\boldsymbol{\Theta}}$, each fuzzy field is then described in discretized form in compliance with Eq. (2.84)

$$\begin{aligned}
\tilde{x}_1(\underline{\theta}) &= \left\{ x_1((\tilde{\underline{s}}_{1,1}, \underline{\theta}_1), x_1((\tilde{\underline{s}}_{1,2}, \underline{\theta}_2), \ldots, x_1((\tilde{\underline{s}}_{1,k_p}, \underline{\theta}_{k_p}) \right\} \\
\tilde{x}_2(\underline{\theta}) &= \left\{ x_2((\tilde{\underline{s}}_{2,1}, \underline{\theta}_1), x_2((\tilde{\underline{s}}_{2,2}, \underline{\theta}_2), \ldots, x_2((\tilde{\underline{s}}_{2,k_p}, \underline{\theta}_{k_p}) \right\} \\
&\bullet \\
&\bullet \\
&\bullet \\
\tilde{x}_w(\underline{\theta}) &= \left\{ x_w((\tilde{\underline{s}}_{w,1}, \underline{\theta}_1), x_w((\tilde{\underline{s}}_{w,2}, \underline{\theta}_2), \ldots, x_w((\tilde{\underline{s}}_{w,k_p}, \underline{\theta}_{k_p}) \right\}.
\end{aligned} \tag{5.58}$$

The bunch parameter vectors in $\tilde{x}_1(\underline{\theta})$ comprise e_1 elements, the vectors in $\tilde{x}_2(\underline{\theta})$ include e_2 elements, and the vectors assigned to $\tilde{x}_w(\underline{\theta})$ contain e_w elements. The w fuzzy fields described at k_p discrete points representing, e.g., finite element nodal points thus comprise a total of

$$n_{f_2} = (e_1 + e_2 + \ldots + e_w) \cdot k_p \tag{5.59}$$

bunch parameters. All fuzzy bunch parameters are convex fuzzy variables.

All fuzzy structural parameters and all fuzzy bunch parameters of interactive as well as of noninteractive fuzzy fields are together introduced into α-level optimization, i.e., a total number of

$$n_e = n_p + n_{f_1} + n_{f_2} \tag{5.60}$$

are to be considered as fuzzy input variables. These are lumped together in the fuzzy vector

$$\begin{aligned}
\tilde{\underline{e}} = (&\tilde{p}_1, \tilde{p}_2, \ldots, \tilde{p}_{n_p}, \\
&\tilde{s}_{1,1}, \ldots, \tilde{s}_{1,n_1}, \ldots, \tilde{s}_{v,1}, \ldots, \tilde{s}_{v,n_v}, \\
&\tilde{\underline{s}}_{1,1}, \ldots, \tilde{\underline{s}}_{1,k_p}, \ldots, \tilde{\underline{s}}_{w,1}, \ldots, \tilde{\underline{s}}_{w,k_p}).
\end{aligned} \tag{5.61}$$

Additionally, p fuzzy model parameters may appear in the uncertain mapping model $\tilde{M}$: $\tilde{M}\,(\tilde{m}_1, \tilde{m}_2, \ldots, \tilde{m}_p)$.

Fuzzy structural analysis yields the fuzzy result vectors $\tilde{\underline{z}} = (\tilde{z}_1, \tilde{z}_2, ..., \tilde{z}_m) \subseteq \underline{\mathbf{Z}}$, it may now be expressed by the relationship

$$\tilde{\underline{z}} = \tilde{M}(\tilde{\underline{e}}). \tag{5.62}$$

When applying α-level optimization all fuzzy variables contained in $\tilde{\underline{e}}$ are subdivided into the same α-levels α_k. For each α-level a crisp subspace $\underline{E}_{\alpha_k}$ possessing n_e dimensions is to be searched. The subspace $\underline{E}_{\alpha_k}$ may be constrained by interaction relationships $I_{i,\alpha}$. After having selected a point in the input subspace $\underline{E}_{\alpha_k}$ the assigned point in the result subspace $\underline{Z}_{\alpha_k}$ is computed with the deterministic fundamental solution from Eq. (5.48). For this purpose each arbitrary finite element code may be used. The optimum points for each α-level are determined with the aid of α-level optimization. The numerical principle for solving Eq. (5.62) by α-level optimization is illustrated in Fig. 5.34.

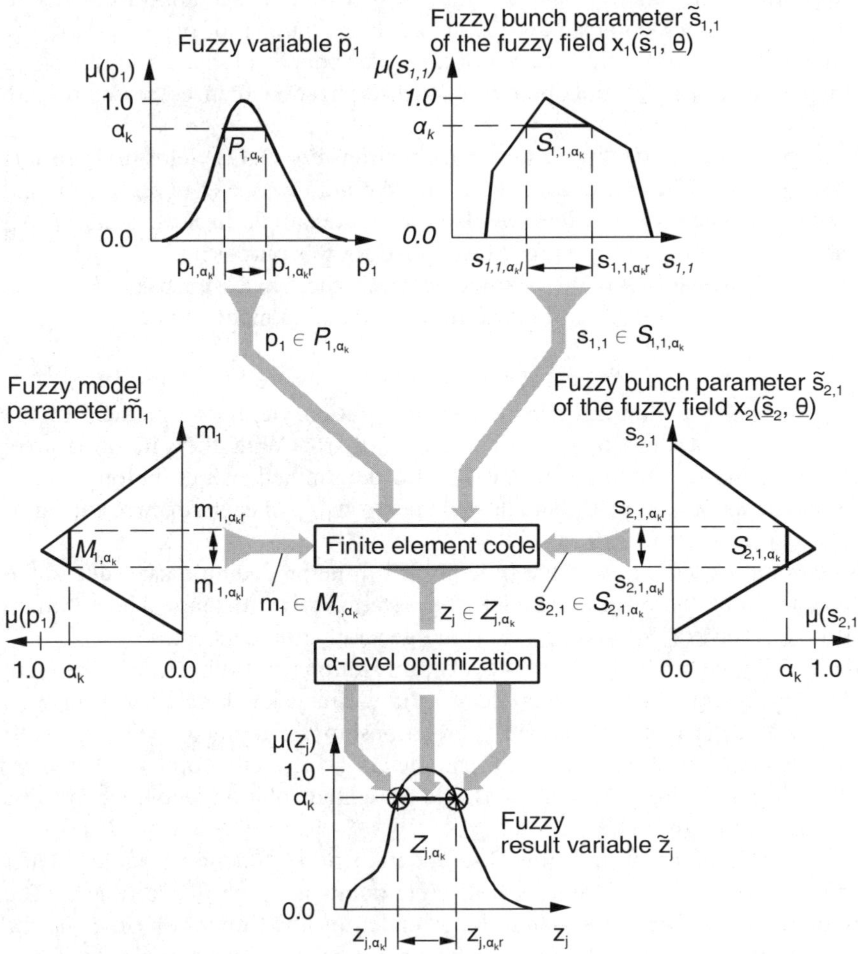

Fig. 5.34. Fuzzy structural analysis, numerical solution by α-level optimization

Solution Technique for Time-dependent Problems. When solving time-dependent problems the fuzzy variables

$$\tilde{p} = \left(\left(x, \mu_p(x)\right) \mid x \in \mathbf{X}\right) \tag{5.63}$$

as well as the stationary and nonstationary fuzzy functions

$$\begin{array}{lll} \tilde{x}(\underline{t}) = x(\tilde{\underline{s}},\underline{\theta},\tau) & \text{(nonstationary in } \underline{\theta} \text{ and } \tau) & \\ \tilde{x}(\underline{t}) = x(\tilde{\underline{s}},\tau) & \text{(nonstationary in } \tau, \text{ stationary in } \underline{\theta}) & (5.64) \\ \tilde{x}(\underline{t}) = x(\tilde{\underline{s}}) & \text{(stationary in } \underline{\theta} \text{ and } \tau) & \end{array}$$

may appear. For the functional values of the fuzzy functions at arbitrary points in time full a priori interaction is next presumed. If the number of fuzzy variables $\tilde{p}$ is designated by n_p and if the number of all scalar bunch parameters $\tilde{s}$ is n_s, each crisp input subspace $\underline{E}_{\alpha_k}$ obtained from α-discretization then possesses $n_p + n_s$ dimensions.

Solving the system of fuzzy differential equations Eq. (5.47) yields fuzzy results $\tilde{\underline{v}}(\underline{\theta}_j, \tau) = \tilde{\underline{z}}(\underline{\theta}_j, \tau)$. These are computed with the aid of α-level optimization and numerical time-step integration (see also the first example in Sect. 5.2.5.1). All available integration techniques may be applied for this purpose.

For $\tau = 0$ a starting point is selected from the input subspace $\underline{E}_{\alpha_k}$. The coordinates of this point are introduced into the mapping model

$$\underline{M} \cdot \ddot{\underline{v}} + \underline{D} \cdot \dot{\underline{v}} + \underline{K} \cdot \underline{v} = \underline{F}. \tag{5.65}$$

At the time point τ_i numerical time-step integration yields one element of the α-level set $\underline{Z}_{\alpha_k}$ of the fuzzy result vector $\tilde{\underline{z}}(\underline{\theta}_j, \tau_i)$. With the aid of α-level optimization the optimum points in $\underline{E}_{\alpha_k}$ are determined, which belong to the largest elements $z_{h,\alpha_k r}$ and the smallest elements $z_{h,\alpha_k l}$ of each fuzzy result variable $\tilde{z}_h(\underline{\theta}_j, \tau_i)$ included in $\tilde{\underline{z}}(\underline{\theta}_j, \tau_i)$.

For computing $\tilde{\underline{z}}(\underline{\theta}_j, \tau_{i+1})$ at the time point τ_{i+1} the procedure must – due to the interaction within the mapping model – be restarted at $\tau = 0$ again. The optimum points in $\underline{E}_{\alpha_k}$ for $\tilde{\underline{z}}(\underline{\theta}_j, \tau_i)$ and $\tilde{\underline{z}}(\underline{\theta}_j, \tau_{i+1})$ are generally not identical.

In Fig. 5.35 the principle of an α-level optimization for solving time-dependent problems is illustrated for the example of a three-dimensional input subspace $\underline{E}_{\alpha_k}$.

If no interaction or an interactive relationship $I_{i,\alpha}$ defined by an a priori interaction exists between the fuzzy parameters at discrete points in space $\underline{\theta}_j$ (j = 1, ..., k_p) and time $\underline{\tau}_i$ (j = 1, ..., t), the presented solution technique is to be enhanced accordingly.

On the basis of point and time discretization fuzzy functional values of the function $x(\tilde{\underline{s}}, \underline{\theta}, \tau)$ are determined at discrete points in space $\underline{\theta}_j$ and time τ_i. The fuzzy fields in Eq. (5.58) are then to be extended by the dimension time and the appropriate fuzzy functional values at the time points $\underline{\tau}_i$. For example, for fuzzy function $x(\tilde{\underline{s}}, \underline{\theta}, \tau)$ the following holds:

$$x(\tilde{\underline{s}},\underline{\theta},\tau) = \left\{ \begin{array}{c} x((\tilde{\underline{s}}_{1,1},\underline{\theta}_1,\tau_1),\ldots,x((\tilde{\underline{s}}_{j,1},\underline{\theta}_j,\tau_1),\ldots,x((\tilde{\underline{s}}_{k_p,1},\underline{\theta}_{k_p},\tau_1) \\ \cdots \\ x((\tilde{\underline{s}}_{1,i},\underline{\theta}_1,\tau_i),\ldots,x((\tilde{\underline{s}}_{j,i},\underline{\theta}_j,\tau_i),\ldots,x((\tilde{\underline{s}}_{k_p,i},\underline{\theta}_{k_p},\tau_i) \\ \cdots \\ x((\tilde{\underline{s}}_{1,t},\underline{\theta}_1,\tau_t),\ldots,x((\tilde{\underline{s}}_{j,t},\underline{\theta}_j,\tau_t),\ldots,x((\tilde{\underline{s}}_{k_p,t},\underline{\theta}_{k_p},\tau_t) \end{array} \right\}. \tag{5.66}$$

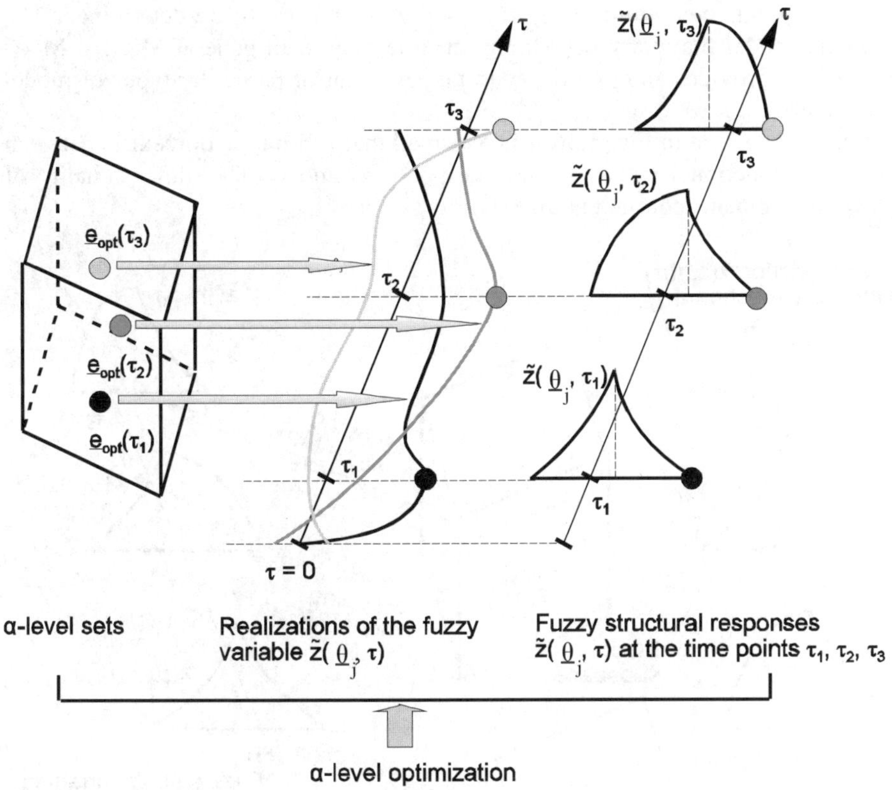

Fig. 5.35. α-level optimization for time-dependent problems – full interaction

As no full interaction exists, after τ_t time steps all fuzzy bunch parameters $\tilde{s}_{1,1},\ldots,\tilde{s}_{k_p,t}$ together form the $k_p\cdot t$-dimensional input subspace $\underline{E}_{\alpha_k}$. The dimensionality of $\underline{E}_{\alpha_k}$ increases from time step to time step. For the example of the fuzzy function $x(\tilde{\underline{s}}, \underline{\theta}, \tau)$ in Eq. (5.66) $\underline{E}_{\alpha_k}$ possesses $1\cdot k_p$ dimensions at time τ_1, $i\cdot k_p$ dimensions at time τ_i, and $t\cdot k_p$ dimensions at time τ_t, provided that k_p discrete points in space are considered. For each time point τ_i the optimum points in the input subspace $\underline{E}_{\alpha_k}$, which belong to the largest elements $z_{h,\alpha_k r}$ and the smallest elements $z_{h,\alpha_k l}$ of each fuzzy result variable $\tilde{\underline{z}}_h(\underline{\theta}_j, \tau_i)$ are to be determined by applying α-level optimization and numerical time-step integration.

In Fig. 5.36 this procedure is shown for the fuzzy variable $\tilde{p}$ and the fuzzy function $x(\tilde{\underline{s}}, \underline{\theta}, \tau)$ forming the input subspace $\underline{E}_0$ assigned to the α-level $\alpha = 0$. At time τ_1 the input subspace $\underline{E}_0(\tau_1)$ is two-dimensional. By applying the mapping model $M(\tau_1)$ the extreme values $z_{\alpha=0\,l}(\tau_1)$ and $z_{\alpha=0\,r}(\tau_1)$ of the fuzzy result variable $\tilde{\underline{z}}_h(\underline{\theta}_j, \tau_i)$ are computed.

At the time point τ_2 $\underline{E}_0(\tau_2)$ possesses three dimensions. Searching in $\underline{E}_0(\tau_2)$ the optimum points belonging to $z_{\alpha=0\,l}(\tau_2)$ and $z_{\alpha=0\,r}(\tau_2)$ are to be determined. The mapping model may vary depending on time τ, i.e., in general $M(\tau_1) \neq M(\tau_2)$ holds. This provides an opportunity to take account of particular types of model uncertainty.

For the example in Fig. 5.36 it is assumed that full interaction exists between the fuzzy functional values at the time points τ_2 and τ_3. The dimensionality of $\underline{E}_0(\tau_3)$ thus remains constant at time τ_3.

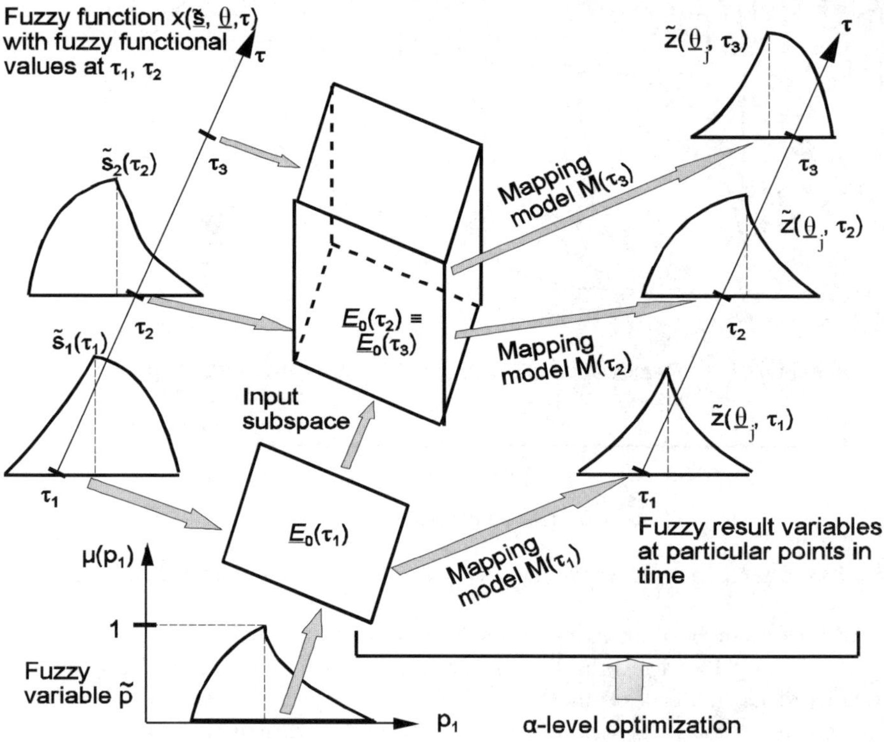

Fig. 5.36. α-level optimization for time-dependent problems – no interaction

5.2.5 Application of Fuzzy Structural Analysis

The developed algorithms for fuzzy structural analysis may be applied to almost arbitrary problems in structural engineering. Geometrical and physical nonlinearities and arbitrary loading processes, e.g., concerning statical or dynamic short-term and long-term behavior, may thereby be accounted for. Uncertain damage processes and lifetime predictions may also be considered [69, 121, 122].

5.2.5.1 Multistory Frame, Linear Dynamic Analysis – Initial Value Problem

The multistory frame shown in Fig. 5.37 is investigated for a dynamic loading. A geometrically and physically nonlinear analysis is carried out. When modeling the system it is assumed that the extensional stiffness of all bars and the bending stiffness of the horizontal bars of the frame are very high in comparison with the bending stiffness of the vertical bars. Hence only the three horizontal displacements v(1), v(2), v(3), the three velocities $\dot{v}(1)$, $\dot{v}(2)$, $\dot{v}(3)$, and the three accelerations $\ddot{v}(1)$, $\ddot{v}(2)$, $\ddot{v}(3)$ of the horizontal bars of the frame are to be considered in the investigation (Fig. 5.37). The masses are modeled as being concentrated in the horizontal bars.

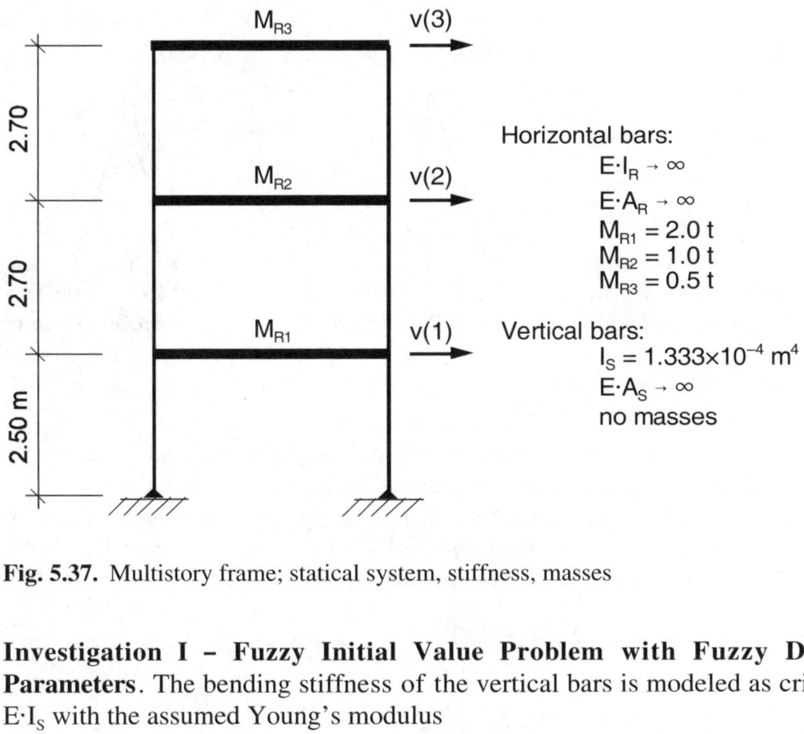

Fig. 5.37. Multistory frame; statical system, stiffness, masses

Investigation I – Fuzzy Initial Value Problem with Fuzzy Damping Parameters. The bending stiffness of the vertical bars is modeled as crisp value $E{\cdot}I_S$ with the assumed Young's modulus

$$E = 3 \cdot 10^7 \text{ kN/m}^2. \tag{5.67}$$

For the dynamic investigation the damping is presumed to be velocity proportional, and the informally uncertain elements d_{ik} of the damping matrix $\underline{D}$ are taken to be fuzzy variables

$$\underline{\tilde{D}} = \begin{bmatrix} \tilde{d}_{11} & 0 & 0 \\ 0 & \tilde{d}_{22} & 0 \\ 0 & 0 & \tilde{d}_{33} \end{bmatrix}. \tag{5.68}$$

The $\tilde{d}_{ii}$ are estimated as fuzzy triangular numbers (Fig. 5.38)

$$\tilde{d}_{11} = <0.045,\ 0.050,\ 0.060>\ \mathrm{kNs/m}, \tag{5.69}$$

$$\tilde{d}_{22} = <0.040,\ 0.050,\ 0.055>\ \mathrm{kNs/m}, \tag{5.70}$$

$$\tilde{d}_{33} = <0.045,\ 0.050,\ 0.060>\ \mathrm{kNs/m}. \tag{5.71}$$

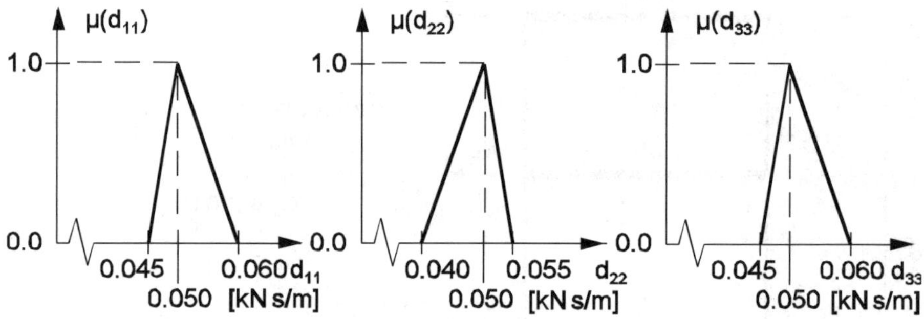

Fig. 5.38. Fuzzy damping parameters $\tilde{d}_{ii}$

The fuzziness of the damping parameters may, e.g., be understood as model uncertainty of the *damping model* for structural analysis.

Starting from a static state the multistory frame is put into vibration by means of prescribed initial values for the displacements $v_0(i)$ and the velocities $\dot{v}_0(i)$ of its horizontal bars. These initial values are assumed to be

$$\underline{v}_0 = (0, 0, 0)\ \mathrm{m}, \tag{5.72}$$

and

$$\underline{\tilde{\dot{v}}}_0 = (0, 0, \tilde{\dot{v}}_0(3))\ \mathrm{m/s}. \tag{5.73}$$

The initial velocity of the upper horizontal bar of the frame in Eq. (5.73) is uncertain, it is introduced into the investigation as fuzzy triangular number (Fig. 5.39)

$$\tilde{\dot{v}}_0(3) = <0.9, 1.0, 1.2>\ \mathrm{m/s}. \tag{5.74}$$

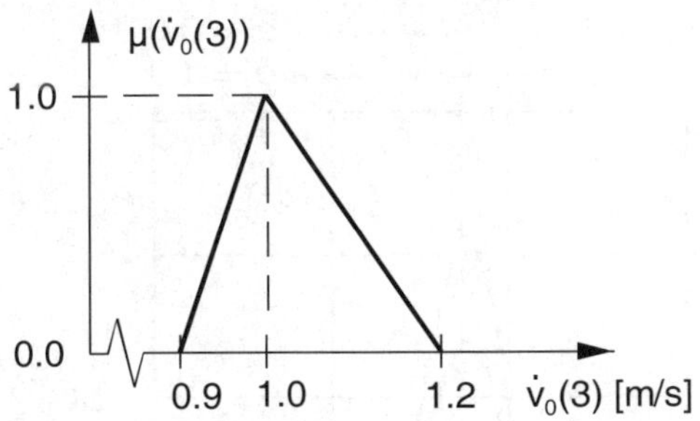

Fig. 5.39. Fuzzy initial velocity $\tilde{\dot{v}}_0(3)$ of the upper horizontal bar of the frame

The fuzziness of $\tilde{\dot{v}}_0(3)$ may, e.g., be interpreted as data uncertainty for the vibration analysis. But it may also be considered as model uncertainty of the model *loading*.

The free, damped vibration of the system is described with the aid of the system of linear, ordinary fuzzy differential equations of first order

$$\frac{d\underline{\tilde{x}}}{dt} = \underline{\dot{\tilde{x}}} = \underline{\tilde{A}} \cdot \underline{\tilde{x}} \; ; \; \underline{\tilde{x}}(t) = \left(\underline{\tilde{v}}(t), \underline{\dot{\tilde{v}}}(t) \right). \tag{5.75}$$

The solution is obtained from a numerical time-step integration with time increments $\Delta t = 0.001$ s. The implicit time-step operator used in this example has been gained from a Padé series [158] in time considering the quadratic terms in both the numerator and the denominator polynomials. It resembles the Gellert operator [59] and is unconditionally numerically stable.

The displacements and velocities of the three horizontal bars of the frame are fuzzy result variables for each point in time $t = t_0 + k \cdot \Delta t$. The mapping of the four-dimensional fuzzy input set onto the six-dimensional fuzzy result set considered here additionally depends on a crisp parameter – the time t (Sect. 5.2.2.2).,

$$\left(\tilde{d}_{11}, \tilde{d}_{22}, \tilde{d}_{33}, \tilde{\dot{v}}_0(3), t \right) \rightarrow \left(\tilde{v}(1), \tilde{v}(2), \tilde{v}(3), \tilde{\dot{v}}(1), \tilde{\dot{v}}(2), \tilde{\dot{v}}(3) \right). \tag{5.76}$$

Monotonicity only exists between $\dot{v}_0(3)$ and the result variables, i.e., the optimum points lie on the boundary but not necessarily in the corners of the input subspace. The mapping according to Eq. (5.76) is not biunique.

For the fuzzy analysis of the initial value problem the α-levels $\alpha_1 = 0.0$ and $\alpha_2 = 1.0$ were chosen. The computation was carried out with the modified evolution strategy and different adjustments of the control parameters (Sect. 5.2.3.4), a reference solution was gained with the aid of a multilayer mesh search technique. For the purpose of comparison the additional solution method described in [19] was applied to the problem. According to [19] it was necessary to extend the system of differential equations from Eq. (5.75) by the three conditions

$$\dot{\tilde{d}}_{ii} = 0 \; ; \; i = 1, \ldots, 3. \tag{5.77}$$

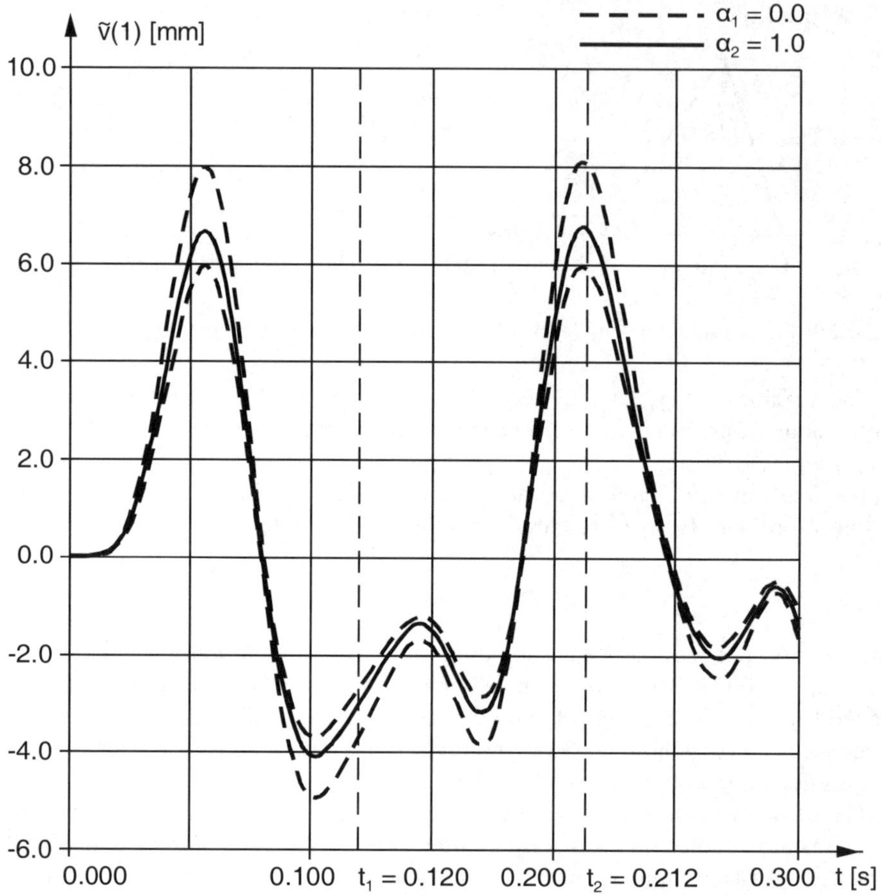

Fig. 5.40. Fuzzy displacement–time dependency for the lower horizontal bar of the frame

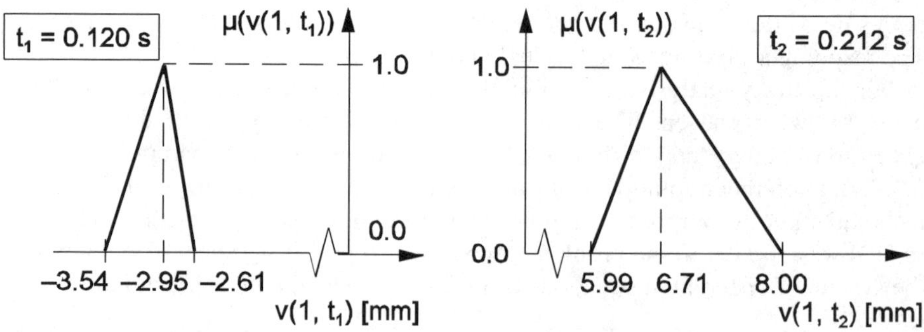

Fig. 5.41. Fuzzy displacement $\tilde{v}(1)$ at time $t_1 = 0.120$ s and $t_2 = 0.212$ s

This leads to the modified mapping

$$\left(\tilde{d}_{11}, \tilde{d}_{22}, \tilde{d}_{33}, \tilde{\dot{v}}_0(3), t\right) \rightarrow \left(\tilde{v}(1), \tilde{v}(2), \tilde{v}(3), \tilde{\dot{v}}(1), \tilde{\dot{v}}(2), \tilde{\dot{v}}(3), \tilde{d}_{11}, \tilde{d}_{22}, \tilde{d}_{33}\right). \tag{5.78}$$

For each value of the crisp parameter t the biuniqueness of the mapping according to Eq. (5.78) is ensured [19]. Boundary points of the result subspace always remain in boundary position for a successively changing time parameter t. The results obtained from the three solution variants are virtually identical.

The numerical time-step integration was carried out for $t = 0$ to $t = 0.300$ s. The fuzzy displacement–time dependency for the lower horizontal bar of the frame is shown in Fig. 5.40, the fuzzy displacements $\tilde{v}(1)$ assigned to the time points $t = t_1 = 0.120$ s and $t = t_2 = 0.212$ s are displayed in Fig. 5.41.

Investigation II – Fuzzy Initial Value Problem with Fuzzy Young's Modulus. The analysis of the free, damped vibration of the multistory frame in Fig. 5.37 is repeated with modified input data.

Owing to informal uncertainty the bending stiffness of the vertical bars is now described by a fuzzy variable. It is calculated from the crisp moment of inertia I_S (Fig. 5.37) and the fuzzy Young's modulus

$$\tilde{E} = <2.7 \cdot 10^7, 3.0 \cdot 10^7, 3.2 \cdot 10^7> \ \text{kN/m}^2, \tag{5.79}$$

which is introduced as fuzzy triangular number (Fig. 5.42).

The elements d_{ik} of the damping matrix $\underline{D}$ are now accounted for as crisp parameters

$$d_{ii} = 0.05 \ \text{kNs/m} \ ; \ i = 1, \ldots, 3, \tag{5.80}$$

$$d_{ik} = 0 \ ; \ i, k = 1, \ldots, 3 \ ; \ i \neq k. \tag{5.81}$$

The initial conditions are prescribed as crisp values. For the displacements $v_0(i)$ of the horizontal bars

$$\underline{v}_0 = (0, 0, 0) \ \text{m} \tag{5.82}$$

is assumed and the velocities $\dot{v}_0(i)$ are taken to be

$$\underline{\dot{v}}_0 = (0, 0, 1{,}0) \ \text{m/s}. \tag{5.83}$$

The numerical solution of the system of fuzzy differential equations according to Eq. (5.75) yields the fuzzy displacements and fuzzy velocities of the horizontal bars of the frame for each point in time $t = t_0 + k \cdot \Delta t$. In the following only the fuzzy displacements $\tilde{v}(1)$, $\tilde{v}(2)$, and $\tilde{v}(3)$ are accounted for as fuzzy results at each time point t. The mapping

$$\left(\tilde{E}, t\right) \rightarrow \left(\tilde{v}(1), \tilde{v}(2), \tilde{v}(3)\right) \tag{5.84}$$

considered is neither monotonic nor biunique, the optimum points may now also lie within the input subspace.

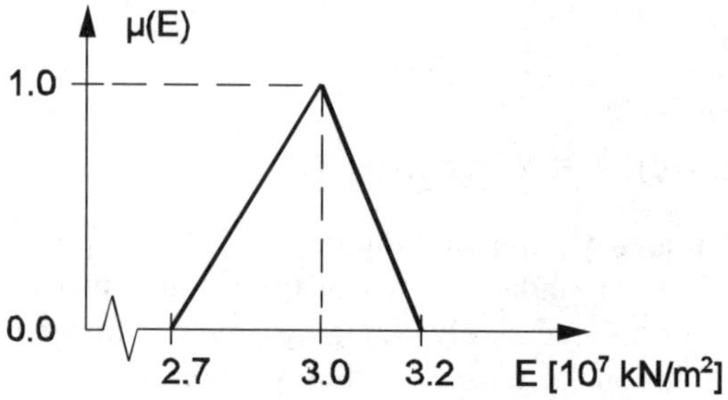

Fig. 5.42. Fuzzy Young's modulus $\tilde{E}$

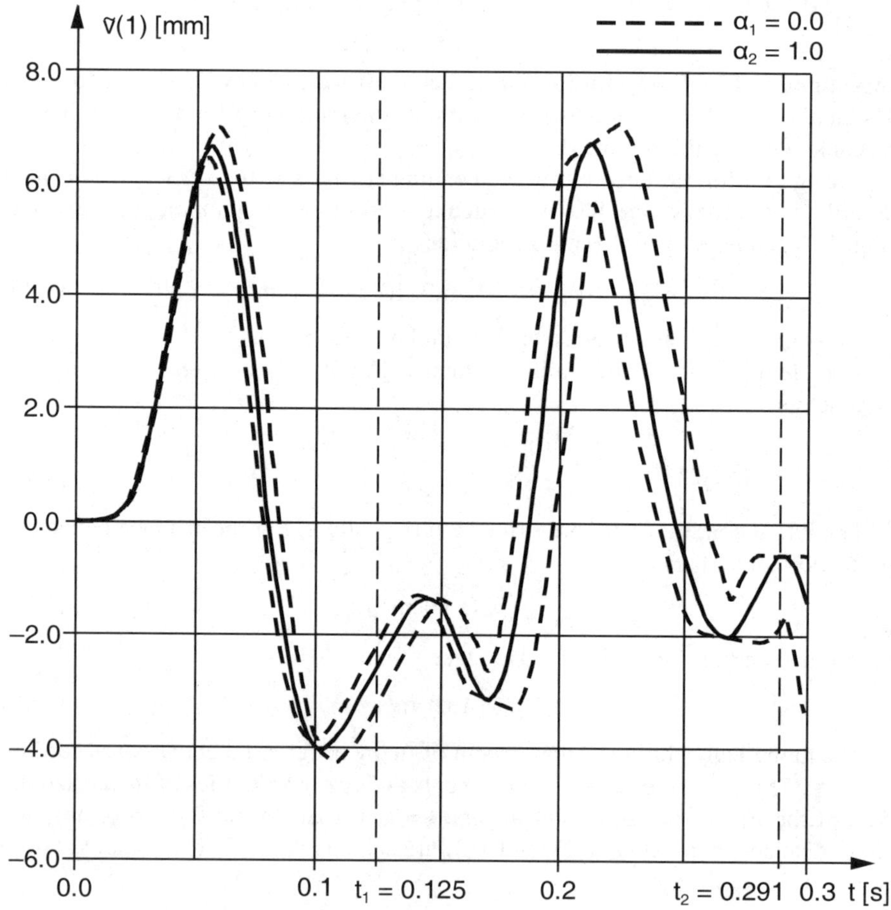

Fig. 5.43. Fuzzy displacement–time dependency for the lower horizontal bar of the frame

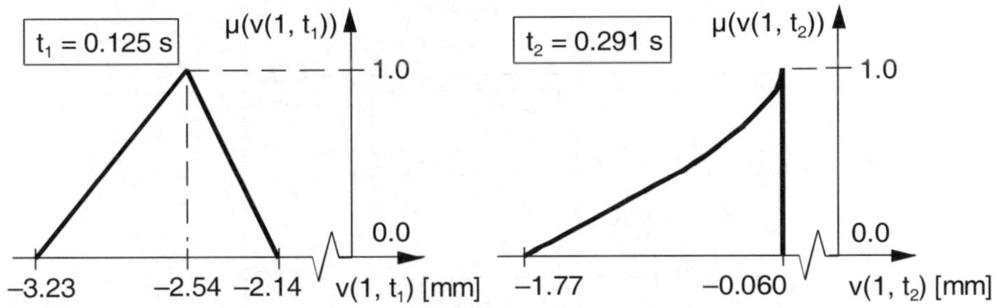

Fig. 5.44. Fuzzy displacement $\tilde{v}(1)$ at time $t_1 = 0.125$ s and $t_2 = 0.291$ s

The fuzzy analysis was carried out with eleven equidistant α-levels $\alpha_1 = 0.0$, $\alpha_2 = 0.1$, ..., $\alpha_{11} = 1.0$. For solving the α-level optimization a modified evolution strategy was applied. Again a multilayer mesh technique was used for comparison purposes. The results from both methods are virtually identical.

Again the time span from t = 0 to t = 0.300 s is considered. The computed fuzzy displacement–time dependency for the lower horizontal bar of the frame is shown in Fig. 5.43 for the α-levels $\alpha_1 = 0.0$ and $\alpha_{11} = 1.0$. The fuzzy displacements $\tilde{v}(1)$ for the points in time $t = t_1 = 0.125$ s and $t = t_2 = 0.291$ s are plotted in Fig. 5.44.

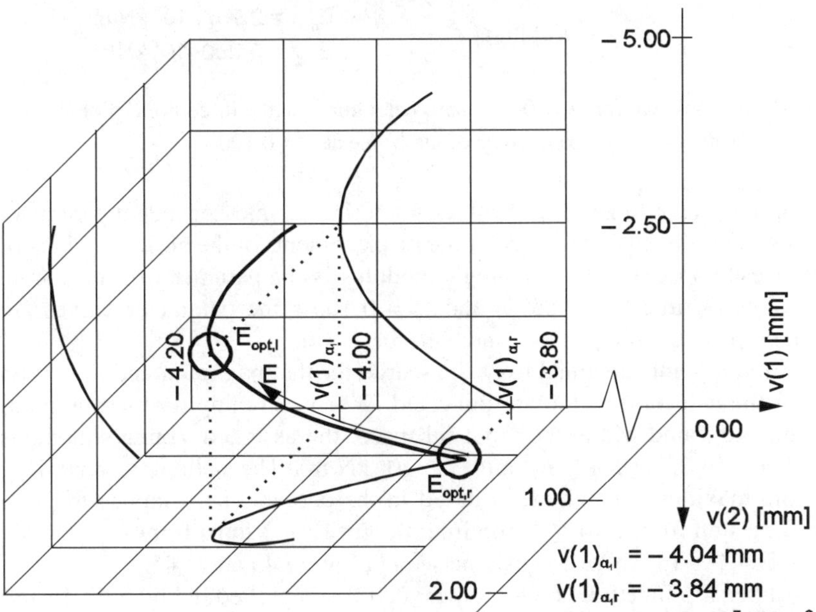

Fig. 5.45. Result subspace for $\alpha_1 = 0$ as a parameter curve in the three-dimensional space of the displacements of the horizontal bars of the frame at $t_3 = 0.099$ s

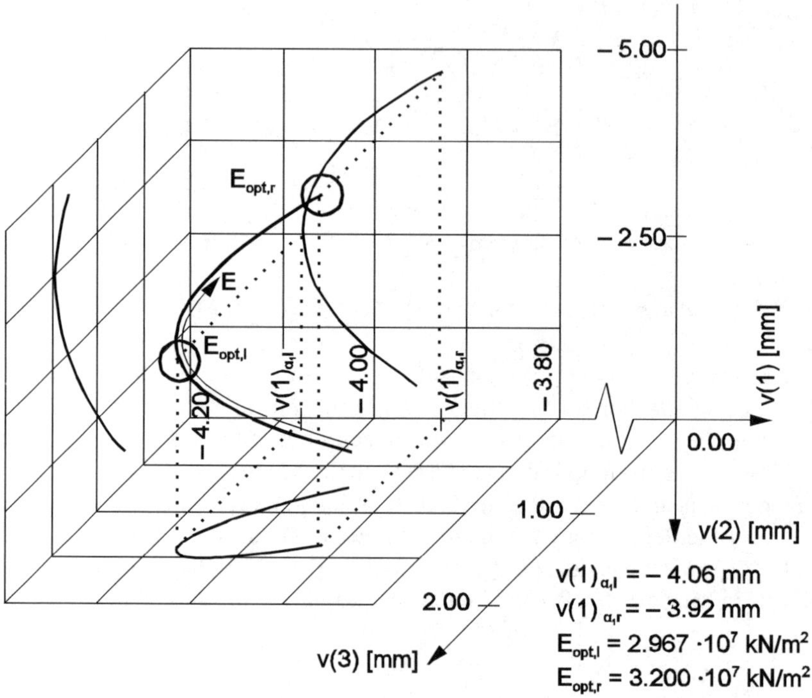

Fig. 5.46. Result subspace for $\alpha_1 = 0$ as a parameter curve in the three-dimensional space of the displacements of the horizontal bars of the frame at $t_4 = 0.100$ s

The mapping according to Eq. (5.84) characterizes a time-dependent parameter curve in the three-dimensional space of the displacements of the horizontal bars of the frame (result space) and the Young's modulus is the parameter of this curve. For the points in time $t_3 = 0.099$ s and $t_4 = 0.100$ s the parameter curves are illustrated graphically in Figs. 5.45 and 5.46 for $\alpha = 0$.

The optimum points obtained from the search for the maximum and minimum of v(1) for the α-level $\alpha_1 = 0.0$ are indicated. When searching for the minimum these points are found within the input subspace, the assigned Young's modulus $E_{opt,l}$ lies between $2.7 \cdot 10^7$ kN/m^2 and $3.2 \cdot 10^7$ kN/m^2. The optimum points $E_{opt,r}$ gained from maximum search are situated in the corners of the input subspace, with the time step from t_3 to t_4 the optimum point $E_{opt,r}$ jumps from $E = 2.7 \cdot 10^7$ kN/m^2 to $E = 3.2 \cdot 10^7$ kN/m^2, as may be seen in Figs. 5.45 and 5.46.

The membership function $\mu(v(1))$ for $t_4 = 0.100$ s is plotted in Fig. 5.47. Its left branch possesses a vertical part, which is obtained in consequence of the position of the optimum points $E_{opt,l}$ from the minimum search within the input subspace.

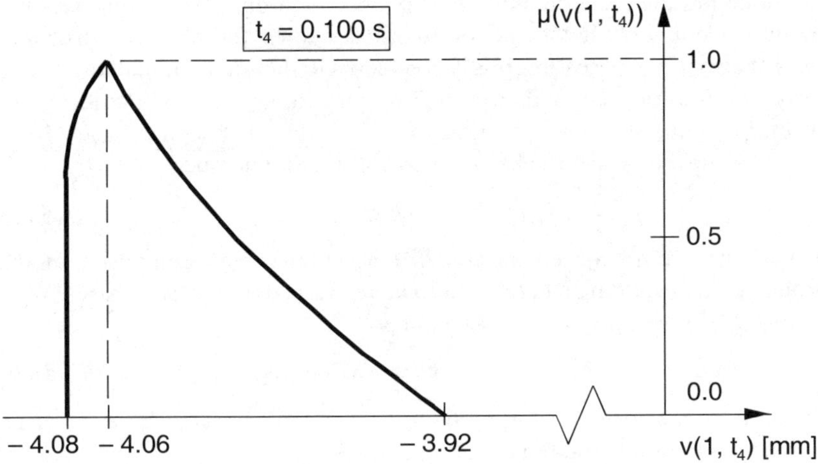

Fig. 5.47. Fuzzy displacement $\tilde{v}(1)$ at time $t_4 = 0.100$ s

5.2.5.2 Steel Frame, Geometrically Nonlinear Statical Analysis

The plane frame consisting of rolled steel shapes (Fig. 5.48) is analyzed under the given load considering geometrical nonlinearities [114].

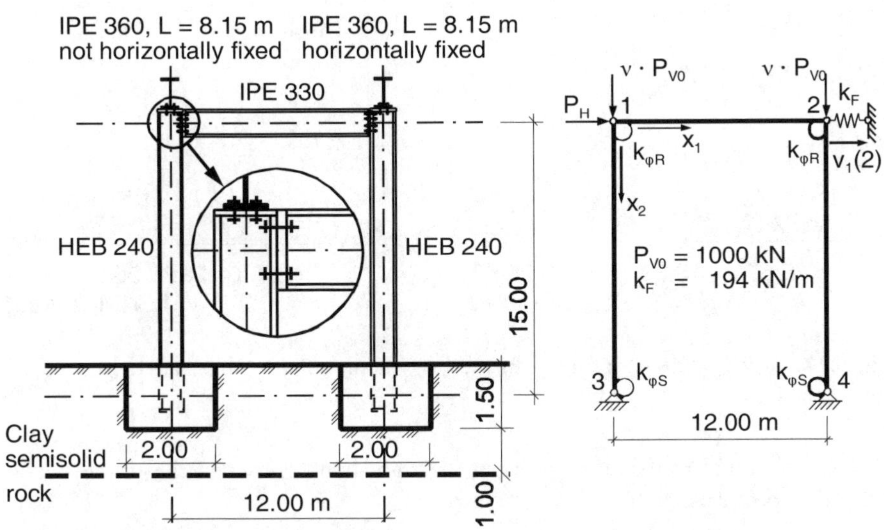

Fig. 5.48. Plane frame consisting of rolled steel shapes; construction sketch, statical system, and loading

When modeling the structure and the support conditions informal uncertainty is taken into account with the aid of fuzzy variables for physical structural parameters. The column bases are rigidly connected to the square foundations. The characteristic values of the foundation soil are not available as real numbers. For the semisolid clay the stiffness is assumed to take values between $E_S = 5$ MN/m^2 and $E_S = 20$ MN/m^2. It is specified as the fuzzy triangular number

$$\tilde{E}_S = <5, 10, 20> \quad \text{MN/m}^2. \tag{5.85}$$

The foundations are idealized as linear elastic elements embedded in the foundation soil, thus yielding a fuzzy rotational restraint of the column bases. With $\tilde{E}_S$ from Eq. (5.85) the fuzzy triangular number

$$\tilde{k}_{\varphi S} = <6.67, 13.33, 26.67> \quad \text{MN m/rad} \tag{5.86}$$

is obtained. It is presumed that the conditions at both column bases may be characterized by the same rotational spring stiffness.

The endplate shear connections at the corners of the frame are modeled as rotational springs and introduced into the analysis as fuzzy triangular numbers (Fig. 5.49)

$$\tilde{k}_{\varphi R} = <20.0, 23.4, 25.0> \quad \text{MN m/rad}. \tag{5.87}$$

The rotational spring stiffness is assumed to be the same at both corners of the frame.

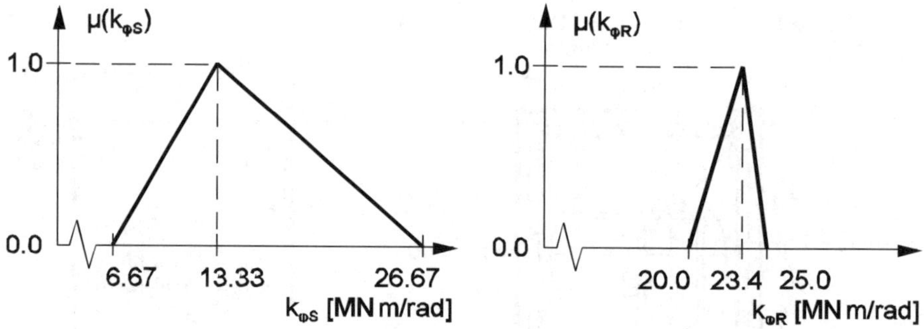

Fig. 5.49. Fuzzy rotational spring stiffnesses $\tilde{k}_{\varphi S}$ at the column bases and $\tilde{k}_{\varphi R}$ at the corners of the frame

The horizontal translation spring at the right-hand corner of the frame is considered to be a crisp value

$$k_F = 194 \ \text{kN/m}. \tag{5.88}$$

The stiffness k_F is determined by the rolled steel shape IPE 360 on the right, running orthogonal to the frame plane.

Investigation I – Linearized, Ideal Fuzzy Bifurcation Load. The linearized, ideal bifurcation load $\nu_{ki} \cdot P_{V0}$ is computed based on second-order elasticity theory presupposing small displacements. For this purpose the horizontal load is taken to be $P_H = 0$. The longitudinal force state $\underline{S}_0$ resulting from the load P_{V0} is computed with the aid of first-order elasticity theory and then proportionally increased by the load factor ν. If the coefficient matrix (system stiffness matrix) of the equation system for computing the unknown node displacement components becomes singular and thus the displacements approach infinity, the linearized bifurcation load $\nu_{ki} \cdot P_{V0}$ is then found with the assigned value ν_{ki}.

Consideration of the fuzzy variables $\tilde{k}_{\varphi S}$ and $\tilde{k}_{\varphi R}$ in the system stiffness leads to the transcendental fuzzy eigenvalue problem

$$\det\left(\underline{\tilde{K}}(\tilde{\nu}_{ki} \cdot \underline{\tilde{S}}_0)\right) = 0, \tag{5.89}$$

the evaluation of which yields the linearized fuzzy bifurcation load $\tilde{\nu}_{ki} \cdot P_{V0}$.

The nonlinear mapping

$$\left(\tilde{k}_{\varphi S}, \tilde{k}_{\varphi R}\right) \to \left(\tilde{\nu}_{ki}\right) \tag{5.90}$$

is monotonic, i.e., the optimum points lie in the corners of the two-dimensional input subspace.

The analysis was carried out with eleven equidistant α-levels. For determining the optimum points the modified evolution strategy was applied. The fuzzy result from the bifurcation load investigation is shown in Fig. 5.50.

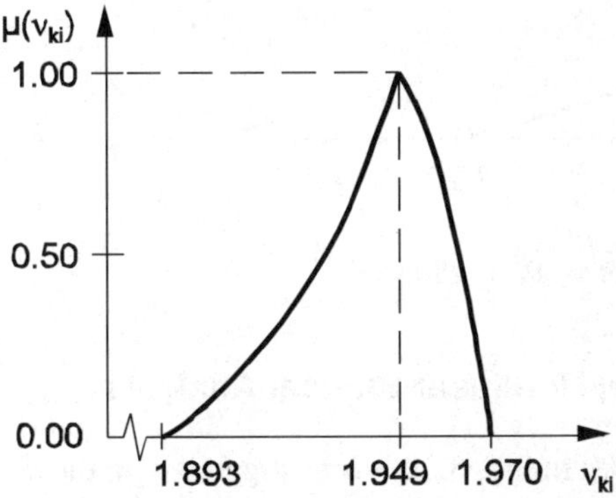

Fig. 5.50. Linearized fuzzy bifurcation load $\tilde{\nu}_{ki} \cdot P_{V0}$

Investigation II – Fuzzy Displacement. The horizontal displacement $v_1(2)$ of the right-hand frame corner is to be computed for a defined loading $\nu_1 \cdot P_{V0}$. For this purpose the deterministic fundamental solution after [116] and [133] is applied. The system behavior is analyzed on the basis of second-order elasticity theory considering large displacements and moderate rotations, physical nonlinearities are not taken into account. The load is incrementally increased.

For achieving an initial horizontal displacement of the system the horizontal load $P_H = 1$ kN is applied. The vertical load P_{V0} is then increased with the load factor ν from zero up to

$$\nu \cdot P_{V0} = 1800 \text{ kN}. \tag{5.91}$$

The fuzzy rotational spring stiffnesses $\tilde{k}_{\varphi S}$ and $\tilde{k}_{\varphi R}$ are mapped onto the fuzzy nodal displacement $\tilde{v}_1(2)$

$$\left(\tilde{k}_{\varphi S}, \tilde{k}_{\varphi R}\right) \to \left(\tilde{v}_1(2)\right). \tag{5.92}$$

The mapping is monotonic, the optimum points are corner points of the input subspace.

For computing the fuzzy result eleven equidistant α-levels were used, the α-level optimization was solved with the aid of the modified evolution strategy. The control parameters were adjusted according to the proposal in Sect. 5.2.3.4.

The fuzzy nodal displacement $\tilde{v}_1(2)$ is shown in Fig. 5.51.

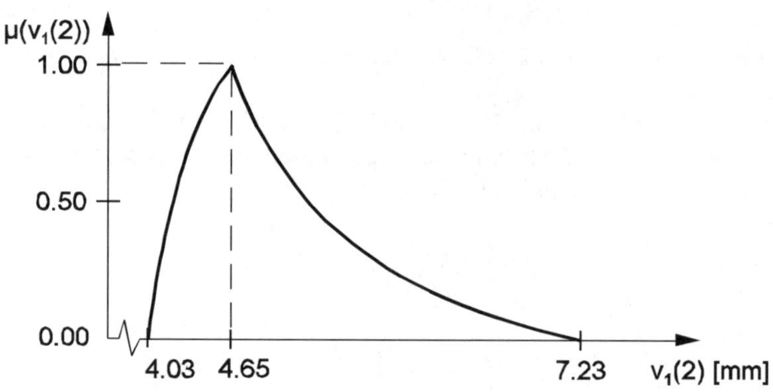

Fig. 5.51. Fuzzy nodal displacement $\tilde{v}_1(2)$ for $\nu \cdot P_{V0} = 1800$ kN

5.2.5.3 Reinforced-concrete Frame, Nonlinear Statical Analysis

The plane reinforced-concrete frame shown in Fig. 5.52 is investigated with the aid of the geometrically and physically nonlinear analysis algorithm after [116] and [133]. The effects of different fuzzy variables are considered [118].

Concerning the deterministic fundamental model the system is described using three bars. Fifty integration increments are chosen for each bar and each cross section is subdivided into 60 layers. The geometrically and physically nonlinear analysis is carried out using the material laws for reinforcement steel and concrete after Oetes. Tension stiffening and the effects of stirrup reinforcement are accounted for in the concrete material law [133].

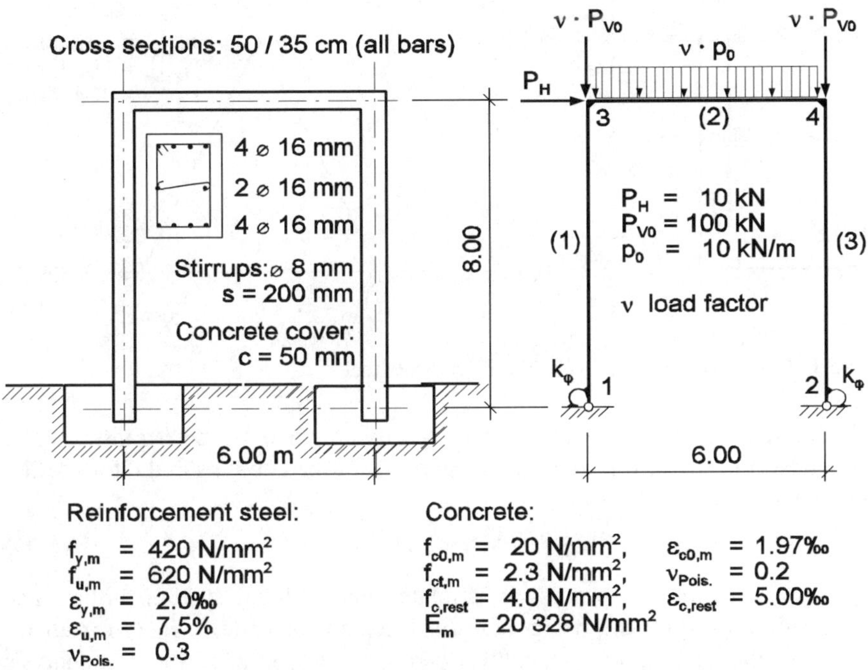

Fig. 5.52. Reinforced-concrete frame (plane); structure, cross sections, materials, statical system with loading

The simulated loading process is comprised of deadload, horizontal load P_H, vertical nodal loads $P_V = \nu \cdot P_{V0}$ and the line load $p = \nu \cdot p_0$. After applying deadload the horizontal load P_H is introduced; P_V and p are then simultaneously increased incrementally using the load factor ν.

Investigation I – Fuzzy System Behavior in Consequence of Fuzzy Arrangement of the Reinforcement Steel. When modeling the structure it is necessary to define the arrangement of the reinforcement steel. The distances h_1, h_2, and h_3 (Fig. 5.53) between the cross-sectional boundaries and the position of the reinforcement steel are prescribed at each end of the bars. In the horizontal bar of the frame the reinforcement arrangement is also specified in the middle of the bar. The distances h_1, h_2, and h_3 are considered to be independent of each other; the reinforcement arrangements at the various prescribed locations are also independent of each other. Assuming a tolerance of ±5 mm in the laying of the reinforcement steel, h_1, h_2, and h_3 become fuzzy variables. For the case in question a simple fuzzification of the values h_i is applied, Fig. 5.53 shows the fuzzy triangular numbers $\tilde{h}_i$. Thus a total of 21 fuzzy variables enter the investigation.

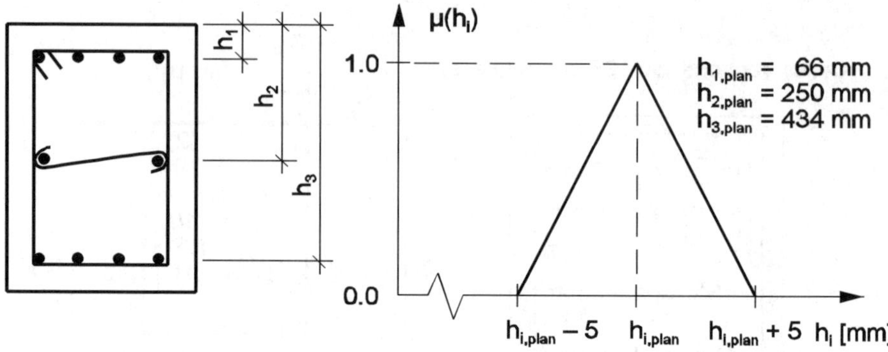

Fig. 5.53. Fuzzy arrangement of the reinforcement steel

The remaining structural parameters are assumed to be deterministic. The rotational spring stiffness k_φ relating to the fixing of the columns in the foundation soil is specified by

$$k_\varphi = 5 \text{ MNm/rad}. \tag{5.93}$$

The described loading process is simulated up to global system failure. The fuzzy result values are the horizontal displacement of node 3 $\tilde{v}_H(3)$ for all increments of the loading process and the load factor $\tilde{v}_{g+p}$(g+p = structural analysis under consideration of geometrical and physical nonlinearities) for the fuzzy failure load.

The computation was carried out with the aid of the modified evolution for the three α-levels $\alpha_1 = 0.00$, $\alpha_2 = 0.50$, and $\alpha_3 = 1.00$. The fuzzy load–displacement dependency for $\tilde{v}_H(3)$ and the result for $\tilde{v}_{g+p}$ are presented in Fig. 5.54.

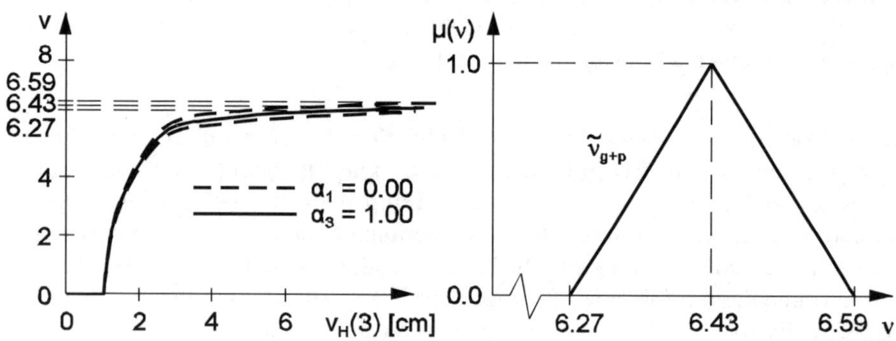

Fig. 5.54. Fuzzy load–displacement dependency for $\tilde{v}_H(3)$ and $\alpha_1 = 0.00$ and $\alpha_3 = 1.00$; load factor $\tilde{v}_{g+p}$ specifying the fuzzy failure load for global system failure according to the geometrically and physically nonlinear investigation

The obtained fuzzy failure load may, e.g., be used for a simple safety verification based on load level without determining a safety measure. If the actual load factor is known as deterministic value $extg\_\nu = 6.30$, the inequality

$$extg\_\nu = 6.30 \le <6.27,\ 6.43,\ 6.59> = \tilde{\nu}_{g+p} \tag{5.94}$$

containing $\tilde{\nu}_{g+p}$ approximated by a fuzzy triangular number is then only partially satisfied. Without further evaluation the structure is assessed as being not safe. In contrast to the latter a common structural analysis based on the assumption of the reinforcement in the planned position yields the deterministic failure load characterized by $\nu_{g+p} = 6.43$. The verification criterion

$$extg\_\nu = 6.30 \le 6.43 = \nu_{g+p} \tag{5.95}$$

is fulfilled, the influence of the tolerance when laying the reinforcement steel remains concealed.

Investigation II – Fuzzy Displacement in Consequence of Fuzzy Loading and Fuzzy Boundary Conditions. The investigation of the frame is repeated with altered structural parameters. The reinforcement arrangement is deterministically assumed to be in the planned position. The fuzzy input variables are now the load factor $extg\_\tilde{\nu}$ and the rotational spring stiffness $\tilde{k}_\varphi$. These informally uncertain structural parameters (Fig. 5.55) are modeled as fuzzy triangular numbers

$$extg\_\tilde{\nu} = <5.5,\ 5.9,\ 6.7>, \tag{5.96}$$

and

$$\tilde{k}_\varphi = <5,\ 9,\ 13>\ \text{MN m/rad}. \tag{5.97}$$

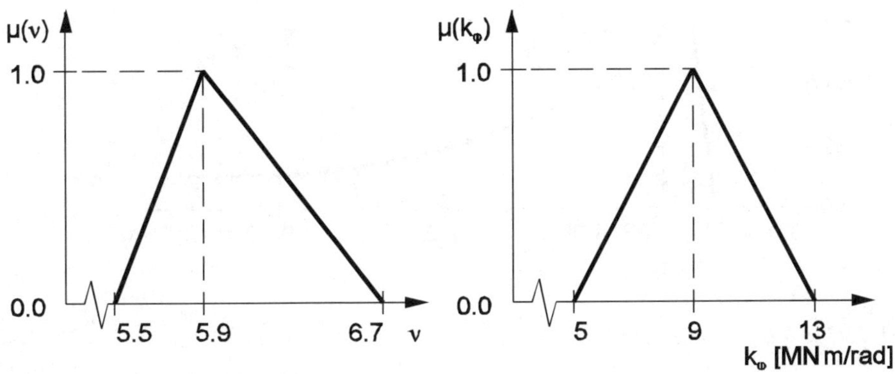

Fig. 5.55. Fuzzy input variables load factor $extg\_\tilde{\nu}$ and rotational spring stiffness $\tilde{k}_\varphi$

The fuzzy result is the horizontal displacement of node 3 $\tilde{v}_H(3)$. The investigation was carried out for the α-levels $\alpha_1 = 0.00$, $\alpha_2 = 0.144$, $\alpha_3 = 0.25$, $\alpha_4 = 0.50$, $\alpha_5 = 0.75$, and $\alpha_6 = 1.00$. The fuzzy load–displacement dependency for $\alpha_1 = 0.00$ and $\alpha_6 = 1.00$ is shown in Fig. 5.56, and Fig. 5.57 shows the fuzzy result $\tilde{v}_H(3)$.

For $\alpha_1 = 0.00$ global system failure already occurs before the attainment of $v_{\alpha_l r} = 6.7$; the search for maximum $v_H(3)$ yields the result $v_H(3)_{\alpha_l r} \rightarrow \infty$ on this α-level.

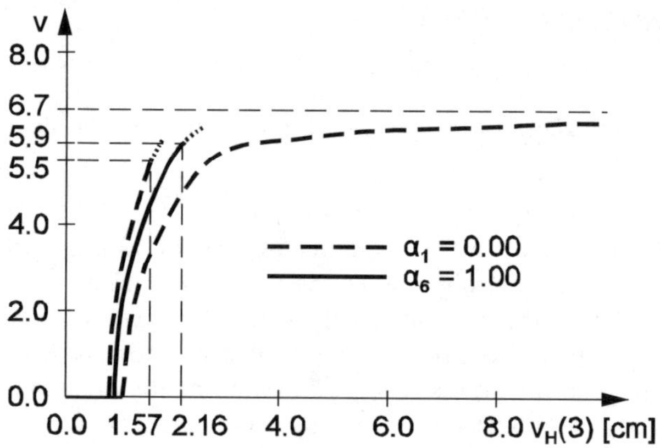

Fig. 5.56. Fuzzy load–displacement dependency for $\tilde{v}_H(3)$ and $\alpha_1 = 0.00$ and $\alpha_6 = 1.00$

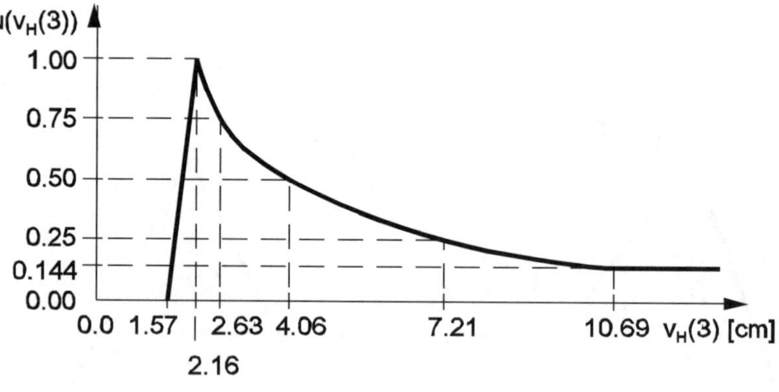

Fig. 5.57. Fuzzy result $\tilde{v}_H(3)$

Investigation III – Fuzzy Displacement in Consequence of Fuzzy Boundary Conditions, Influence of the Deterministic Fundamental Solution. The load factor $\tilde{v}_{g+p}$ of the fuzzy failure load for global system failure (for geometrically and physically nonlinear behavior) is computed as the fuzzy result variable. In contrast to investigation II, only the rotational spring stiffness according to Eq. (5.97) enters the analysis as a fuzzy input variable. The loads P_V and p are incrementally increased with ν until global system failure is attained. For the purpose of comparison the quasicrisp failure load represented by ν_p, under exclusive consideration of physical nonlinearities, and the fuzzy failure load represented by $\tilde{\nu}_g$, under exclusive consideration of geometrical nonlinearities, are computed. The results for $\tilde{\nu}_{g+p}$, ν_p, and $\tilde{\nu}_g$ are compared in Fig. 5.58. This example demonstrates the decisive influence of the deterministic fundamental solution on the quality of the results. Realistic prognoses concerning structural behavior are closely related to capable simulation algorithms.

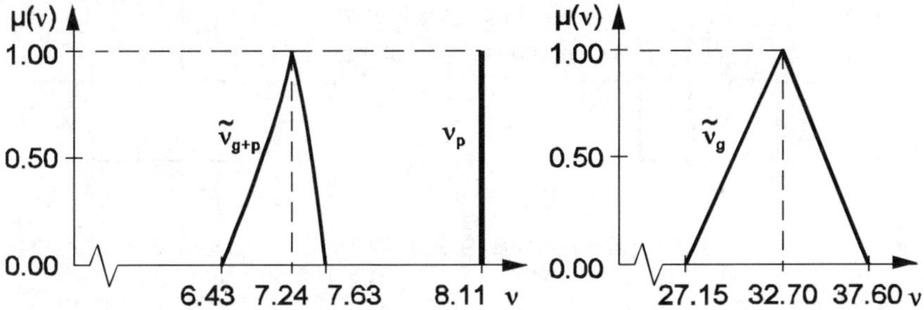

Fig. 5.58. Load factors $\tilde{\nu}_{g+p}$, ν_p, and $\tilde{\nu}_g$ of the fuzzy failure load for global system failure; considered nonlinearities: g = geometrical, p = physical, g+p = geometrical and physical

5.2.5.4 Prestressed Reinforced Concrete Frame, Nonlinear Dynamic Analysis

The plane reinforced-concrete frame with a prestressed horizontal bar according to Fig. 5.59 is investigated under dynamic transient loading [118]. For the dynamic fuzzy structural analysis the geometrically and physically nonlinear simulation algorithm after [116, 133] is applied as the deterministic fundamental solution.

The system is modeled using three bars, 49 integration increments are chosen for the horizontal bar and 40 integration increments for the columns. Each of the cross sections is subdivided into 60 layers. The geometrically and physically nonlinear analysis is carried out using the material laws after Ma, Bertero and Meskouris, Krätzig according to [113] and [133] for reinforcement steel and concrete. Contact forces associated with fracture closure and the effects of stirrup reinforcement are accounted for in the concrete material law; tension stiffening is neglected in this case.

The structure is constructed from prefabricated parts and the frame corners are rigidly connected on site. The fixture of the columns in the foundation soil is modeled by means of linear-elastic rotational springs (Fig. 5.60).

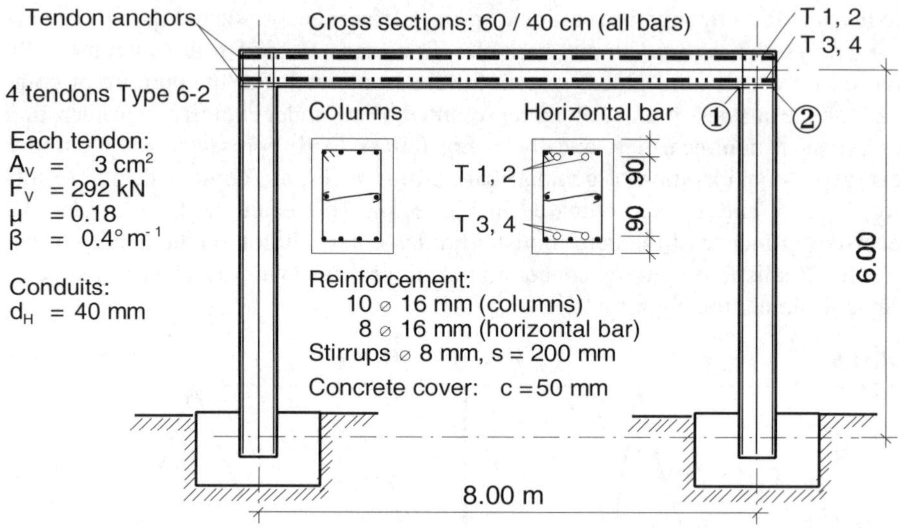

Concrete: C 35/45	Reinforcement steel: S 500	Prestressing steel: S 1570/1770
$f_{c0,m}$ = 43 N/mm^2	$f_{y,m}$ = 551 N/mm^2	E = 195 000 N/mm^2
$\varepsilon_{c0,m}$ = 2.89‰	$f_{u,m}$ = 667 N/mm^2	(remains in the linear range)
$f_{ct,m}$ = 3.8 N/mm^2	$\varepsilon_{y,m}$ = 2.62‰	
E_m = 29 807N/mm^2	$\varepsilon_{u,m}$ = 7.0%	
v = 0.2	v = 0.3	

Fig. 5.59. Reinforced concrete frame (plane); structure, cross sections, materials

The simulated loading process, including system modification, is comprised of the following components:

- Simultaneous prestressing of all tendons in the horizontal bar according to the specified prestressing force without the effects of deadload, grouting of the conduits
- Application of the deadload of the columns, hinged connection of the columns and the horizontal bar at the frame corners, and application of the deadload of the horizontal bar
- Transformation of the hinged joints at the frame corners into rigid connections
- Application of additional translational mass at the frame corners and along the horizontal bar (statical loads P_V and $p_{R,V}$ in Fig. 5.60)
- Introduction of dynamic loading ($P_H(t)$, $p_{R,H}(t)$, and $p_{S,H}(t)$ in Fig.5.60) due to the horizontal acceleration according to the normalized load-time function

The stiffness of the rotational springs (the same at each column base) and the horizontal acceleration are modeled by the fuzzy triangular numbers $\tilde{k}_\varphi$ and $\tilde{a}$ due to informal uncertainty (Fig. 5.61). For the purpose of determining a suitable time increment Δt for the numerical time-step integration the first three natural angular frequencies for the linear case and k_φ = 9.0 MN m/rad were computed to be

$\omega_1 = 14.48\ s^{-1}$, $\omega_2 = 68.43\ s^{-1}$, and $\omega_3 = 108.64\ s^{-1}$. It is assumed that the prescribed loading excites the system exclusively at the first natural frequency. The time increment for the numerical integration is therefore chosen to be $\Delta t = 0.025$ s.

The time dependency of the horizontal fuzzy displacement $\tilde{v}_H(t)$ of the left-hand frame corner up to $t = 2.5$ s is plotted in Fig. 5.62. For the purpose of comparison the results of a linear deterministic investigation for $\mu(v_H(t)) = 1.0$ are also plotted.

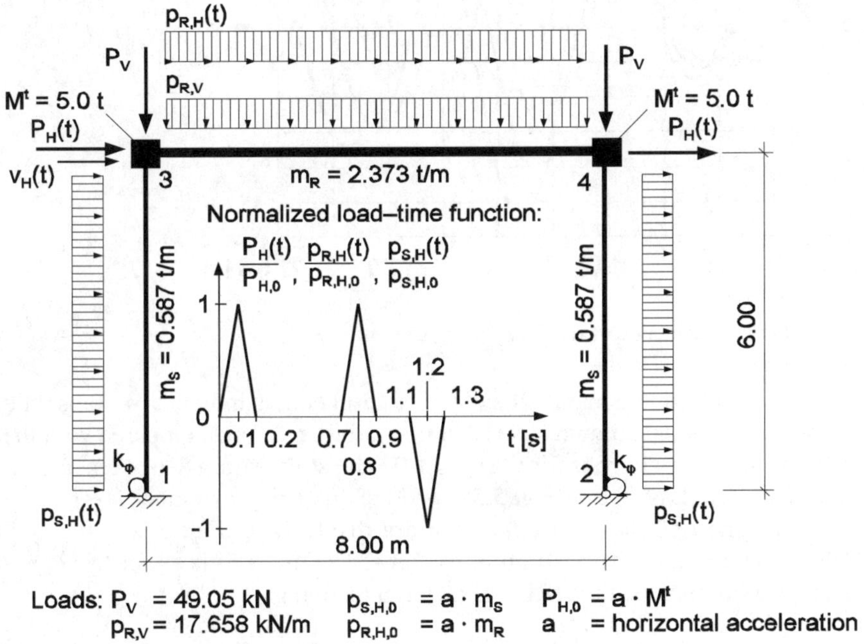

Fig. 5.60. Statical system (final state) with statical and dynamic loading

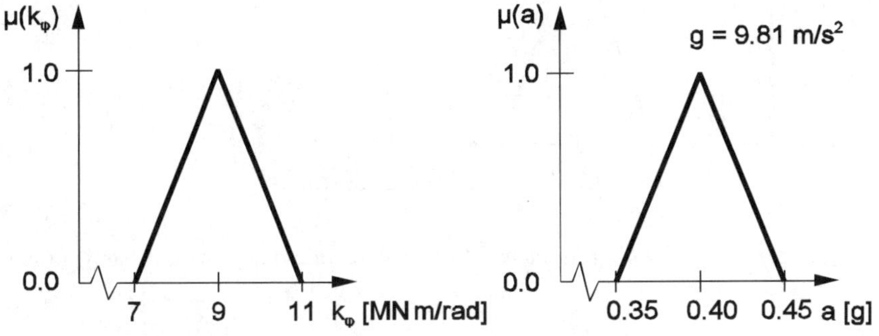

Fig. 5.61. Fuzzy input variables $\tilde{k}_\varphi$ and $\tilde{a}$

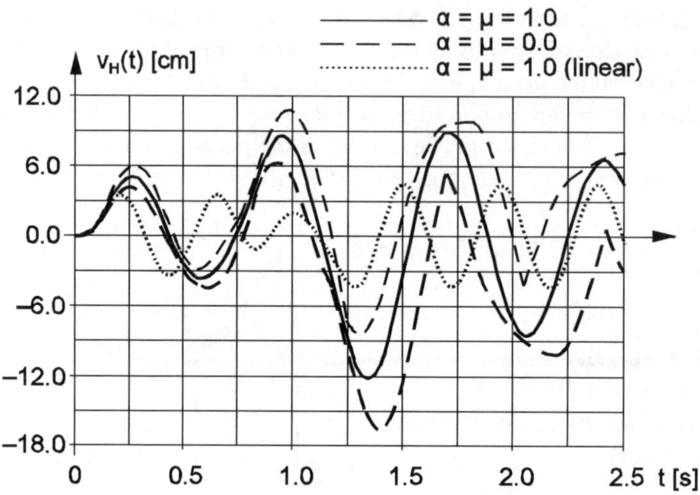

Fig. 5.62. Time dependency of the fuzzy displacement $\tilde{v}_H(t)$

The fuzzy result for the magnitude of the largest bending moment $\tilde{M}_S$ (absolute value) at the right-hand column base is plotted in Fig. 5.63 and compared with the result from the linear analysis for $\mu(M_S) = 1.0$. The stress-strain dependencies for the inner concrete layer (① in Fig. 5.59) and the outer reinforcement layer (② in Fig. 5.59) for the right-hand column head are displayed in Fig. 5.64 for $k_\varphi = 7$ MN m/rad and a = 0.45 g (μ = 0) up to t = 2.5 s. The material laws for hysteresis material behavior include the effects of material damping.

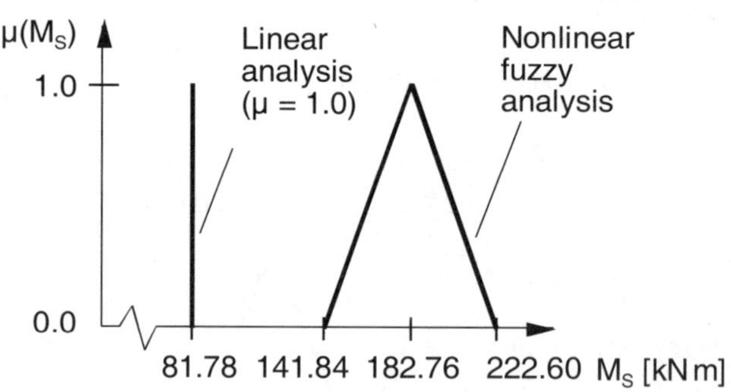

Fig. 5.63. Largest fuzzy bending moment $\tilde{M}_S$ (absolute value) at the right-hand column base

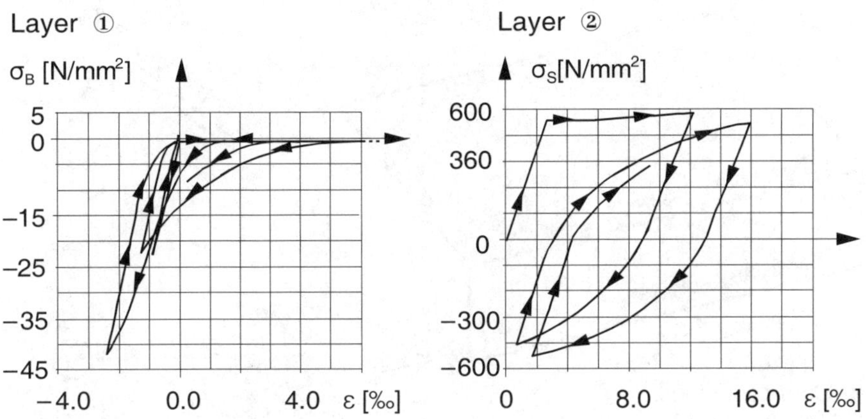

Fig. 5.64. Stress-strain dependencies for concrete (layer ①) and steel (layer ②)

5.2.5.5 Reinforced-concrete Folded-plate Structure, Fuzzy Finite Element Method, Physically Nonlinear Analysis

The reinforced-concrete folded-plate structure shown in Fig. 5.65 is analyzed under consideration of the governing nonlinearities of reinforced concrete and data uncertainty represented by fuzziness [120]. Fuzziness is accounted for in the superficial load $\tilde{p}$, the concrete compressive strength $\tilde{\beta}_R$, and the position of the reinforcement $\tilde{h}_1$. The extension for every plane of the folded-plate structure in $\mathbb{R}^2$ is crisp. Taking advantage of the symmetric properties in relation to two planes the system is meshed using 48 finite elements. Each element, with an overall thickness of 0.1 m, was modeled using seven equidistant concrete layers, with two smeared mesh reinforcement layers each on the upper and lower surfaces (Fig. 5.65). The concrete material law according to Kupfer [95] and Link [102] and a bilinear material law for reinforcement steel were applied. Tensile cracks in the concrete were accounted for in each element on a layer-to-layer basis according to the concept of smeared fixed cracks. The mapping model thus represents a physically nonlinear algorithm. The superficial load characterized by fuzziness was increased incrementally up to the prescribed service load. Selected result values are observed at this stage of loading. Owing to the fuzziness of the uncertain input variables, the result variables are also obtained as fuzzy variables.

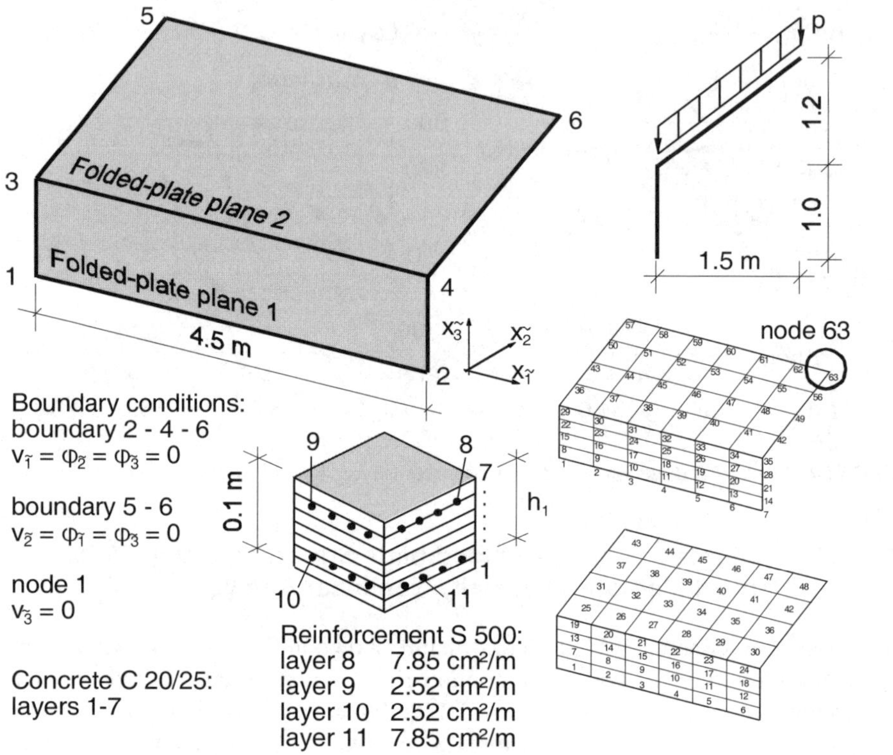

Fig. 5.65. Folded-plate structure; geometry and finite element model

Investigation I – Fuzzy Displacement in Consequence of Fuzzy Loading and Fuzzy Material Behavior. The superficial load $\tilde{p}$ and the concrete compressive strength $\tilde{\beta}_R$ enter the analysis as fuzzy structural parameters. The superficial load on the folded-plate plane 2 is approximated as a linear fuzzy field with the aid of fuzzy interpolation nodes at the corner points 3, 5, and 6. The fuzzy field thus depends on the three bunch parameters $\tilde{s}_1 = \tilde{p}(3)$, $\tilde{s}_2 = \tilde{p}(5)$, $\tilde{s}_3 = \tilde{p}(6)$

$$x_1(\underline{\tilde{s}}, \underline{\theta}) = x_1(\tilde{s}_1, \tilde{s}_2, \tilde{s}_3, \underline{\theta})\,. \tag{5.98}$$

Interaction between the fuzzy bunch parameters does not exist and their membership functions are chosen to be equal (Fig. 5.65). A crisp function (trajectory) $x_1(\underline{s}, \underline{\theta})$ from the bunch of functions $x_1(\underline{\tilde{s}}, \underline{\theta})$ is shown in Fig. 5.66 for one α-level and three values $s_1 \in S_{1,\alpha}$, $s_2 \in S_{2,\alpha}$, and $s_3 \in S_{3,\alpha}$. The uncertain concrete compressive strength is accounted for with a stationary fuzzy field. Only one bunch parameter $\tilde{s}_4 = \tilde{\beta}_r$ is sufficient for describing this fuzzy field. This bunch parameter is specified by the membership function $\mu(\beta_R)$ (Fig. 5.67). Applying α-level optimization a four-dimensional input subspace $\underline{E}_{\alpha_k}$ is to be searched for each α-level.

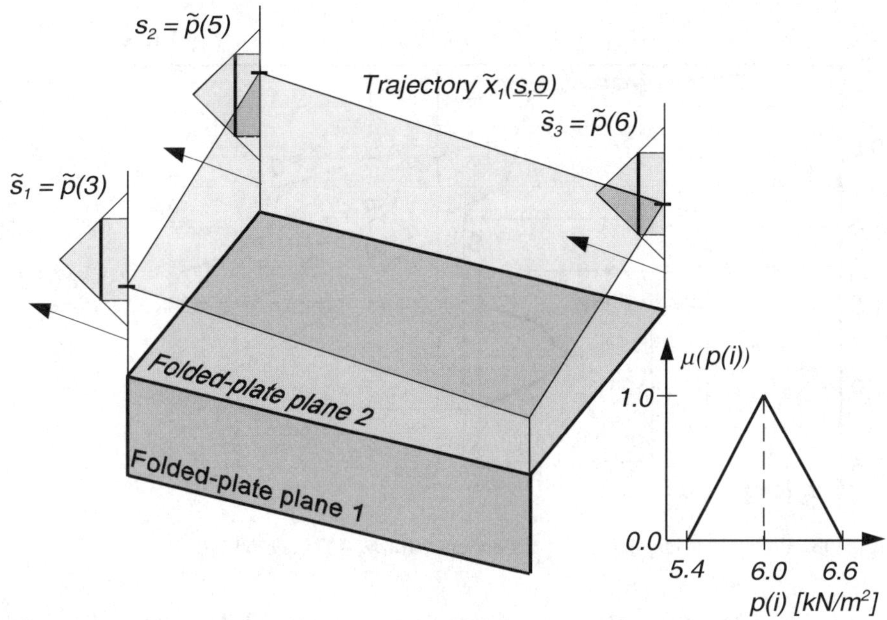

Fig. 5.66. Fuzzy field $x_1(\tilde{\underline{s}},\underline{\theta})$ for the superficial load $\tilde{p}$; trajectory $x_1(\underline{s}, \underline{\theta})$, fuzzy bunch parameters $\tilde{s}_1 = \tilde{p}(3)$, $\tilde{s}_2 = \tilde{p}(5)$, $\tilde{s}_3 = \tilde{p}(6)$

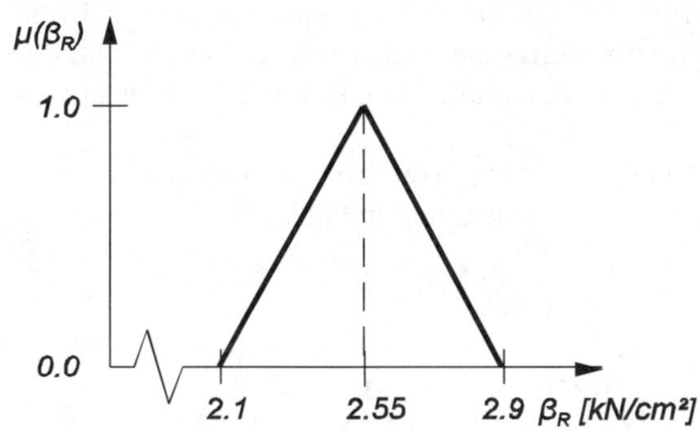

Fig. 5.67. Fuzzy concrete compressive strength $\tilde{\beta}_R$

As an example of the results the fuzzy load-displacement dependency for the displacement $v_{\tilde{3}}$ of node 63 under incremental load increase up to the service load ($\nu = 1$) is shown in Fig. 5.68.

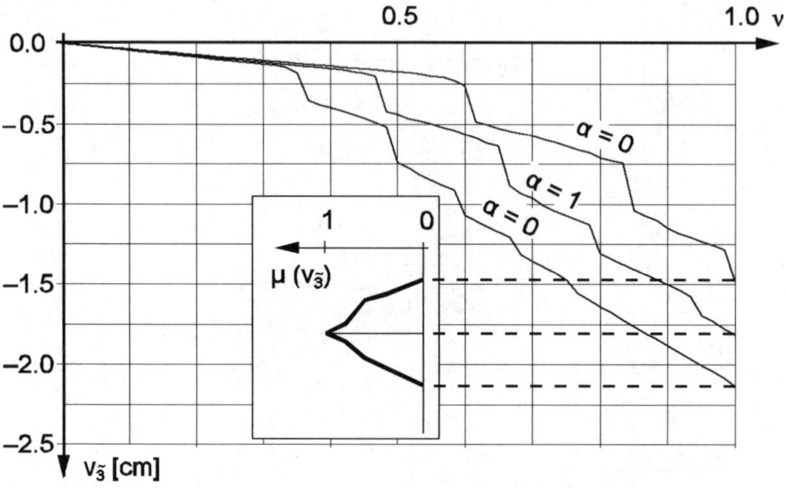

Fig. 5.68. Fuzzy load-displacement dependency for $v_{\tilde{3}}$ of node 63

Investigation II – Fuzzy Displacement in Consequence of Fuzzy Loading and Fuzzy Arrangement of the Reinforcement. The fuzzy structural parameters in this case are the position $\tilde{h}_1$ of reinforcement layer 11 and the superficial load $\tilde{p}$. The spatial configuration of these fuzzy structural parameters is to be accounted for with stationary fuzzy fields. Two fuzzy bunch parameters $\tilde{s}_1 = \tilde{h}_1$ and $\tilde{s}_2 = \tilde{p}$ are thus required, whose membership functions are plotted in Fig. 5.69. In this investigation the concrete compressive strength is taken to be a deterministic value $\beta_R = 2.55$ kN/m^2.

Representing a characteristic result, again the fuzzy load-displacement dependency for the displacement $v_{\tilde{3}}$ of node 63 is illustrated in Fig. 5.70.

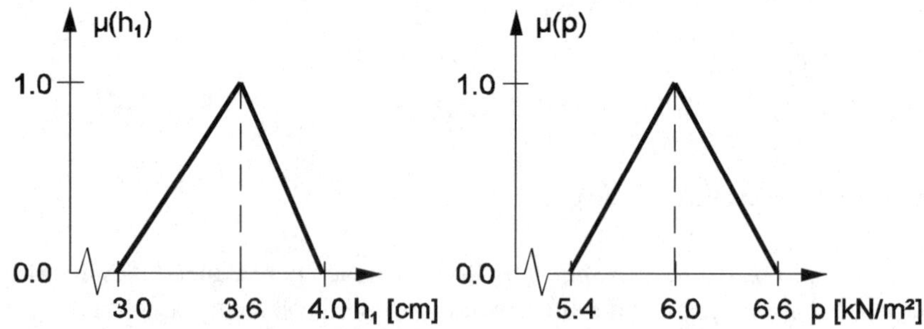

Fig. 5.69. Fuzzy bunch parameters and $\tilde{s}_1 = \tilde{h}_1$ and $\tilde{s}_2 = \tilde{p}$

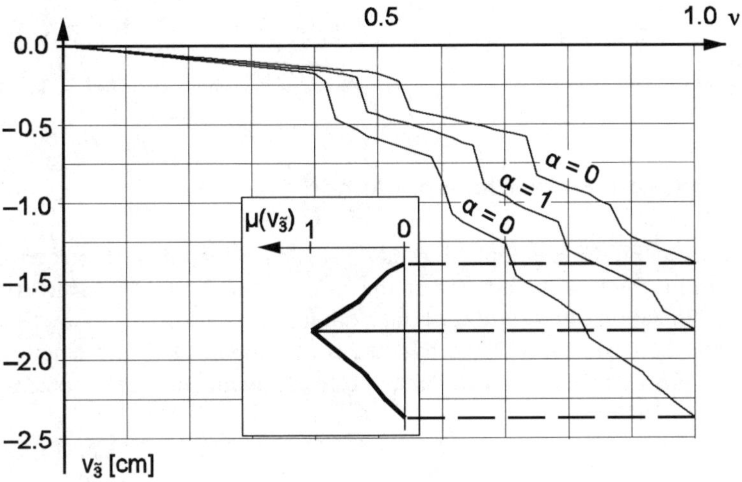

Fig. 5.70. Fuzzy load-displacement dependency for $v_{\tilde{3}}$ of node 63

5.3 Fuzzy Stochastic Structural Analysis [1]

If the uncertainty of the input parameters of a structural analysis is described with the aid of fuzzy random functions, the following problem is then to be solved for a crisp mapping model

$$\mathbf{F}\colon \tilde{\underline{X}}(\underline{t}) \rightarrow \tilde{\underline{Z}}(\underline{t}). \tag{5.99}$$

Applying Eq. (5.99) the fuzzy random functions $\tilde{\underline{X}}(\underline{t})$ (structural input parameters) are mapped onto the fuzzy random functions $\tilde{\underline{Z}}(\underline{t})$ (structural responses). As fuzzy vectors and real random vectors are special cases of fuzzy random functions, these uncertainty models are also accounted for with Eq. (5.99).

The mapping according to Eq. (5.99) is the symbolic representation of a fuzzy stochastic structural analysis. For the deterministic fundamental solution arbitrary algorithms of structural analysis may be applied as already explained for fuzzy structural analysis.

If the mechanical behavior of a structure may be described on the basis of the nonlinear differential equation

$$\tilde{m}_t(.) \cdot \ddot{\tilde{v}} + \tilde{d}_t(.) \cdot \dot{\tilde{v}} + \tilde{k}_t(.) \cdot \tilde{v} = \tilde{f}_t(.) \quad \forall\ \underline{t} \in \mathbf{T}, \tag{5.100}$$

discretization in space then leads to a special form of the fuzzy stochastic structural analysis – the Fuzzy Stochastic Finite Element Method (FSFEM), which may be characterized by the system of nonlinear differential equations

[1] in collaboration with Dipl.-Ing. Jan-Uwe Sickert

$$\underline{\tilde{M}}(.) \cdot \ddot{\tilde{\underline{v}}} + \underline{\tilde{D}}(.) \cdot \dot{\tilde{\underline{v}}} + \underline{\tilde{K}}(.) \cdot \tilde{\underline{v}} = \underline{\tilde{F}}(.) \tag{5.101}$$

for the overall system.

5.3.1 Fuzzy Stochastic Finite Element Method

The Fuzzy Stochastic Finite Element Method (FSFEM) is based on the representation of uncertain structural parameters in the space of fuzzy bunch parameters and an α-level optimization of these fuzzy bunch parameters.

In the following the basic form of the FSFEM is developed for fuzzy random fields. In the case of time dependency this may be formally enhanced for considering fuzzy random functions.

Representation of Uncertain Structural Input Parameters in the Space of Fuzzy Bunch Parameters. It is assumed that the uncertainty of the structural parameters is accounted for with the aid of fuzzy random fields, real random fields, and fuzzy fields. Moreover, fuzzy random vectors, real random vectors, and fuzzy vectors may be given additionally at discrete points $\underline{\theta}$ in the parameter space $\underline{\boldsymbol{\Theta}}$.

Regarding fuzzy random fields an appropriate representation in closed or discrete form is required. Owing to the characterization of a fuzzy random field as an evaluated set of original functions according to Eqs. (2.214) and (2.215) all available representations developed for real-valued random fields, like, e.g., covariance decomposition, Karhunen–Loéve expansion, or polynomial chaos expansion, may be used as a basis for representing fuzzy random fields. Some basic information concerning representation and processing methods for real random fields is given in [171]. In view of applying a general, nonlinear numerical analysis in terms of the deterministic fundamental solution only a discrete representation of fuzzy random fields is considered subsequently. The discretization may thereby also be realized starting from closed-form representations. A variety of discrete representations of real random fields is, e.g., presented in [167] and [62].

The discretization of fuzzy random fields is explained for the example of point discretization methods, which are applicable in a particularly simple way. According to Eq. (2.243) each fuzzy random field is the set of the fuzzy random vectors at the points $\underline{\theta}$ in the parameter space $\underline{\boldsymbol{\Theta}}$. For each point $\underline{\theta}_i$ the fuzzy random vector $\underline{\tilde{X}}_{\theta_i} = \underline{\tilde{X}}(\underline{\theta}_i)$ belonging to the fuzzy random field $\underline{\tilde{X}}(\underline{\theta})$ is known.

When applying the midpoint method for the discretization in space, the midpoints of the finite elements, e.g., defined by the center of gravity, are taken to be the parameter points $\underline{\theta}_i$. According to the nodal-point method the parameter points $\underline{\theta}_i$ lie on the nodal points of the finite elements. The obvious disadvantage is the dependance of the chosen finite element mesh. These methods are particularly practicable in the case of strongly correlated fuzzy random fields.

At the points $\underline{\theta}_i$ the fuzzy probability distribution functions and parameters of the fuzzy random vectors $\tilde{\underline{X}}_{\theta_i} = \tilde{\underline{X}}(\underline{\theta}_i)$ are known.

In the *one-dimensional case* these represent, e.g., the fuzzy probability distribution function $\tilde{F}_{\theta_i}(x) = \tilde{F}(x, \underline{\theta}_i)$ with the fuzzy expected value $\tilde{m}_X(\theta_i) = E[\tilde{X}(\underline{\theta}_i)]$ and fuzzy variance $\tilde{\sigma}_X{}^2(\theta_i) = D^2[\tilde{X}(\underline{\theta}_i)]$ of the fuzzy random field.

The dependency between fuzzy random variables $\tilde{X}(\underline{\theta}_1)$ and $\tilde{X}(\underline{\theta}_2)$ at different points $\underline{\theta}_1$ and $\underline{\theta}_2$ is characterized by the fuzzy covariance (Eq. (2.235))

$$\tilde{K}(\underline{\theta}_1, \underline{\theta}_2) = \operatorname{Cov}[\tilde{X}(\underline{\theta}_1), \tilde{X}(\underline{\theta}_2)] = E[(\tilde{X}(\underline{\theta}_1) - \tilde{m}_X(\underline{\theta}_1)) \cdot (\tilde{X}(\underline{\theta}_2) - \tilde{m}_X(\underline{\theta}_2))]. \quad (5.102)$$

This model may also be applied for describing the dependency between fuzzy random variables representing different dimensions of a fuzzy random vector.

For a stationary fuzzy random field

$$\tilde{K}(\underline{\theta}_1, \underline{\theta}_2) = \tilde{\sigma}_X{}^2 \cdot \tilde{k}_X(L_{12}) \quad (5.103)$$

follows from Eq. (2.241) with $\tilde{\sigma}_X{}^2$ being the fuzzy variance of the fuzzy random field. The fuzzy covariance is a fuzzy function that only depends on the distance $\|\underline{\delta}\| = \|\underline{\theta}_2 - \underline{\theta}_1\| = L_{12}$. The dependency of the fuzzy covariance $\tilde{K}(\underline{\theta}_1, \underline{\theta}_2)$ on the distance L_{12} is accounted for with the fuzzy function $\tilde{k}_X(L_{12})$. This fuzzy function represents an assumption for the correlation function $\tilde{\varrho}(\underline{\theta}_1, \underline{\theta}_2)$ according to Eq. (2.236). The linear decrease of $\tilde{k}_X(L_{12})$ illustrated in Fig. 5.71 may be described by

$$\tilde{k}_X(L_{12}) = k_X(\tilde{L}_X, L_{12}) = \begin{cases} 1 - \dfrac{L_{12}}{\tilde{L}_X} & \text{if } L_{12} \in [0, \tilde{L}_X] \\ 0 & \text{otherwise} \end{cases} . \quad (5.104)$$

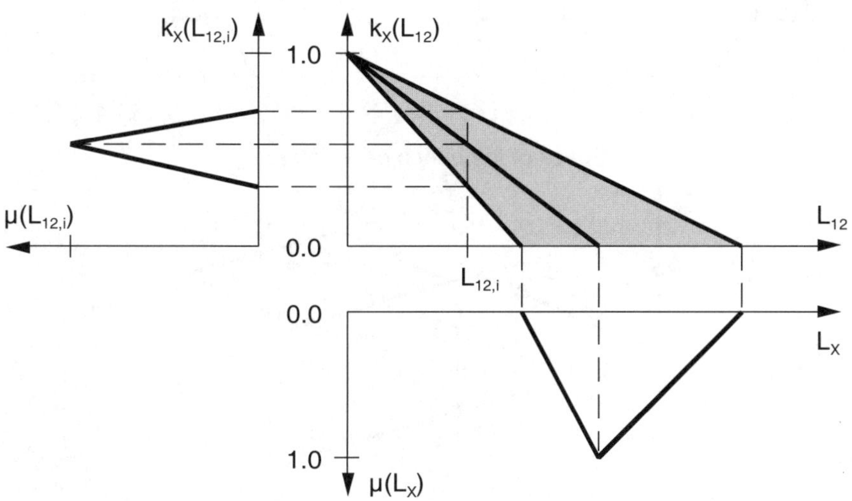

Fig. 5.71. Fuzzy function $\tilde{k}_X(L_{12})$ for describing fuzzy correlation of fuzzy random fields

The fuzzy correlation length $\tilde{L}_X$ is thereby taken into consideration as the fuzzy bunch parameter $\tilde{s}_i = \tilde{L}_X$. When prescribing a fuzzy correlation length instead of a deterministic correlation length, the uncertainty of this parameter may be accounted for.

In consequence of the discretization of the fuzzy random field $\underline{\tilde{X}}(\underline{\theta})$ the fuzzy random vectors $\underline{\tilde{X}}_{\theta_i} = \underline{\tilde{X}}(\underline{\theta}_i) \mid i = 1,...,p_1$ together with their fuzzy probability distribution functions $\tilde{F}_{\theta_i}(\underline{x}) = \tilde{F}(\underline{x},\underline{\theta}_i) = F(\underline{\tilde{s}}_i,\underline{x},\underline{\theta}_i)$ are determined at p_1 points in the parameter space $\boldsymbol{\Theta}$. The total number of all fuzzy bunch parameters $\tilde{s}_1,...,\tilde{s}_{n_1}$ in the fuzzy bunch parameter vectors $\underline{\tilde{s}}_i \mid i = 1,...,p_1$ is denoted by n_1.

Additionally, existing fuzzy fields are discretized according to Sect. 2.1.11.2. This leads to the fuzzy vectors $\underline{x}(\underline{\tilde{s}}_i,\underline{\theta}_i) \mid i = 1,...,p_2$ at p_2 points with n_2 fuzzy bunch parameters $\tilde{s}_1,...,\tilde{s}_{n_2}$ in total, which represent fuzzy input variables.

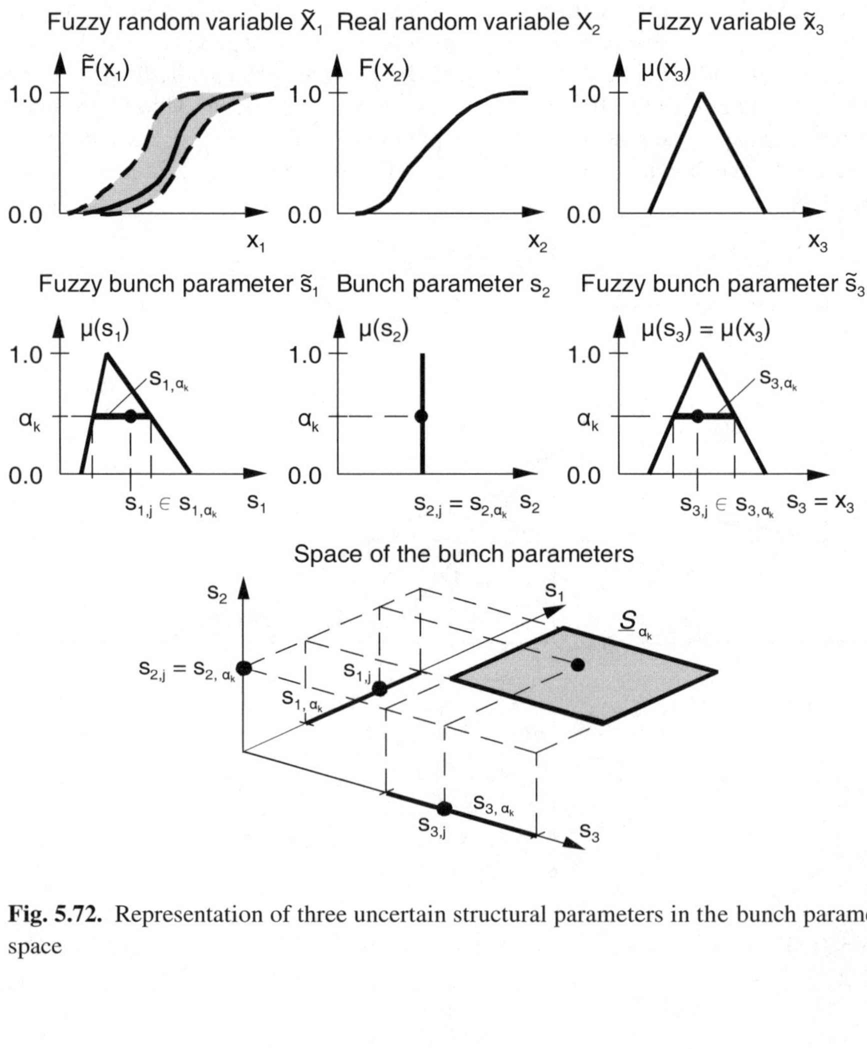

Fig. 5.72. Representation of three uncertain structural parameters in the bunch parameter space

If, furthermore, real random vectors characterized by $F(\underline{x}, \underline{\theta}_i)$ are to be considered at p_3 points, the parameters of their probability distribution functions may then be interpreted as being real bunch parameters. The total number of these bunch parameters $s_1, ..., s_{n_3}$ is given by n_3.

Finally, n_4 fuzzy bunch parameters $\tilde{s}_1, ..., \tilde{s}_{n_4}$ may appear from taking account of fuzzy covariance.

Combining all discretization results, a total of $n = n_1 + n_2 + n_3 + n_4$ bunch parameters $\tilde{s}_i$ are considered in the FSFEM analysis. These bunch parameters are combined in the fuzzy bunch parameter vector $\underline{\tilde{s}}_i$.

The fuzzy bunch parameters all together constitute the space of the bunch parameters, in which the uncertain structural input parameters are completely described. A three-dimensional bunch parameter space is illustrated in Fig. 5.72. The three bunch parameters are assigned to the fuzzy random variable $\tilde{X}_1$ with its fuzzy probability distribution function $\tilde{F}(x_1) = F(\tilde{s}_1, x_1)$, the real random variable X_2 with its probability distribution function $F(x_2) = F(s_2, x_2)$, and the fuzzy variable $\tilde{x}_3 = \tilde{s}_3$ with the membership function $\mu(x_3 = s_3)$.

An α-discretization of the fuzzy random variable $\tilde{X}_1$ yields the α-level set S_{1,α_k} representing the interval s_{1,α_k}. Each crisp element $s_{1,j}$ from this α-level set specifies precisely one original $X_{1,j}$ of $\tilde{X}_1$.

α-level Optimization in the Space of the Fuzzy Bunch Parameters. It is now intended to compute the structural responses $\underline{\tilde{Z}}(\underline{\theta})$ according to Eq.(5.99) as fuzzy random vectors $\underline{\tilde{Z}}_{\theta_r} = \underline{\tilde{Z}}(\underline{\theta}_r) \mid r = 1, ..., q_1$ with the fuzzy probability distribution functions $\tilde{F}_{\theta_r}(\underline{z}) = \tilde{F}(\underline{z}, \underline{\theta}_r) = F(\underline{\tilde{\sigma}}_r, \underline{z}, \underline{\theta}_r)$ at q_1 points $\underline{\theta}_r$ in the parameter space $\boldsymbol{\Theta}$. For this purpose q_1 fuzzy bunch parameter vectors $\underline{\tilde{\sigma}}_r$ are to be determined, which comprise a total of m_1 bunch parameters $\tilde{\sigma}_1, ..., \tilde{\sigma}_{m_1}$. The m_1 fuzzy bunch parameters are combined in the fuzzy vector $\underline{\tilde{\sigma}}$. The fuzzy stochastic structural analysis characterized by Eq. (5.99) has thus been transformed into the mapping

$$\mathbf{F}: \underline{\tilde{s}} \rightarrow \underline{\tilde{\sigma}}. \tag{5.105}$$

The mapping model may be denoted in the form

$$(\sigma_1, ..., \sigma_{m_1}) = f(s_1, ..., s_n). \tag{5.106}$$

On this basis Eq. (5.105) may be treated with the aid of α-level optimization (Sect. 5.2.3) in compliance with the procedure for fuzzy structural analysis.

The mapping model is to be established by a probabilistic structural analysis. Associated with a finite element approach this leads to the Stochastic Finite Element Method (SFEM).

An α-discretization of the fuzzy bunch parameters $\tilde{s}_1, ..., \tilde{s}_{n_1}$ belonging to fuzzy probability distribution functions $F(\underline{\tilde{s}}_i, \underline{x}, \underline{\theta}_i)$ yields the α-level sets $S_{1,\alpha_k}, ..., S_{n_1,\alpha_k}$ for the level α_k (Fig. 5.73). These α-level sets together with the α-level sets $S_{h,\alpha_k} \mid h = n_1 + 1, ..., n$ form the n-dimensional crisp subspace $\underline{S}_{\alpha_k}$. If one element is selected from each α-level set, one crisp point $\underline{s}$ is then obtained in the subspace $\underline{S}_{\alpha_k}$.

With each set of crisp elements $s_{1,j} \in S_{1,\alpha_k}, \ldots, s_{n_1,j} \in S_{n_1,\alpha_k}$ constituting the vector $\underline{s}_j \in \underline{S}_{\alpha_k}$ precisely one trajectory $F_{\theta_i,j}(\underline{x}) \in F_{\theta_i,\alpha_k}(\underline{x}) \mid i = 1,\ldots,p_1$ with the membership value $\mu(F_{\theta_i,j}(\underline{x})) = \mu(\underline{s}_j) \geq \alpha_k$ is simultaneously selected from each of the p_1 fuzzy probability distribution functions $\tilde{F}_{\theta_i}(\underline{x})$ (Fig. 5.74). The trajectories $F_{\theta_i,j}(\underline{x})$ are real-valued probability distribution functions. Each α-function set $F_{\theta_i,\alpha_k}(\underline{x})$ comprises all trajectories of the fuzzy probability distribution function $\tilde{F}_{\theta_i}(\underline{x})$ at the point $\underline{\theta}_i \in \underline{\Theta}$ for the level α_k.

Having selected one crisp point $\underline{s}_j$ from the subspace $\underline{S}_{\alpha_k}$ one real probability distribution function (trajectory) is known for each fuzzy random vector $\underline{\tilde{X}}_{\theta_i} = \underline{\tilde{X}}(\underline{\theta}_i)$. Moreover, one element from the respective α-level set (for the same level α_k) of each bunch parameter belonging to the fuzzy vectors, fuzzy fields, real random vectors, and fuzzy covariances is to be selected. Based on a finite element approach one stochastic structural analysis may now be carried out for the crisp bunch parameter vector $\underline{s}_j \in \underline{S}_{\alpha_k}$ defined.

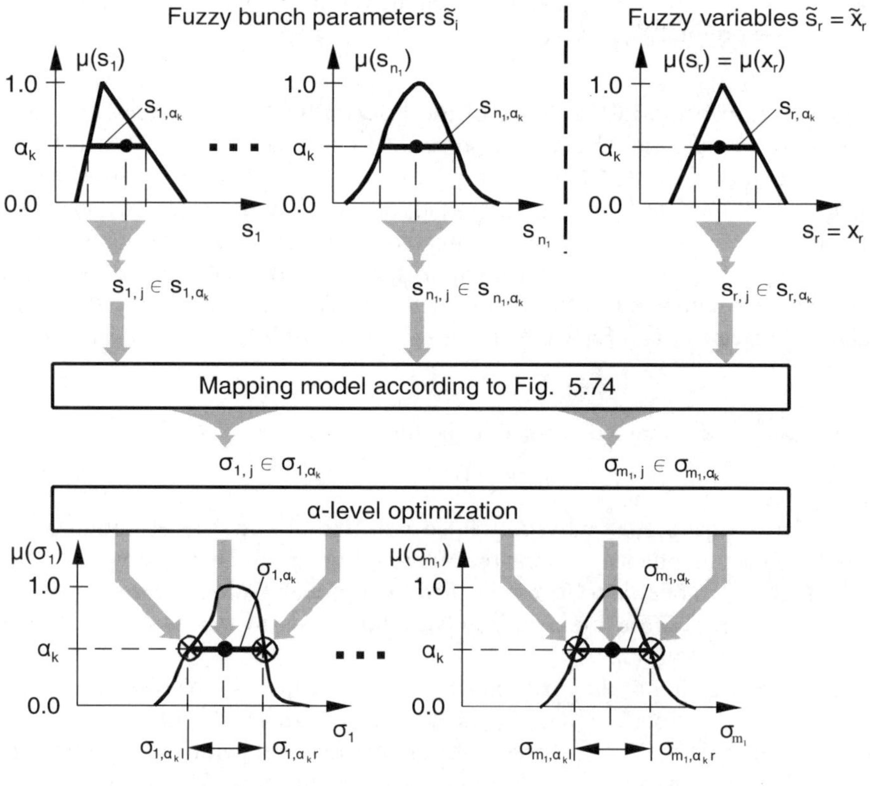

Fig. 5.73. α-level optimization and FSFEM

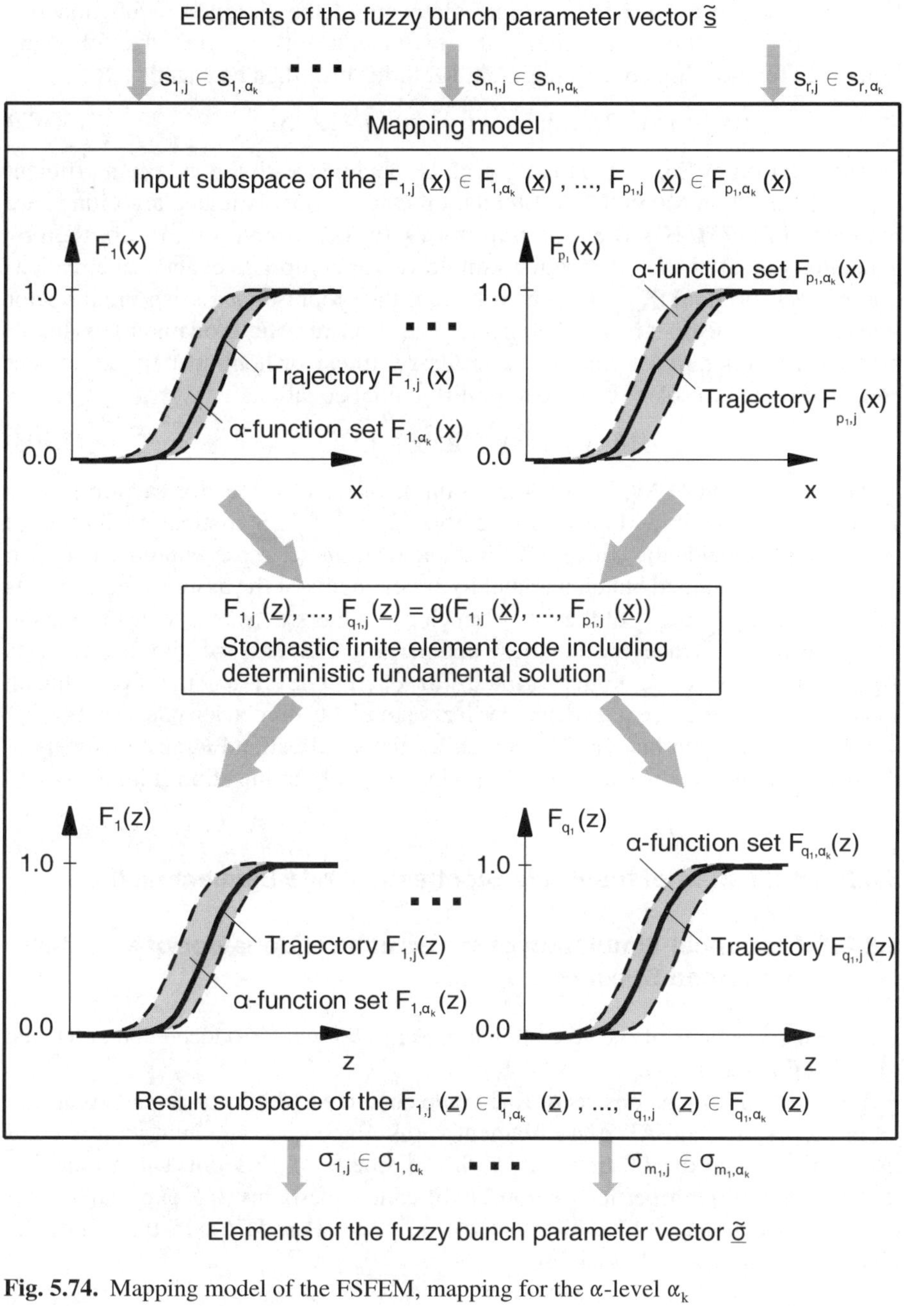

Fig. 5.74. Mapping model of the FSFEM, mapping for the α-level α_k

The trajectories of the fuzzy probability distribution functions $F(\underline{\tilde{\sigma}}_r, \underline{z}, \underline{\theta}_r)$ of the result vectors $\underline{\tilde{Z}}_{\theta_r} = \underline{\tilde{Z}}(\underline{\theta}_r) \mid r = 1, \ldots, q_1$ are designated by $F_{\theta_r, j}(\underline{z}) \mid r = 1, \ldots, p_1$.

For each defined α-level α_k these are elements of the assigned α-function sets $F_{\theta_r,j}(\underline{z}) \in F_{\theta_r,\alpha_k}(\underline{z})$. For determining the α-function sets $F_{\theta_r,\alpha_k}(\underline{z})$ the following functional relationship concerning the trajectories may then be stated:

$$\left(F_{\theta_r,j}(\underline{z}) \mid r = 1,\ldots,q_1\right) = g\left(F_{\theta_i,j}(\underline{x}) \mid i = 1,\ldots,p_1\right). \tag{5.107}$$

The solution of Eq. (5.107) may, e.g., be obtained with the aid of an efficient approach based on Monte Carlo Simulation (MCS). Sophisticated algorithms are presented in [173]. Based on the trajectories $F_{\theta_i,j}(\underline{x})$ sample vectors are thereby generated one after the other. Each sample vector comprises exactly one realization of each original $\underline{X}_{\theta_i,j}$, respectively, and thus represents a crisp input vector for one deterministic structural analysis. The deterministic fundamental solution required for this purpose may be an arbitrary, linear or nonlinear finite element code that is able to solve the system of differential equations

$$\underline{M} \cdot \ddot{\underline{v}} + \underline{D} \cdot \dot{\underline{v}} + \underline{K} \cdot \underline{v} = \underline{F}. \tag{5.108}$$

The application of MCS results in a sample of result values for each trajectory $F_{\theta_r,j}(\underline{z})$ of the fuzzy random result vectors $\tilde{\underline{Z}}_{\theta_r} = \tilde{\underline{Z}}(\underline{\theta}_r)$. Statistical evaluation of these samples yields the trajectories in bunch parameter representation. For each α-level α_k the obtained bunch parameters are elements of the assigned α-level sets $\sigma_1 \in \sigma_{1,\alpha_k},\ldots,\sigma_{m_1} \in \sigma_{m_1,\alpha_k}$ of the fuzzy bunch parameters $\tilde{\sigma}_1,\ldots,\tilde{\sigma}_{m_1}$ constituting the fuzzy vector $\tilde{\underline{\sigma}}$. Once the smallest and largest elements of the α-level sets $\sigma_{1,\alpha_k},\ldots,\sigma_{m_1,\alpha_k}$ have been determined for each α-level α_k, the fuzzy bunch parameter vectors $\tilde{\underline{\sigma}}_1,\ldots,\tilde{\underline{\sigma}}_{q_1}$ and hence the fuzzy probability distribution functions $\tilde{F}(\underline{z},\underline{\theta}_r) = F(\tilde{\underline{\sigma}}_r,\underline{z},\underline{\theta}_r)$ are then known. The search for the smallest and largest elements of the bunch parameters is realized by applying α-level optimization (Fig. 5.74).

5.3.2 Application of the Fuzzy Stochastic Finite Element Method

5.3.2.1 Numerical Simulation of the Structural Behavior of a Textile-reinforced Specimen

Eight test specimens of the type shown in Fig. 5.75 were produced and subjected to uniaxial loading [77].

Each test specimen consists of fine grade concrete and reinforcement comprised of 66 parallel-laid AR glass filament yarns. Each glass filament yarn is comprised of a bunch of 800 filaments with a diameter of 13.5 μm. Due to the fact that the material properties exhibit significant uncertainty the experiments for each test specimen yield a different σ – ε curve (Fig. 5.76). In the region IIa all curves exhibit a slight increase.

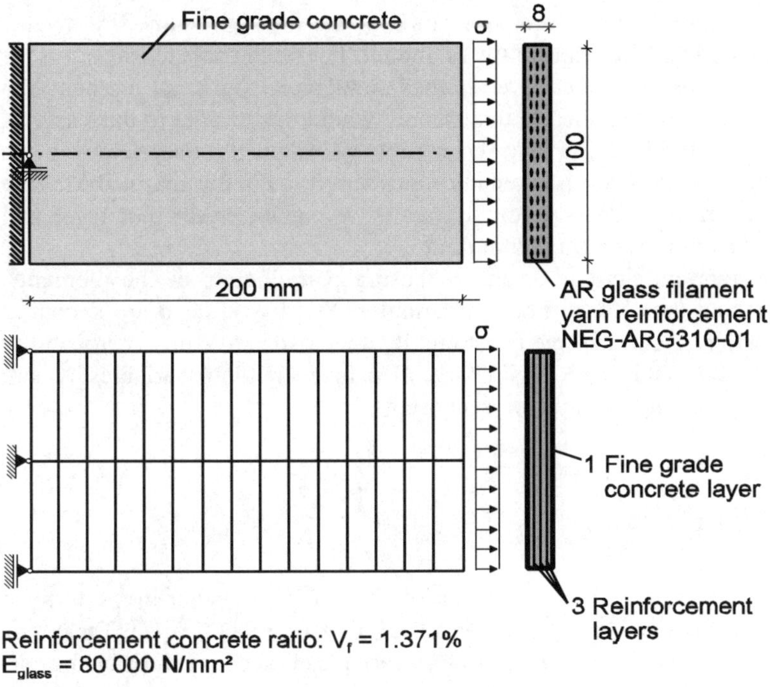

Fig. 5.75. Experimental specimen and finite element model

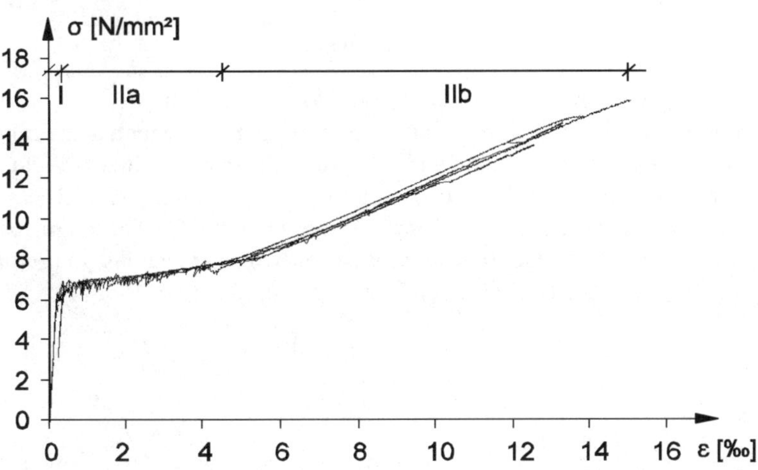

Fig. 5.76. Uncertain experimental stress-strain curve, test series B1-004 [77], NEG-ARG310-01, $V_f = 1.371\%$

If deterministic material parameters for the fine grade concrete and filament yarn are assumed in the numerical simulation of the tests, the same stress value occurs in region I in all elements of the FE model. The sustainable tensile stress for the fine grade concrete is therefore attained simultaneously in all elements. In region IIa the curve determined in this manner would run parallel to the ε axis. A realistic numerical simulation is only possible provided the uncertainty in the sensitive material parameters is taken into consideration. For this reason the tensile strength of the fine grade concrete is treated as an uncertain parameter and modeled as an isotropic fuzzy random field.

The fuzzy random field is discretized using p midpoints of the elements. Accordingly, correlated fuzzy random variables $\tilde{X}_i \mid i = 1, ..., p$ are known at p points. Owing to the assumed stationarity the associated fuzzy probability distribution functions $\tilde{F}_i(x) = F(\underline{\tilde{s}}, x)$ for all p fuzzy random variables $\tilde{X}_i$ are identical. The following formulation is chosen

$$F_i(\tilde{s}, x) = \frac{1}{\sqrt{2\pi}} \int_{-\infty}^{\frac{\ln x - \ln 6.268 + 0.5\tilde{\sigma}_u^2}{\tilde{\sigma}_u}} \exp\left(-\frac{u^2}{2}\right) du . \tag{5.109}$$

The standard deviation is introduced as a fuzzy bunch parameter in the form of the fuzzy triangular number $\tilde{\sigma}_u = \tilde{s}_1 = < 5.995,\ 6.991,\ 7.987 > \cdot 10^{-2}$. The correlation between the discretized fuzzy random variables is accounted for by the correlation function

$$\tilde{k}_X(L_{12}) = \begin{cases} 1 - \dfrac{L_{12}}{\tilde{L}_X} & \text{if } L_{12} \in [0, \tilde{L}_X] \\ 0 & \text{otherwise} \end{cases} , \tag{5.110}$$

with the fuzzy correlation length $\tilde{L}_X = \tilde{s}_2 = <100,\ 200,\ 300>$ mm.

Under consideration of nonlinear material properties the Fuzzy Stochastic Finite Element Method yields the $\sigma - \varepsilon$ relationship as a one-dimensional fuzzy random field for σ depending on prescribed ε. According to the definition Eq. (2.243) each functional value represents a fuzzy random variable (Fig. 5.77). The empirical fuzzy probability distribution functions are plotted for two strain values. In region IIa the numerically determined curves exhibit the required positive slope.

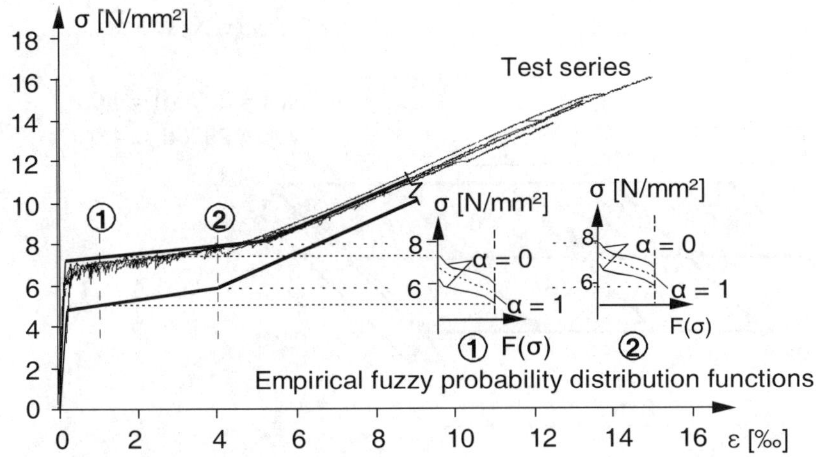

Fig. 5.77. Numerically determined $\sigma - \varepsilon$ curve as one-dimensional fuzzy random field in comparison with the test series B1-004 [77]

5.3.2.2 Reinforced-concrete Plate, Physically Nonlinear Analysis

The uniaxial reinforced-concrete plate (Fig. 5.78) is analyzed under consideration of data uncertainty and the governing nonlinearities of reinforced concrete [180]. A physically nonlinear deterministic fundamental solution on the basis of mixed hybrid finite elements with assumed stress distribution is used. Endochronic material laws are applied to the concrete and reinforcement steel. Tensile cracks in the concrete are accounted for in each element on a layer-to-layer basis according to the concept of smeared fixed cracks. Each finite element consists of 12 concrete layers (of 0.01 m thickness each) and two uniaxial smeared reinforcement layers.

The loading process (Fig. 5.79) consists of the deadload, a superficial load, and a nodal load P = 1 kN in the center of the plate.

Uncertain input parameters are both the time-dependent, superficial load and the concrete compressive strength in the whole plate. These are modeled as special fuzzy random functions.

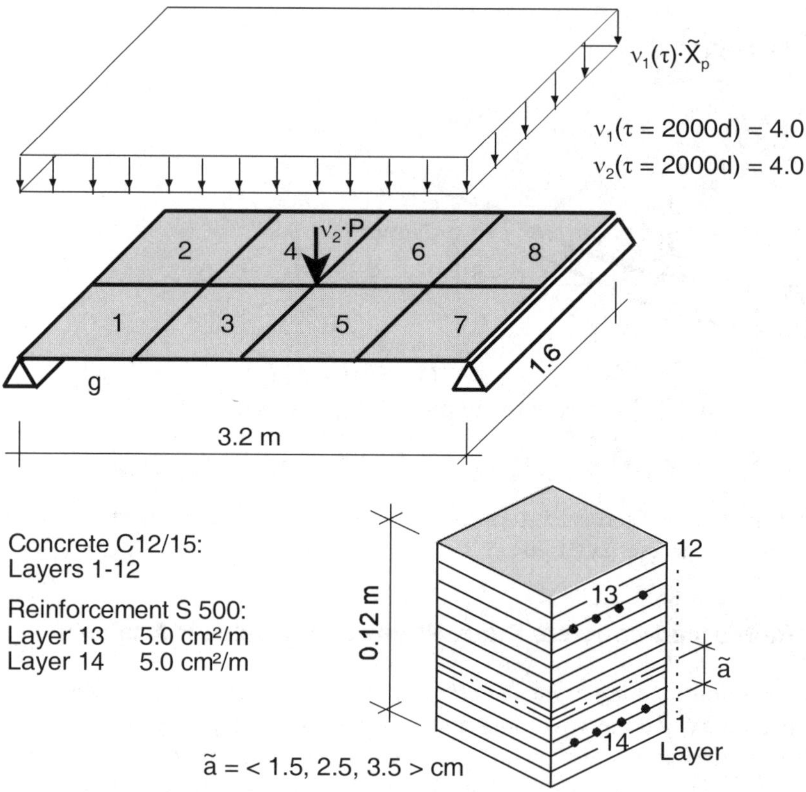

Fig. 5.78. Reinforced-concrete plate; geometry and finite element model

Investigation I – Fuzzy Random Crack State. The investigation aims at the determination of the fuzzy random crack state under service load. The fuzzy random superficial load is described with the aid of the product $\nu_1(\tau) \cdot \tilde{X}_p$ consisting of the crisp, time-dependent load factor $\nu_1(\tau)$ and the time-invariant fuzzy random variable $\tilde{X}_p$ with perfect correlation and full interaction in time. The latter is considered to be Gumbel distributed with the fuzzy expected value $\tilde{m}_X = E[\tilde{X}_p]$ (Fig. 5.79) and the crisp standard deviation $\sigma_X = 0.1$ kN/m². The associated fuzzy probability distribution function is formulated in bunch parameter representation

$$F(\tilde{\underline{s}}, x_p) = \exp\bigl(-\exp\bigl(-s_1\,(x_p - \tilde{s}_2)\bigr)\bigr), \tag{5.111}$$

with

$$s_1 = \frac{1.28255}{\sigma_X}, \tag{5.112}$$

and

$$\tilde{s}_2 = E(\tilde{X}_p) - \frac{0.577216}{s_1}. \tag{5.113}$$

The fuzzy expected value $E[\tilde{X}_p]$ is given in the form of the fuzzy triangular number

$$\tilde{m}_X = E(\tilde{X}_p) = <0.9,\ 1.0,\ 1.1>\ \mathrm{kN/m^2}\,. \tag{5.114}$$

The concrete compressive strength β_c is modeled as a perfectly correlated, Gaussian random field $\tilde{X}_\beta$ with the expected value $E[\tilde{X}_\beta] = 20\ \mathrm{N/mm^2}$ and a coefficient of variation $v = 0.10$. The concrete tensile strength β_t is calculated from the compressive strength with an endochronic concrete material law and under consideration of the strain velocity. Between the concrete compressive strength β_c and the concrete tensile strength β_t a relationship exists in the form

$$\beta_t = f(\beta_c)\,. \tag{5.115}$$

Therefore the tensile strength is also a random variable. Because of the perfect correlation the introduction of a random variable that mirrors the random fluctuations of all elements at once is sufficient.

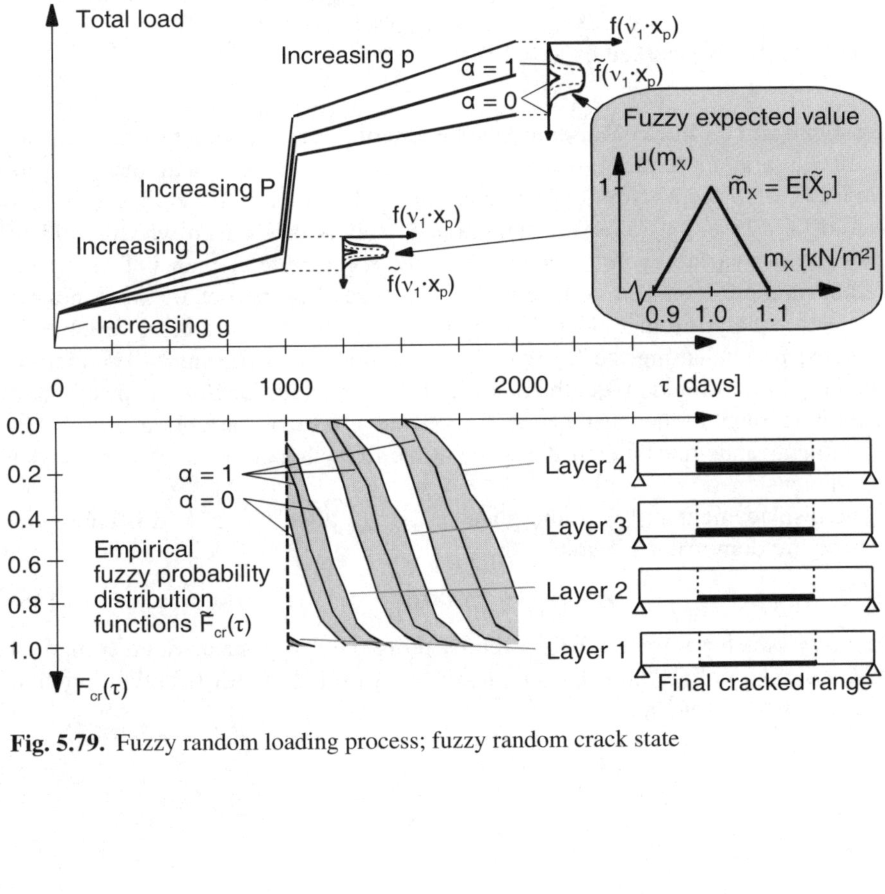

Fig. 5.79. Fuzzy random loading process; fuzzy random crack state

The time is increased incrementally up to τ = 2000 days. At each time increment the crack state is computed with the aid of the Fuzzy Stochastic Finite Element Method. The solution of Eq. (5.107) is obtained from Monte Carlo simulation. Each trajectory $F_{cr,\theta_l,j}(\tau)$ is computed on the basis of a sample consisting of 100 sample elements. The statistical evaluation yields the trajectories $F_{cr,\theta_l,j}(\tau)$ as empirical probability distribution functions and the assigned fuzzy expected value as fuzzy bunch parameter.

The empirical fuzzy probability distributions $\tilde{F}_{cr}(\tau)$ that indicate the fuzzy probability with which the first cracks occur in the concrete layers depending on time τ are shown in Fig. 5.79. The first cracks in layer 1 already occur when applying the nodal load. After 2000 days the lower three layers are cracked in one direction for all simulated realizations, whereas layer 4 still remained uncracked for some realizations.

In consequence of the multiplicative combination of $v_1(\tau)$ and $\tilde{X}_p$ the uncertainty of the loading process as well as the uncertainty of the crack state increases in time. This may be recognized by means of the growing distance between the empirical fuzzy probability distribution functions bounding the α-level $\alpha = 0$. A common probabilistic analysis, which does not consider the fuzziness of the expected value $\tilde{m}_X$, only yields one crisp empirical distribution function for each layer, respectively (marked by $\alpha = 1$ in Fig. 5.79).

Investigation II – Fuzzy Random Displacement. Uncertain input parameters are the fuzzy random loading process according to Fig. 5.79 and the concrete compressive strength. Now the latter is modeled as a partially correlated fuzzy random field in extension of investigation I. This is discretized with the aid of fuzzy random variables in the centroids of the eight finite elements. This discretization method is chosen with regard to the special properties of the nonlinear finite element algorithm applied. The correlation of the fuzzy random variables is computed by evaluating the fuzzy correlation function according to Eq. (5.104) with the fuzzy correlation length $\tilde{L}_X = \langle 2, 4, 10 \rangle$ m. Additionally, the position of the reinforcement in the tensile zone is modeled as a fuzzy variable $\tilde{a}$ (Fig. 5.78), i.e., without randomness. Thus the space of the fuzzy bunch parameters possesses three dimensions.

The displacement $v(\underline{\theta}_m)$ in the plate center $\underline{\theta}_m$ after τ = 2000 d is chosen as the fuzzy random result variable

$$\tilde{Z} = \left(\theta_m, \tau = 2000\,\mathrm{d}\right). \tag{5.116}$$

The fuzzy bunch parameter $\tilde{\bar{v}}(\underline{\theta}_m)$ representing the fuzzy mean value computed for $\tilde{Z}$ is shown in Fig. 5.80. The empirical fuzzy probability distribution function of $\tilde{Z}$ is plotted in Fig. 5.81.

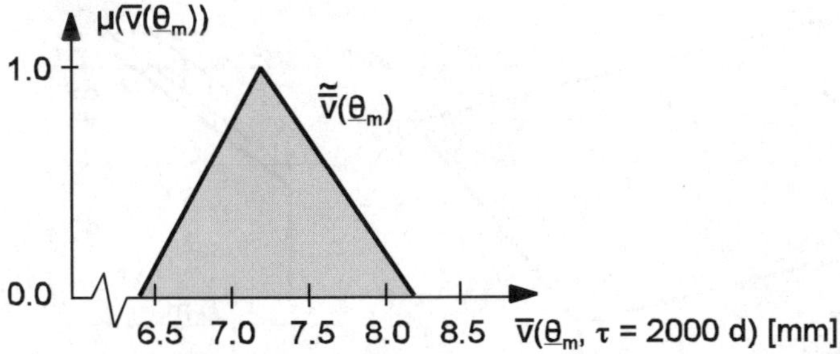

Fig. 5.80. Fuzzy mean value of the fuzzy random displacement

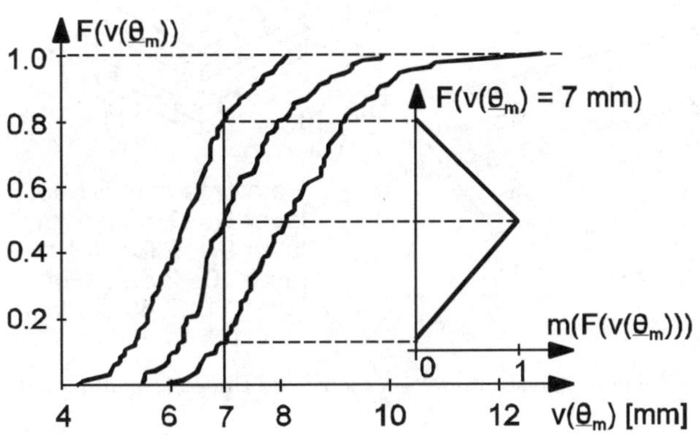

Fig. 5.81. Empirical fuzzy probability distribution function of the fuzzy random displacement

5.3.2.3 Reinforced-concrete Folded-plate Structure, Physically Nonlinear Analysis

The reinforced-concrete folded-plate structure shown in Fig. 5.82 is computed under consideration of the governing nonlinearities of reinforced concrete, of a random superficial load, and fuzzy randomness associated with the concrete compressive strength [123, 125] (Fig. 5.83).

The dimensions for every plane of the folded-plate structure in $\mathbb{R}^2$ are crisp. The system is meshed using 48 hybrid finite elements with assumed stress distribution. Each element, with an overall thickness of 0.1 m, was modeled using seven equidistant concrete layers, with two smeared mesh reinforcement layers each on the upper and lower surfaces (Fig. 5.83).

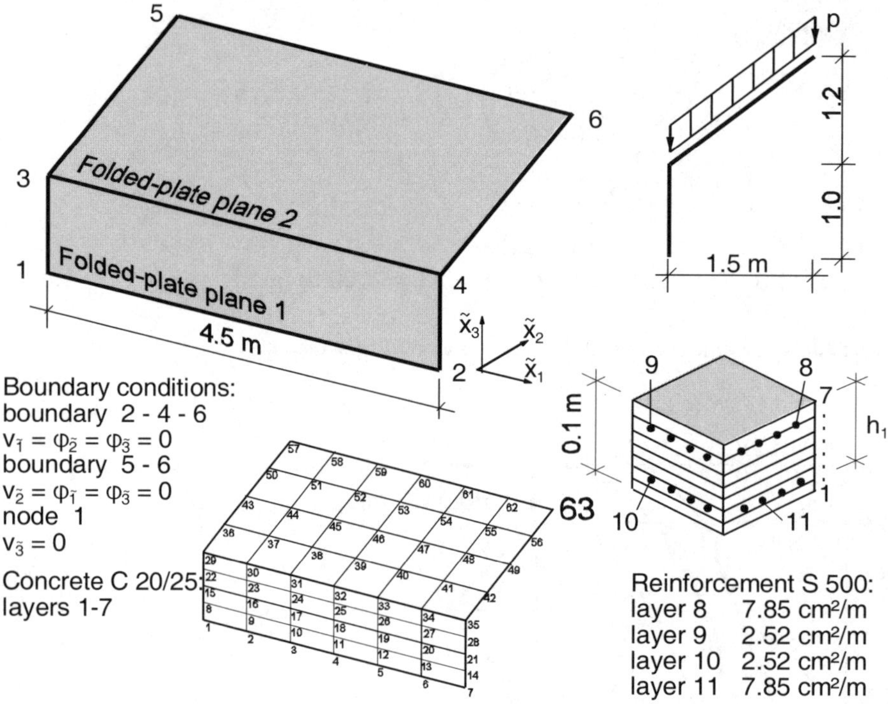

Fig. 5.82. Reinforced-concrete folded-plate structure; geometry and finite element model

A nonlinear concrete material law and a bilinear material law for reinforcement steel were applied. Tensile cracks in the concrete were accounted for in each element on a layer-to-layer basis according to the concept of smeared fixed cracks. The superficial load with randomness was increased incrementally up to the service load. Selected result values were computed at this stage of loading. Under consideration of the fuzzy randomness and randomness of the uncertain input variables, the result variables are also fuzzy random variables.

The fuzzy randomness of the concrete compressive strength is modeled using a correlated, isotropic fuzzy random field in $\mathbb{R}^2$ of the planes of the folded-plate structure. The correlation of the discretized variables of a folded-plate plane is taken into account by prescribing the fuzzy correlation function from Eq. (5.110) with the fuzzy correlation length $\tilde{L}_X$ shown in Fig. 5.83. One realization of the correlated fuzzy random field for the concrete compressive strength is illustrated graphically in Fig. 5.84. According to Eqs. (2.212) and (2.213) each realization is a fuzzy function. The discretized fuzzy random variables of different folded-plate planes are not correlated. The fuzzy expected value and the crisp standard deviation (Table 5.2) are the same for every discrete fuzzy random variable.

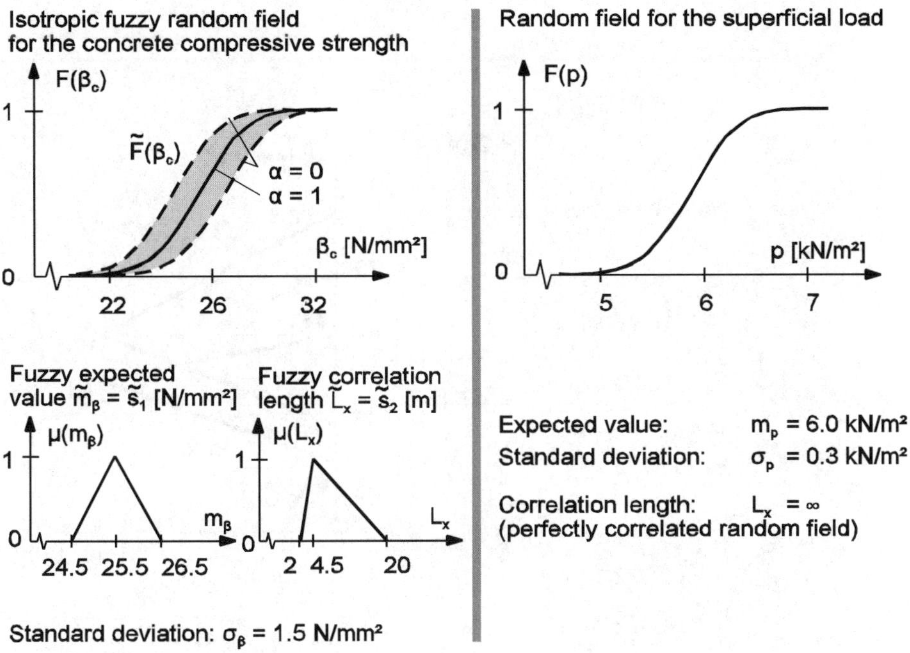

Fig. 5.83. Uncertain input variables

The fuzzy random field is discretized in the centroids of the finite elements. This discretization variant is chosen under consideration of the special properties of the applied nonlinear finite element algorithm. The principal stress criterion (crack criterion) is evaluated in the element centroid, i.e., the discretization point.

The following relationship between the tensile and compressive strength of concrete is adopted

$$\tilde{\beta}_t = 0.092 \cdot \tilde{\beta}_c \,. \tag{5.117}$$

Therefore a perfect correlation is assumed between the compressive strength and the tensile strength as well as between the strengths of the elements of the layers in a finite element.

The superficial load is modeled as a real-valued, perfectly correlated random field. The parameters are listed in Table 5.2.

For the numerical analysis with the Fuzzy Stochastic Finite Element Method the α-levels $\alpha = 0$ and $\alpha = 1$ are considered. The displacement v of node 63 in the direction of x_3 is selected as a representative result from the computation. The approximated membership function of the fuzzy mean value and the empirical fuzzy probability distribution function of the displacement are shown in Fig. 5.85.

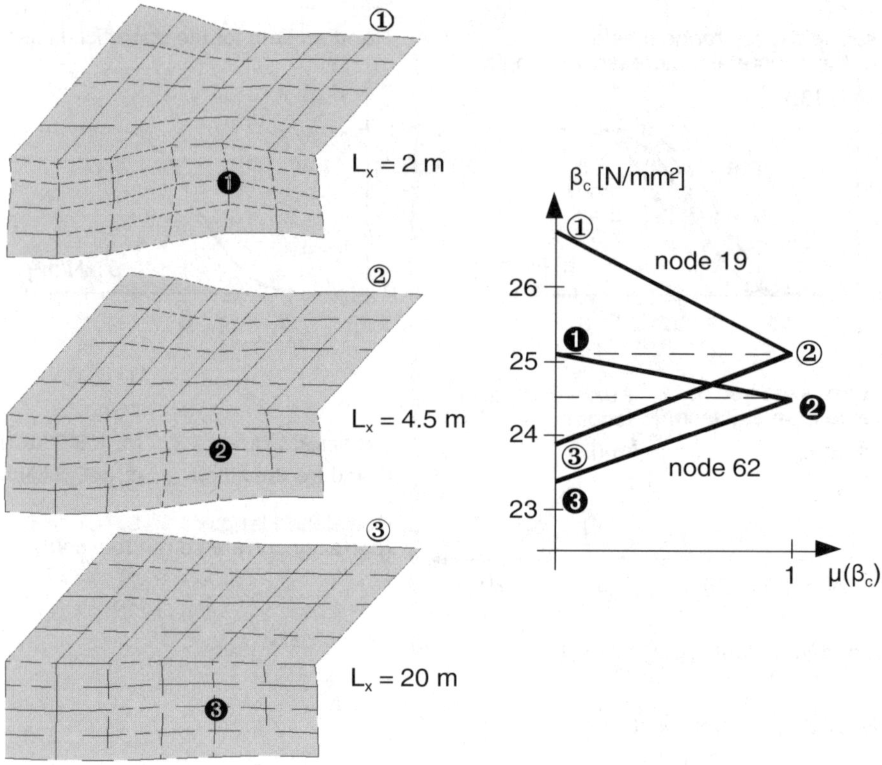

Fig. 5.84. Realization of the fuzzy random field for the concrete compressive strength

Table 5.2. Parameters of the (fuzzy) random fields

	Concrete compressive strength β_c [N/mm²]	Superficial load p [kN/m²]
Model	Fuzzy random function	Random function
Type of distribution	Normal distribution	Normal distribution
Expected value	< 24.5, 25.5, 26.5 >	6.0
Standard deviation	1.5	0.3
Correlation length	< 2.0, 4.5, 20.0 >	∞

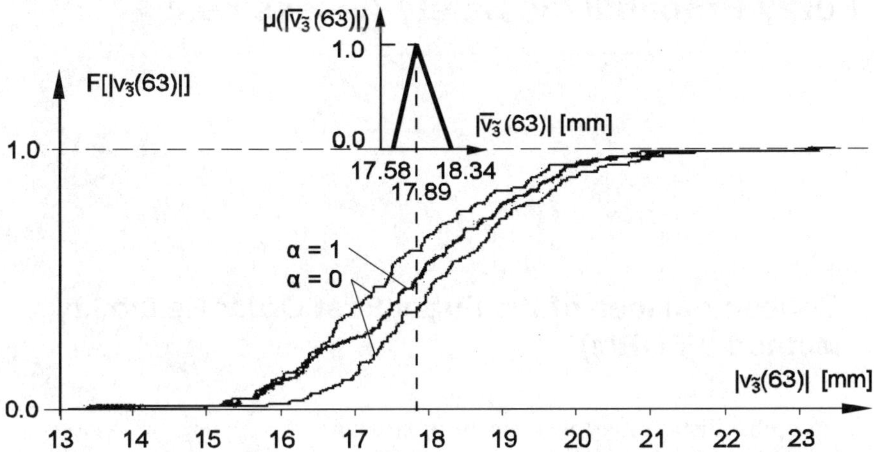

Fig. 5.85. Fuzzy mean value and empirical fuzzy probability distribution function of the displacement

6 Fuzzy Probabilistic Safety Assessment

6.1 Conceptual Idea of the Fuzzy First Order Reliability Method (FFORM)

In fuzzy probabilistic safety assessment of structures fuzzy random vectors are introduced for the mathematical description of the uncertainty characteristic fuzzy randomness. The failure and survival of a structure may be assessed with the aid of the probability measure for fuzzy random vectors, which has been introduced in Sect. 2.3.1.2; the fuzzy failure probability and fuzzy survival probability are to be determined.

As the fuzzy probability $\tilde{P}(\underline{A}_i)$ has been defined as being the set of the probabilities $P_j(\underline{A}_i)$ of all originals $\underline{X}_j$ of the fuzzy random vector $\tilde{\underline{X}}$, the failure probability and survival probability are to be calculated for each original. This is possible by applying algorithms of the First Order Reliability Method [107, 166, 182, 189, 190].

For each (one-dimensional) fuzzy random variable $\tilde{X}$ that is taken to be a fuzzy probabilistic basic variable a fuzzy probability distribution function $\tilde{F}(x)$ must be known. Since $\tilde{F}(x)$ represents the set of the probability distribution functions $F_j(x)$ (trajectories) of all originals X_j, a virtually infinite number of originals are to considered for the numerical solution. The parameters of the fuzzy probability distributions are thus represented by α-discretization and the governing originals are determined with the aid of α-level optimization.

Once one original from each fuzzy probabilistic basic variable is known, the assigned failure probability may be calculated, e.g., by means of the First Order Reliability Method. This yields one element P_f with the membership value $\mu(P_f)$ from the fuzzy set $\tilde{P}_f$ representing the fuzzy failure probability. The consideration of all originals of the fuzzy random variables requires the enhancement of the First Order Reliability Method to develop the Fuzzy First Order Reliability Method (FFORM) [115, 119, 124] (Fig. 6.1).

In FFORM, data uncertainty is accounted for with fuzzy random variables. With the α-level $\alpha = 1$ real random variables are included as a special case. The fuzziness of the fuzzy random variables leads to the joint fuzzy probability density function $\tilde{f}(\underline{x})$ in the original space of the fuzzy probabilistic basic variables.

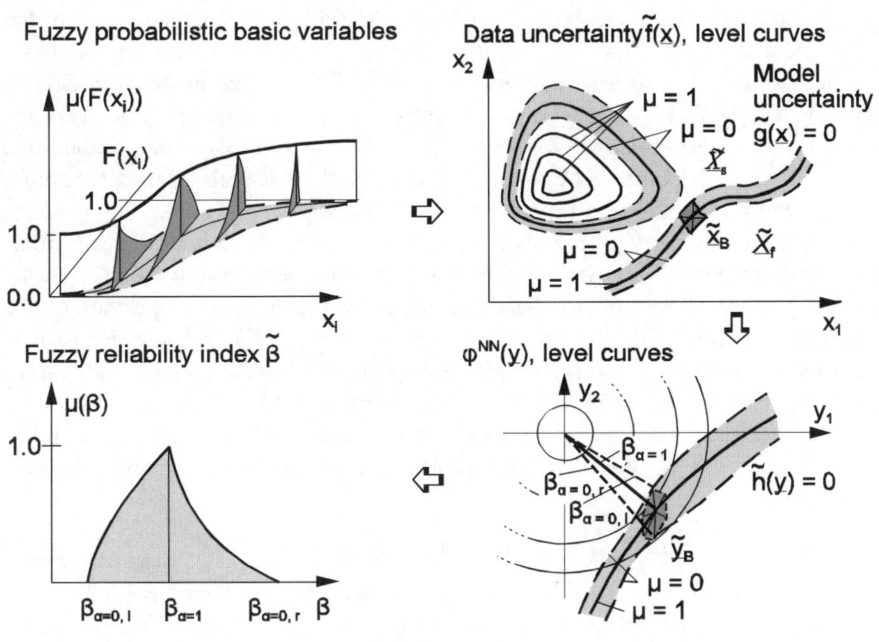

Fig. 6.1. Idea of the fuzzy probabilistic safety assessment with FFORM

The original space of the fuzzy probabilistic basic variables is subdivided into a survival region and a failure region by the limit state surface. The additional consideration of model uncertainty as fuzziness yields a fuzzy limit state surface separating a fuzzy survival and a fuzzy failure region. The fuzzy survival region is then described by the fuzzy set $\tilde{\underline{X}}_s$ and the fuzzy failure region by the fuzzy set $\tilde{\underline{X}}_f$.

By integrating the joint fuzzy probability density function $\tilde{f}(\underline{x})$ on the fuzzy failure region $\tilde{\underline{X}}_f$ the fuzzy failure probability $\tilde{P}_f$ may be computed. Based on Fuzzy First Order Reliability Method it is possible to determine a fuzzy design point $\tilde{\underline{x}}_B$ and a fuzzy reliability index $\tilde{\beta}$.

6.2 Original Space of the Fuzzy Probabilistic Basic Variables

6.2.1 Fuzzy Probabilistic Basic Variables and Joint Fuzzy Probability Density Function

The original space of the basic variables (x-space) is constituted by n fuzzy probabilistic basic variables $\tilde{X}_i$, which are represented in a Cartesian coordinate system with n axes x_i.

Each fuzzy probabilistic basic variable is described by its fuzzy probability density function $\tilde{f}_i(x_i)$. The parameters of the $\tilde{f}_i(x_i)$ are fuzzy numbers, and as a rule they are coupled by interactive dependencies. The fuzzy probability density functions $\tilde{f}(x_i)$ of all $\tilde{X}_i$ are lumped together on the original space to form the *joint fuzzy probability density function* $\tilde{f}(\underline{x})$, which assigns one fuzzy functional value $\tilde{f}(\underline{x})$ to each point in the x-space, respectively. The joint fuzzy probability density function comprises all combinations of trajectories $f_{i,j}(x_i)$ belonging to the originals $X_{i,j}$ of the fuzzy probabilistic basic variables $\tilde{X}_i$. Each combination yields one crisp joint probability density function $f(\underline{x})$. The membership values μ of the functional values $f(\underline{x}) \in \tilde{f}(\underline{x})$ are determined by the max-min operator of the extension principle. *The set of all trajectories (crisp joint probability density functions) established from the originals together with their membership values represents the joint fuzzy probability density function* $\tilde{f}(\underline{x})$.

Using the fuzzy numbers $\tilde{p}_t(\tilde{X}_i)$ for the parameters of the fuzzy probability distributions of the $\tilde{X}_i$, the joint fuzzy probability density function is the result of the mapping

$$\left\{\tilde{p}_t(\tilde{X}_i)\,;\, i = 1, \ldots, n\,;\, t = 1, \ldots, r_i\right\} \rightarrow \tilde{f}(\underline{x})\,. \qquad (6.1)$$

Owing to the close relation or partial equivalence between fuzzy parameters of fuzzy random vectors (Sect. 2.3.1.4) and fuzzy functional parameters of fuzzy probability distributions the denotation $\tilde{p}_t(\tilde{\underline{X}})$ is here used for both.

An example of a joint fuzzy probability density function $\tilde{f}(\underline{x})$ formed by two fuzzy random variables is shown in Fig. 6.2. Level curves belonging to constant functional values $\tilde{f}(\underline{x}) = c$ are plotted for the membership level $\mu(f(\underline{x})) = 1$. Additionally, the support, i.e., the bounding of the membership level $\mu(f(\underline{x})) = 0$ is illustrated for the lowest level curve.

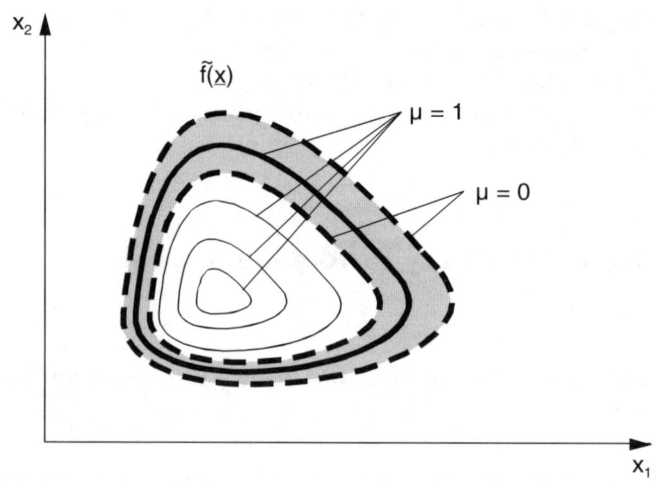

Fig. 6.2. Joint fuzzy probability density function $\tilde{f}(\underline{x})$, level curves for constant $\tilde{f}(\underline{x}) = c$

If all originals of different fuzzy probabilistic basic variables are stochastically independent of each other, the multiplication theorem of probability theory may be applied for determining the joint fuzzy probability density function

$$\tilde{f}(\underline{x}) = \tilde{f}_1(x_1) \cdot \tilde{f}_2(x_2) \cdot \ldots \cdot \tilde{f}_i(x_i) \cdot \ldots \cdot \tilde{f}_n(x_n) . \qquad (6.2)$$

This is to be carried out on an original-by-original basis.

Dependencies between the $\tilde{X}_i$ may be taken into account original-by-original in compliance with the treatment of real random variables, e.g., with the aid of a correlation matrix. Such dependencies are not considered in the following. These are not relevant for developing the new safety assessment method and may easily be included in the FFORM algorithm (Sect. 6.6).

Real-valued random variables together with fuzzy random variables may be introduced as basic variables, these are simultaneously accounted for. In contrast to fuzzy random variables, real random variables possess only one original. For the special case that all basic variables possess only one original, the joint fuzzy probability density function $\tilde{f}(\underline{x})$ reduces to the real-valued joint probability density function $f(\underline{x})$.

6.2.2 Fuzzy Limit State Surface

The limit state surface is specified by the computational model. Uncertain computational models with fuzzy model parameters (Sect. 1.1) result in a *fuzzy limit state surface* $\tilde{g}(\underline{x}) = 0$ in the original space of the basic variables. The space of the fuzzy probabilistic basic variables is subdivided into a fuzzy survival region $\tilde{\underline{X}}_s$ and a fuzzy failure region $\tilde{\underline{X}}_f$ by the fuzzy limit state surface. The fuzzy function $\tilde{g}(\underline{x}) = 0$ may be expressed in the form

$$\left(\tilde{g}(\underline{x}) = 0\right) = \left\{\left(g(\underline{x}) = 0, \mu(g(\underline{x}) = 0)\right) \mid \underline{x} \in \mathbf{X}\right\} . \qquad (6.3)$$

A fuzzy limit state surface $\tilde{g}(x_1, x_2) = 0$ on $\mathbb{R}^2$ is shown in Fig. 6.3. The $g(\underline{x}) = 0$ form a bunch of functions with the membership values $\mu(g(\underline{x}) = 0)$; these are elements of the fuzzy set $\tilde{g}(\underline{x}) = 0$ and represent crisp limit state surfaces.

In order to compute the elements $g(\underline{x}) = 0$ it is necessary to discretize the fuzzy model parameters, i.e., selection of an α-level and selection of elements from the α-level sets. By this means, possible values of the fuzzy model parameters are defined. These values serve as input data to a (nonlinear) analysis algorithm with which the crisp limit state surface $g(\underline{x}) = 0$ may be computed. The respective analysis algorithm is referred to as the deterministic fundamental solution. The quality of the deterministic fundamental solution has a decisive influence on the results of the safety assessment; thus the system behavior of the structure has to be realistically, numerically simulated.

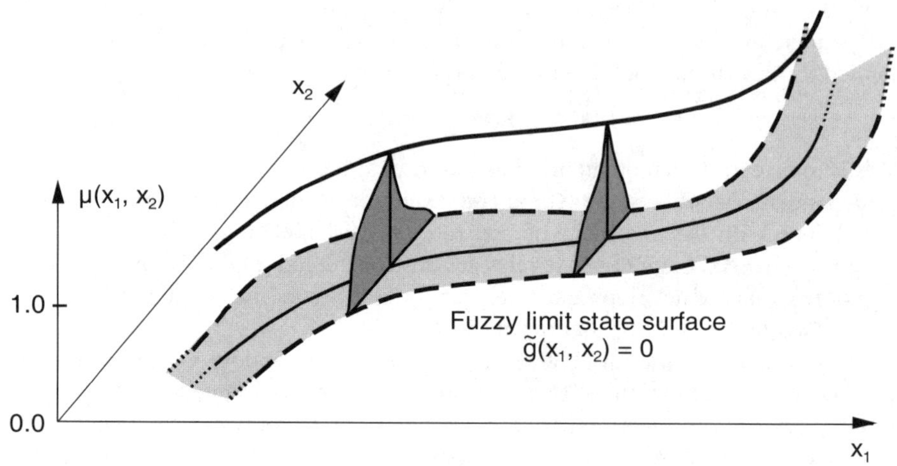

Fig. 6.3. Fuzzy limit state surface $\tilde{g}(\underline{x}) = 0$ for two basic variables

In general, a sophisticated nonlinear structural analysis algorithm is applied as deterministic fundamental solution, like, e.g., the numerical simulation of the behavior of plane reinforced-concrete bar structures presented in [116] and [133]. Each single application of the fundamental solution yields one crisp limit state point. The computation of several crisp limit state points (for the same, retained values from the fuzzy model parameters) leads to *one* numerically approximated crisp limit state surface, which represents *one* element $g(\underline{x}) = 0$ of the fuzzy set $\tilde{g}(\underline{x}) = 0$.

The assessment of the points $\underline{x}$ in the space of the fuzzy probabilistic basic variables regarding failure and survival is carried out using membership functions. The fuzzy survival region $\tilde{\underline{X}}_s$ is characterized by the membership function

$$\mu(\underline{x}_s) = \mu(g(\underline{x}) > 0) = \begin{cases} 1 & \forall\, \underline{x} \mid g(\underline{x})_{\alpha=1} > 0 \\ \mu(g(\underline{x}) = 0 + \delta) \quad \text{otherwise} & \mid \delta \to +0 \end{cases}, \tag{6.4}$$

and the membership function of the fuzzy failure region $\tilde{\underline{X}}_f$ is given by

$$\mu(\underline{x}_f) = \mu(g(\underline{x}) \le 0) = \begin{cases} 1 & \forall\, \underline{x} \mid g(\underline{x})_{\alpha=1} \le 0 \\ \mu(g(\underline{x}) = 0) & \text{otherwise} \end{cases}. \tag{6.5}$$

The membership functions determined by Eqs. (6.4) and (6.5) for assessing failure and survival are plotted in Fig. 6.4 for the two-dimensional case.

The membership values $\mu(\underline{x}_s)$ and $\mu(\underline{x}_f)$ assess the propositions $\underline{x} \in \tilde{\underline{X}}_s$ and $\underline{x} \in \tilde{\underline{X}}_f$ for all points $\underline{x} \in \mathbf{X}$. The survival and failure regions overlap in the region of the fuzzy limit state surface $\tilde{g}(\underline{x}) = 0$.

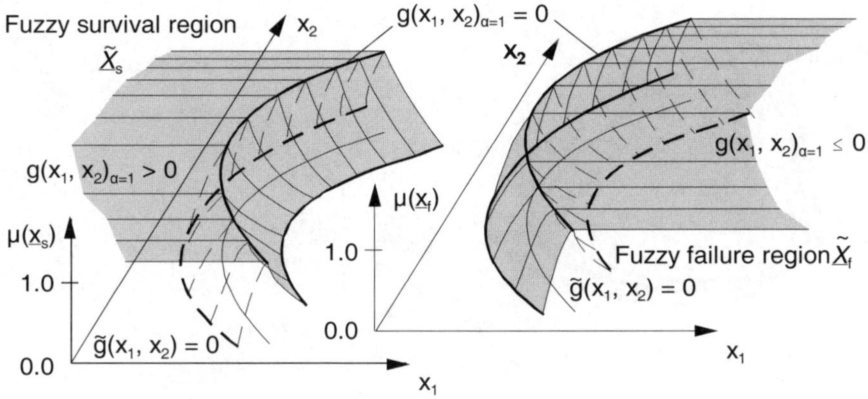

Fig. 6.4. Fuzzy limit state surface $\tilde{g}(\underline{x}) = 0$, fuzzy survival region $\underline{\tilde{X}}_s$, and fuzzy failure region $\underline{\tilde{X}}_f$

6.2.3 Fuzzy Design Point

In the original space of the basic variables the joint fuzzy probability density function $\tilde{f}(\underline{x})$ is plotted together with the fuzzy limit state surface $\tilde{g}(\underline{x}) = 0$. The sought fuzzy failure probability $\tilde{P}_f$ is obtained by integrating $\tilde{f}(\underline{x})$ over the fuzzy failure region $\underline{\tilde{X}}_f$ with $\tilde{g}(\underline{x}) \leq 0$. Analogous to the First Order Reliability Method, this integration is carried out in the space of the standard normal distribution (y-space) and leads to the fuzzy design point and fuzzy reliability index.

The search is for that point on the limit state surface for which the joint fuzzy probability density function takes its maximum functional value, this is the most probable failure point. The fuzzy design point $\underline{\tilde{x}}_B$ indicates that combination of values of the basic variables $\tilde{X}_i$ that leads to failure with the highest probability. The influence of the individual basic variables on the failure event is recognizable. The assigned failure mode is the governing one. The fuzzy design point $\underline{\tilde{x}}_B$ may also be used for the determination of partial safety factors.

Different combinations of uncertain input and model parameters result in various shapes of the fuzzy design point $\underline{\tilde{x}}_B$.

Case I. If no model uncertainty (i.e., only data uncertainty) enters the investigation, the fuzzy design point then lies on a crisp limit state surface $g(\underline{x}) = 0$. Uncertainty with the characteristic fuzziness exists exclusively in the joint fuzzy probability density function $\tilde{f}(\underline{x})$. This fuzziness becomes visible in the fuzzy design point. It may only appear in the direction of the limit state surface, i.e., $\underline{\tilde{x}}_B$ does not possess fuzziness perpendicular to $g(\underline{x}) = 0$. This situation is shown in Fig. 6.5 for two fuzzy probabilistic basic variables. The fuzzy design point $\underline{\tilde{x}}_B$ is plotted depending on the coordinate s, which runs along $g(\underline{x}) = 0$.

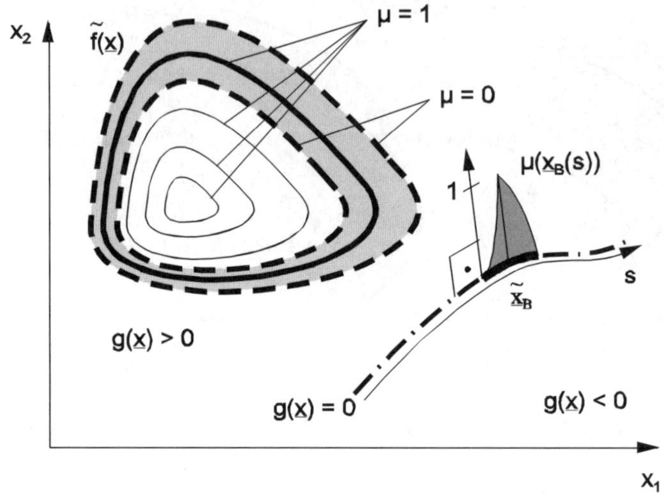

Fig. 6.5. Fuzzy design point $\tilde{\underline{x}}_B$ on a crisp limit state surface $g(\underline{x}) = 0$, no model uncertainty

For $g(\underline{x}) = 0$ each element $f(\underline{x})$ of the joint fuzzy probability density function, i.e., each combination of originals of the n fuzzy probabilistic basic variables $\tilde{X}_i$ yields one (crisp) element $\underline{x}_B$ of the fuzzy design point with the membership value $\mu(\underline{x}_B)$. These elements together with their membership values constitute the fuzzy set $\tilde{\underline{x}}_B$ representing the fuzzy design point. The fuzzy design point is the result of the mapping of the r_i fuzzy parameters $\tilde{p}_t$ of the fuzzy probability distributions of the $\tilde{X}_i$ onto the coordinates of the design point

$$\left\{\tilde{p}_t(\tilde{X}_i); i = 1, ..., n; t = 1, ..., r_i\right\} \to \tilde{\underline{x}}_B . \quad (6.6)$$

Case II. Model uncertainty with the characteristic fuzziness leads to the fuzzy limit state surface $\tilde{g}(\underline{x}) = 0$; the fuzzy design point $\tilde{\underline{x}}_B$, in general, additionally exhibits uncertainty perpendicular to the limit state surface (Fig. 6.6). This fuzziness is described by means of the coordinate t. For each element $\underline{x}_B$ of the fuzzy design point $\tilde{\underline{x}}_B$ the coordinate t runs in the direction of the gradient of that element $f(\underline{x})$ of the joint fuzzy probability density function $\tilde{f}(\underline{x})$ that belongs to $\underline{x}_B$. The coordinate s runs along the assigned element $g(\underline{x}) = 0$ of the fuzzy limit state surface $\tilde{g}(\underline{x}) = 0$; s and t are perpendicular to one another. As a rule, model uncertainty also causes a portion of the fuzziness of $\tilde{\underline{x}}_B$ in the direction of s. Only in special cases may the fuzziness of $\tilde{\underline{x}}_B$ be separated into data uncertainty in direction s and model uncertainty in direction t.

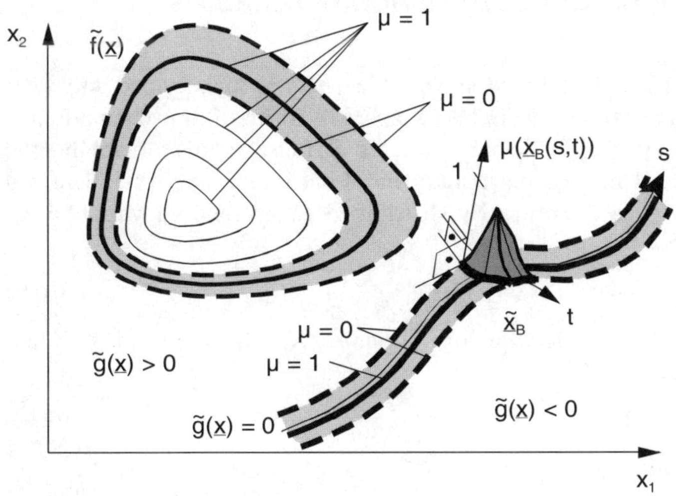

Fig. 6.6. Fuzzy design point $\tilde{\underline{x}}_B$ on a fuzzy limit state surface $\tilde{g}(\underline{x}) = 0$ caused by model uncertainty

If model uncertainty appears, the mapping according to Eq. (6.6) is to be extended by taking account of q additional fuzzy model parameters $\tilde{m}_j$

$$\left\{ \tilde{p}_t(\tilde{X}_i), \tilde{m}_j ; i = 1, ..., n ; t = 1, ..., r_i ; j = 1, ..., q \right\} \rightarrow \tilde{\underline{x}}_B . \tag{6.7}$$

Each combination of elements of the fuzzy model parameters $\tilde{m}_j$ yields precisely one element of the fuzzy limit state surface $\tilde{g}(\underline{x}) = 0$ and each combination of elements of the fuzzy parameters $\tilde{p}_t(\tilde{X}_i)$ yields precisely one element (trajectory) of the joint fuzzy probability density function $\tilde{f}(\underline{x})$. *The evaluation of all combinations of elements of* $\tilde{g}(\underline{x}) = 0$ *and the originals of the* $\tilde{X}_i$ *(i.e., the trajectories* $f(\underline{x})$ *of* $\tilde{f}(\underline{x})$*) yields the elements of the fuzzy design point* $\tilde{\underline{x}}_B$ *with the membership values* $\mu(\underline{x}_B)$.

Case III. If data uncertainty may be described by real random variables alone, i.e., the joint fuzzy probability density function possesses only one element, the mapping in Eq. (6.7) then reduces to

$$\left\{ \tilde{m}_j ; j = 1, ..., q \right\} \rightarrow \tilde{\underline{x}}_B . \tag{6.8}$$

Although no fuzzy random variables enter the investigation in this case, the computation of the fuzzy design point may be executed as a special case of the procedure described.

For all cases discussed, interaction between all involved fuzzy variables is to be accounted for when determining $\tilde{\underline{x}}_B$.

6.3 Transformation of Fuzzy Random Variables

Fuzzy random variables, like real random variables, may be transformed into other (fuzzy) random variables. For FFORM a special form of transformation is necessary: every fuzzy probabilistic basic variable $\tilde{X}_i$ must be transformed into the standard normal space. This requires assignment of one (real-valued) standardized random variable Y_i to each arbitrarily distributed fuzzy random variable $\tilde{X}_i$, respectively,

$$\tilde{X}_i \rightarrow Y_i . \tag{6.9}$$

When considering real-valued random variables X, the values $x \in \mathbf{X}$ are mapped onto the values $y \in \mathbf{Y}$

$$x \rightarrow y . \tag{6.10}$$

The transformation function

$$y = t(x) \tag{6.11}$$

describes the relationship between x and y, and t(x) is required to be differentiable and monotonic. The mapping in Eq. (6.10) is reversible, the assignment

$$x \leftrightarrow y \tag{6.12}$$

is biunique. Regarding the transformation

$$X \rightarrow Y \tag{6.13}$$

the following holds for the functional values of the probability distribution functions of X and Y

$$F(x) = \Phi^{NN}(y) . \tag{6.14}$$

With the aid of Eq. (6.14) the transformation function t(x) may also be designated by

$$y = \Phi^{NN^{-1}}(F(x)) \tag{6.15}$$

with $\Phi^{NN^{-1}}$ being the inverse function of the probability distribution function Φ^{NN} of the standard normal distribution.

For considering fuzzy random variables $\tilde{X}$ as fuzzy sets of real-valued random variables X Eqs. (6.10) to (6.15) are enhanced.

The fuzzy random variable $\tilde{X}_i$ has the fuzzy probability distribution function $\tilde{F}(x)$; the standard normal distribution $\Phi^{NN}(y)$ for the new random variable Y_i is defined as a crisp probability distribution function, however. For this reason, the transformation relationship between $\tilde{X}_i$ and Y_i contains uncertainty with the characteristic fuzziness. The uncertain transformation of crisp values $x \in \mathbf{X}$ leads to fuzzy variables $\tilde{y} \in \mathbf{Y}$

$$x \tilde{\rightarrow} \tilde{y} . \tag{6.16}$$

Hence the transformation function is a fuzzy function

$$\tilde{y} = \tilde{t}(x)\,. \tag{6.17}$$

With the aid of the distribution functions for $\tilde{X}$ and Y the fuzzy transformation function $\tilde{t}(x)$ may be denoted by (Fig. 6.7)

$$\tilde{y} = \Phi^{NN^{-1}}(\tilde{F}(x))\,. \tag{6.18}$$

The transformation of the fuzzy random variable $\tilde{X}$ into the standardized random variable Y is realized by transforming all originals of $\tilde{X}$. Each original is mapped onto Y by means of the transformation function according to Eq. (6.11) or (6.15). The set of the transformation functions for all originals together with the membership values of these originals constitutes the fuzzy transformation function $\tilde{t}(x)$.

Differentiability and monotonicity are complied with for each element t(x) of the fuzzy transformation function $\tilde{t}(x)$. Though biuniqueness of the mapping is ensured for the individual originals, it does not apply to the fuzzy transformation according to Eq. (6.16). The inverse transformation of a crisp value $y \in \mathbf{Y}$, which may also represent an element of a fuzzy variable $\tilde{y}$, may only be carried out if the assigned original is known. Otherwise, starting from a value y or $\tilde{y}$ this leads to a fuzzy value $\tilde{x}$ containing *all possible* values x that might belong to y or $\tilde{y}$.

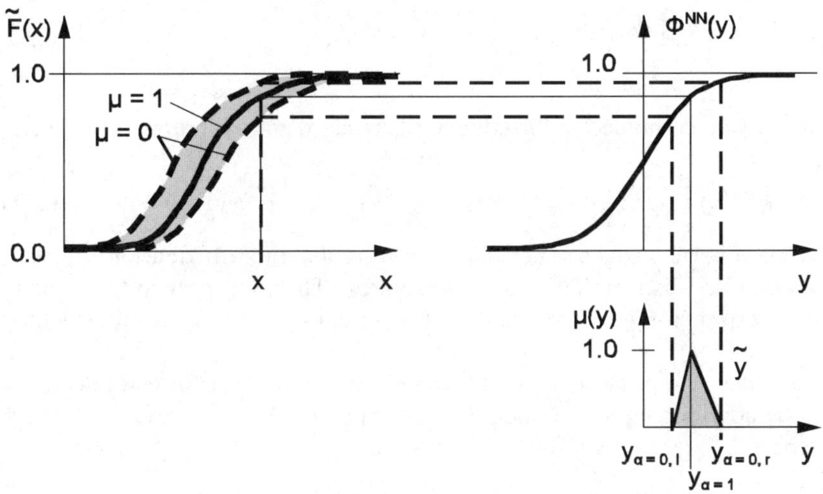

Fig. 6.7. Transformation of fuzzy random variables into standardized random variables

6.4 Standard Normal Space of the Normalized Basic Variables

The standard normal space (y-space) is constituted by n standardized basic variables Y_i. These result from the transformation of the n fuzzy probabilistic basic - variables $\tilde{X}_i$ according to Eq. (6.9). The Y_i are represented in a Cartesian coordinate system with n axes y_i. At each point $\underline{y}$ the standard normal joint probability density function $\varphi^{NN}(\underline{y})$ possesses a crisp functional value.

6.4.1 Standard Normal Joint Probability Density Function and Fuzzy Limit State Surface

The transformation of the fuzzy probabilistic basic variables $\tilde{X}_i$ with Eq. (6.9) leads to the standardized basic variables Y_i with the probability density functions

$$\varphi^{NN}(y_i) = \varphi^{NN}(y) = \frac{1}{\sqrt{2\pi}} e^{-\frac{1}{2}y^2} , \qquad (6.19)$$

and the probability distribution functions

$$\Phi^{NN}(y_i) = \Phi^{NN}(y) = \int_{t=-\infty}^{t=y} \varphi^{NN}(t)\,dt . \qquad (6.20)$$

The $\varphi^{NN}(y_i)$ are combined to form the joint standard normal probability density function

$$\varphi^{NN}(\underline{y}) = \varphi^{NN}(y_1) \cdot \varphi^{NN}(y_2) \cdot \ldots \cdot \varphi^{NN}(y_i) \cdot \ldots \cdot \varphi^{NN}(y_n) \qquad (6.21)$$

in the standard normal space. The standardized probability distribution is a real-valued probability distribution without fuzziness. The joint probability density function is shown in Fig. 6.8 by means of level curves for the two-dimensional case.

The coordinates x_i of the points $\underline{x}$ from the x-space are transformed into fuzzy values $\tilde{y}_i$ by applying Eq. (6.16), i.e., each crisp point $\underline{x}$ becomes a fuzzy point $\tilde{\underline{y}}$ in the y-space

$$\underline{x} \rightrightarrows \tilde{\underline{y}} . \qquad (6.22)$$

The application of the transformation according to Eq. (6.22), in particular to the limit state points $\underline{x}$ or $\tilde{\underline{x}}$, always leads to fuzzy limit state points $\tilde{\underline{y}}$ in the standard normal space consequently. For this reason, the limit state surface in y-space is always obtained as fuzzy limit state surface $\tilde{h}(\underline{y}) = 0$ (Fig. 6.8). Even a crisp limit state surface $g(\underline{x}) = 0$ becomes a fuzzy limit state surface $\tilde{h}(\underline{y}) = 0$ in the y-space

$$g(\underline{x}) = 0 \rightrightarrows \tilde{h}(\underline{y}) = 0 , \qquad (6.23)$$

$$\tilde{g}(\underline{x}) = 0 \rightarrow \tilde{h}(\underline{y}) = 0\,. \tag{6.24}$$

In the standard normal space the survival region is obtained as fuzzy set $\tilde{\underline{Y}}_s$ of all points $\underline{y}$ for which $\tilde{h}(\underline{y}) > 0$ applies. The failure region $\tilde{\underline{Y}}_f$ is characterized by $\tilde{h}(\underline{y}) \leq 0$. Regarding the membership functions of $\tilde{\underline{Y}}_s$ und $\tilde{\underline{Y}}_f$ the following holds in compliance with Eqs. (6.4) and (6.5)

$$\mu(\underline{y}_s) = \mu(h(\underline{y}) > 0) = \begin{cases} 1 & \forall\, \underline{x} \mid h(\underline{y})_{\alpha=1} > 0 \\ \mu(h(\underline{y}) = 0 + \delta) & \text{otherwise} \mid \delta \rightarrow +0 \end{cases}, \tag{6.25}$$

and

$$\mu(\underline{y}_f) = \mu(h(\underline{y}) \leq 0) = \begin{cases} 1 & \forall\, \underline{y} \mid h(\underline{y})_{\alpha=1} \leq 0 \\ \mu(h(\underline{y}) = 0) & \text{otherwise} \end{cases}. \tag{6.26}$$

These assess the propositions $\underline{y} \in \tilde{\underline{Y}}_s$ and $\underline{y} \in \tilde{\underline{Y}}_f$ for all points $\underline{y} \in \underline{\mathbf{Y}}$ of the standard normal space. The fuzzy limit state surface $\tilde{h}(\underline{y}) = 0$ represents the overlapping region of $\tilde{\underline{Y}}_s$ and $\tilde{\underline{Y}}_f$.

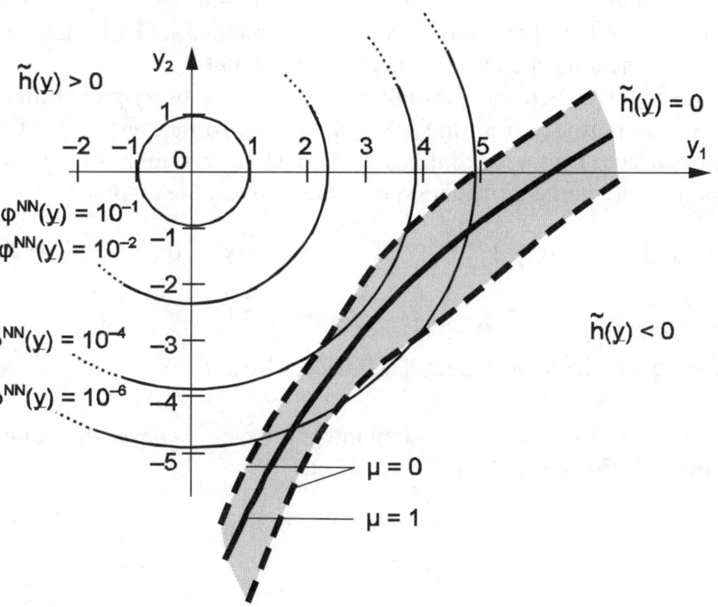

Fig. 6.8. Joint standard normal probability density function $\varphi^{NN}(\underline{y})$ (level curves) and fuzzy limit state surface $\tilde{h}(\underline{y}) = 0$

In the y-space the uncertainty appears separated according to its characteristics. The joint standard normal distribution exclusively describes uncertainty with the characteristic randomness, whereas the fuzzy limit state surface $\tilde{h}(\underline{y}) = 0$ only possesses uncertainty with the characteristic fuzziness. In the transformation from the x-space into the y-space the uncertainty of the structural parameters (data and model uncertainty) is separated according to the characteristics fuzziness and randomness.

The randomness of the fuzzy probabilistic basic variables $\tilde{X}_i$ is completely transferred to the standardized basic variables Y_i, i.e., to the joint standard normal distribution, with the transformation from Eq. (6.9). The function $\varphi^{NN}(\underline{y})$ only describes data uncertainty.

The whole fuzziness of the $\tilde{X}_i$ moves to the coordinates of all points transformed into the y-space, i.e., to limit state points, as may be recognized from Eq. (6.18). When transforming the crisp limit state surface $g(\underline{x}) = 0$ according to Eq. (6.23), each element (trajectory) of the joint fuzzy probability density function $\tilde{f}(\underline{x})$ yields precisely one element $h(\underline{y}) = 0$ of the fuzzy limit state surface $\tilde{h}(\underline{y}) = 0$. The fuzzy limit state surface $\tilde{h}(\underline{y}) = 0$ possesses data uncertainty.

The model uncertainty in the fuzzy limit state surface $\tilde{g}(\underline{x}) = 0$ remains in the limit state when transformed according to Eq. (6.24) and reappears in $\tilde{h}(\underline{y}) = 0$. In this mapping all elements of $\tilde{g}(\underline{x}) = 0$ are transformed into the y-space with Eq. (6.23), i.e., each element $g(\underline{x}) = 0$ of $\tilde{g}(\underline{x}) = 0$ is combined with all elements (trajectories) of $\tilde{f}(\underline{x})$ one after the other in this transformation.

The fuzziness in $\tilde{h}(\underline{y}) = 0$ therefore partially results from the fuzzy probabilistic basic variables $\tilde{X}_i$ and partially from the fuzzy limit state surface $\tilde{g}(\underline{x}) = 0$. The fuzzy limit state surface $\tilde{h}(\underline{y}) = 0$ contains both data uncertainty and model uncertainty with the characteristic fuzziness, it is the result of the mapping

$$\left\{\tilde{p}_t(\tilde{X}_i), \tilde{m}_j ; i = 1,...,n ; t = 1,...,r_i ; j = 1,...,q\right\} \rightarrow \tilde{h}(\underline{y}) = 0 . \tag{6.27}$$

6.4.2 Fuzzy Design Point and Fuzzy Reliability Index

The fuzzy failure probability $\tilde{P}_f$ is obtained by integrating $\varphi^{NN}(\underline{y})$ over the failure region $\tilde{\underline{Y}}_f$ with $\tilde{h}(\underline{y}) \leq 0$ in the standard normal space

$$\tilde{P}_f = \tilde{P}(\underline{y} \in \tilde{\underline{Y}}_f ; \underline{y} \notin \tilde{\underline{Y}}_s) = \int\limits_{\underline{y} \in \tilde{\underline{Y}}_f ; \underline{y} \notin \tilde{\underline{Y}}_s} \cdots \int \cdots \int \varphi^{NN}(\underline{y}) \, d\underline{y} . \tag{6.28}$$

In the framework of the First Order Reliability Method Eq. (6.28) is not directly evaluated, instead of this the fuzzy design point $\tilde{\underline{y}}_B$ and fuzzy reliability index $\tilde{\beta}$ are determined (Fig. 6.9).

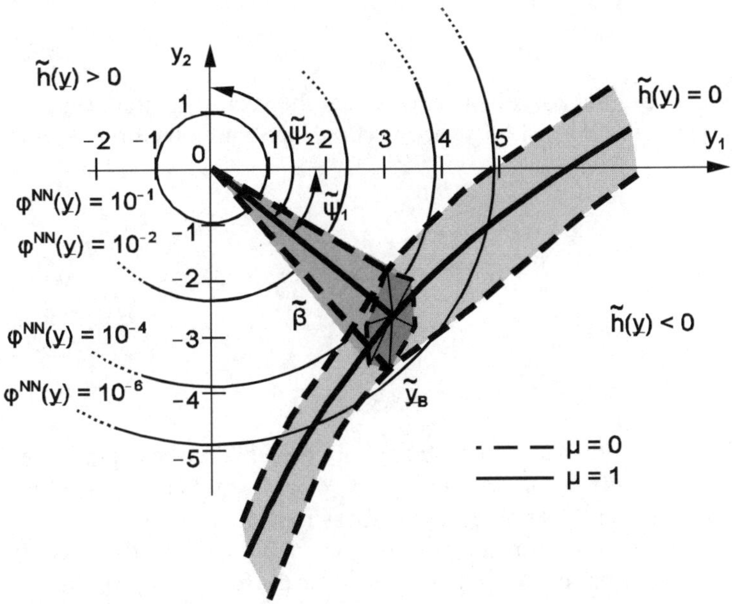

Fig. 6.9. Fuzzy limit state surface $\tilde{h}(\underline{y}) = 0$ with fuzzy design point $\tilde{\underline{y}}_B$ and fuzzy reliability index $\tilde{\beta}$

The fuzzy reliability index $\tilde{\beta}$ – interpreted geometrically – is the shortest distance of the fuzzy limit state surface $\tilde{h}(\underline{y}) = 0$ from the coordinate origin in the standard normal space. The assigned uncertain point on $\tilde{h}(\underline{y}) = 0$ is referred to as fuzzy design point $\tilde{\underline{y}}_B$.

For the conversion of the fuzzy reliability index $\tilde{\beta}$ into the fuzzy failure probability $\tilde{P}_f$ the relationship

$$\tilde{P}_f = \Phi^{NN}(-\tilde{\beta}) \tag{6.29}$$

is used. This yields "exact" values for $\tilde{P}_f$ if the fuzzy limit state surface $\tilde{h}(\underline{y}) = 0$ represents a linear fuzzy function, i.e., if every element of $\tilde{h}(\underline{y}) = 0$ is a linear function. In the case of a nonlinear fuzzy limit state surface $\tilde{h}(\underline{y}) = 0$, the approximation solution

$$\tilde{P}_{f,N} = \Phi^{NN}(-\tilde{\beta}) \approx \tilde{P}_f \tag{6.30}$$

is obtained.

The fuzzy design point $\tilde{\underline{y}}_B$ and fuzzy reliability index $\tilde{\beta}$ are determined by solving the optimization problem

$$\|\tilde{\underline{y}}\| = \sqrt{\sum_{i=1}^{n} \tilde{y}_i^2} \Rightarrow \text{Min}\,. \tag{6.31}$$

As for constraints, $\tilde{\underline{y}} \in (\tilde{h}(\underline{y}) = 0)$ is to be complied with. The objective in Eq. (6.31) is satisfied by $\tilde{\underline{y}} = \tilde{\underline{y}}_B$ for which the following holds

$$\|\tilde{\underline{y}}_B\| = \tilde{\beta} . \tag{6.32}$$

When the angles between the coordinate axes y_i and the vector $\tilde{\underline{y}}_B$ are designated by $\tilde{\psi}_i$ (Fig. 6.9) and the fuzzy sensitivity factors (fuzzy weighting factors)

$$\tilde{\alpha}_i = \cos\,\tilde{\psi}_i \tag{6.33}$$

are introduced, the relationships

$$\tilde{\underline{y}}_B = \tilde{\underline{\alpha}} \cdot \tilde{\beta} \ , \tag{6.34}$$

and

$$\sum_{i=1}^{n} \tilde{\alpha}_i^2 = 1 \tag{6.35}$$

may be stated. The interaction between the fuzzy variables involved must therefore be taken into account. The fuzzy sensitivity factors $\tilde{\alpha}_i$ are a measure of the effect of the fuzzy probabilistic basic variables on the fuzzy reliability index.

For solving the optimization problem according to Eq. (6.31) the generally nonlinear fuzzy limit state surface $\tilde{h}(\underline{y}) = 0$ is linearized in the fuzzy design point $\tilde{\underline{y}}_B$. For this purpose, $\tilde{h}(\underline{y}) = 0$ is expanded in a Taylor series that is broken off with the linear terms. The fuzzy limit state surface is then replaced by the tangential fuzzy hyperplane $\tilde{l}(\underline{y}) = 0$ in $\tilde{\underline{y}}_B$

$$\tilde{l}(\underline{y}) = \tilde{h}(\tilde{\underline{y}}_B) + \sum_{i=1}^{n} \left.\frac{\partial \tilde{h}(\underline{y})}{\partial y_i}\right|_{\tilde{\underline{y}} = \tilde{\underline{y}}_B} \cdot (y_i - \tilde{y}_{Bi}) . \tag{6.36}$$

With the column vector

$$\tilde{\underline{H}}(\tilde{\underline{y}}_B) = \left(\left.\frac{\partial \tilde{h}(\underline{y})}{\partial y_1}\right|_{\tilde{\underline{y}} = \tilde{\underline{y}}_B}, \ldots, \left.\frac{\partial \tilde{h}(\underline{y})}{\partial y_i}\right|_{\tilde{\underline{y}} = \tilde{\underline{y}}_B}, \ldots, \left.\frac{\partial \tilde{h}(\underline{y})}{\partial y_n}\right|_{\tilde{\underline{y}} = \tilde{\underline{y}}_B} \right) \tag{6.37}$$

Eq. (6.36) reads

$$\tilde{l}(\underline{y}) = \tilde{h}(\tilde{\underline{y}}_B) + \tilde{\underline{H}}^T(\tilde{\underline{y}}_B) \cdot (\underline{y} - \tilde{\underline{y}}_B) . \tag{6.38}$$

The fuzzy reliability index $\tilde{\beta}$ is computed as the shortest distance of the tangential fuzzy hyperplane $\tilde{l}(\underline{y}) = 0$ from the coordinate origin. With Eq. (6.38)

$$\tilde{\beta} = \frac{\tilde{h}(\tilde{\underline{y}}_B) - \tilde{\underline{H}}^T(\tilde{\underline{y}}_B) \cdot \tilde{\underline{y}}_B}{\left(\tilde{\underline{H}}^T(\tilde{\underline{y}}_B) \cdot \tilde{\underline{H}}(\tilde{\underline{y}}_B)\right)^{\frac{1}{2}}} \tag{6.39}$$

is obtained. The fuzzy sensitivity factors are computed according to

$$\underline{\tilde{\alpha}} = \frac{-\ \underline{\tilde{H}}(\tilde{\underline{y}}_B)}{\left(\underline{\tilde{H}}^T(\tilde{\underline{y}}_B) \cdot \underline{\tilde{H}}(\tilde{\underline{y}}_B)\right)^{\frac{1}{2}}} \quad . \tag{6.40}$$

The linearization according to Eq. (6.36) requires the fuzzy design point $\tilde{\underline{y}}_B$ to be known. As $\tilde{\underline{y}}_B$ is not available a priori, however, $\tilde{\underline{y}}_B$ and $\tilde{\beta}$ are iteratively determined. At first an assumption is made for $\tilde{\underline{y}}_B$, which is then stepwise improved by applying the iteration rule

$$\overset{[r+1]}{\tilde{\underline{y}}_B} = \underline{\tilde{\alpha}}\left(\overset{[r]}{\tilde{\underline{y}}_B}\right) \cdot \tilde{\beta}\left(\overset{[r]}{\tilde{\underline{y}}_B}\right) \quad , \tag{6.41}$$

with r being the iteration step counter. The terms on the right-hand side of Eq. (6.41) are obtained from Eqs. (6.39) and (6.40). After the completion of the iteration the fuzzy design point $\tilde{\underline{y}}_B$ and fuzzy reliability index $\tilde{\beta}$ are determined.

The numerical computation of $\tilde{\underline{y}}_B$ and $\tilde{\beta}$ is carried out by evaluating all elements of $\tilde{h}(\underline{y}) = 0$, which have been determined with Eq. (6.27). The algorithms of the First Order Reliability Method are therefore applied. For each element of $\tilde{h}(\underline{y}) = 0$ a crisp value β as an element of $\tilde{\beta}$ and a crisp point $\underline{y}_B$ as an element of $\tilde{\underline{y}}_B$ are obtained. The membership values of the elements of $\tilde{h}(\underline{y}) = 0$ are transferred to the values β and $\underline{y}_B$ computed for $\tilde{\beta}$ and $\tilde{\underline{y}}_B$. The set of the elements β together with the membership values $\mu(\beta)$ forms the fuzzy reliability index $\tilde{\beta}$ and the set of the elements $\underline{y}_B$ together with $\mu(\underline{y}_B)$ yields the fuzzy design point $\tilde{\underline{y}}_B$.

With Eq. (6.27) each element of $\tilde{h}(\underline{y}) = 0$ is defined by one combination of elements of all fuzzy variables entering the investigation. The fuzzy parameters of the distributions of the fuzzy probabilistic basic variables and the fuzzy model parameters are mapped onto the fuzzy design point and fuzzy reliability index

$$\left\{\tilde{p}_t(\tilde{X}_i), \tilde{m}_j\, ; i = 1,...,n\, ; t = 1,...,r_i\, ; j = 1,...,q\right\} \rightarrow \tilde{\underline{y}}_B\, , \tag{6.42}$$

$$\left\{\tilde{p}_t(\tilde{X}_i), \tilde{m}_j\, ; i = 1,...,n\, ; t = 1,...,r_i\, ; j = 1,...,q\right\} \rightarrow \tilde{\beta}\, . \tag{6.43}$$

When computing $\tilde{\underline{y}}_B$ and $\tilde{\beta}$ from an evaluation of $\tilde{h}(\underline{y}) = 0$ the interaction between the fuzzy limit state points in standard normal space must be taken into consideration. Special elements of the fuzzy limit state points belong to each element of $\tilde{h}(\underline{y}) = 0$. The elements of the fuzzy limit state points must therefore not be arbitrarily combined and lumped together to form a possible, but not necessarily existent element of $\tilde{h}(\underline{y}) = 0$. Only those elements of $\tilde{h}(\underline{y}) = 0$ may be evaluated that are obtained from the $\tilde{m}_j$ and $\tilde{p}_t(\tilde{X}_i)$ with Eq. (6.27). The "boundaries" of $\tilde{h}(\underline{y}) = 0$, which are shown as dashed lines in Fig. 6.9, do not generally belong to these elements.

The representation of a fuzzy design point $\tilde{\underline{y}}_B$ is shown in Fig. 6.10 for two basic variables; the corresponding fuzzy reliability index $\tilde{\beta}$ is plotted in Fig. 6.11. The interval bounds for the α-level set β_α are obtained for each membership level α as the minimum and maximum distance of the corresponding α-level set of

$\tilde{y}_B$ from the coordinate origin. This relationship is documented graphically in Fig. 6.10 for the membership levels $\alpha = 0$ and $\alpha = 1$.

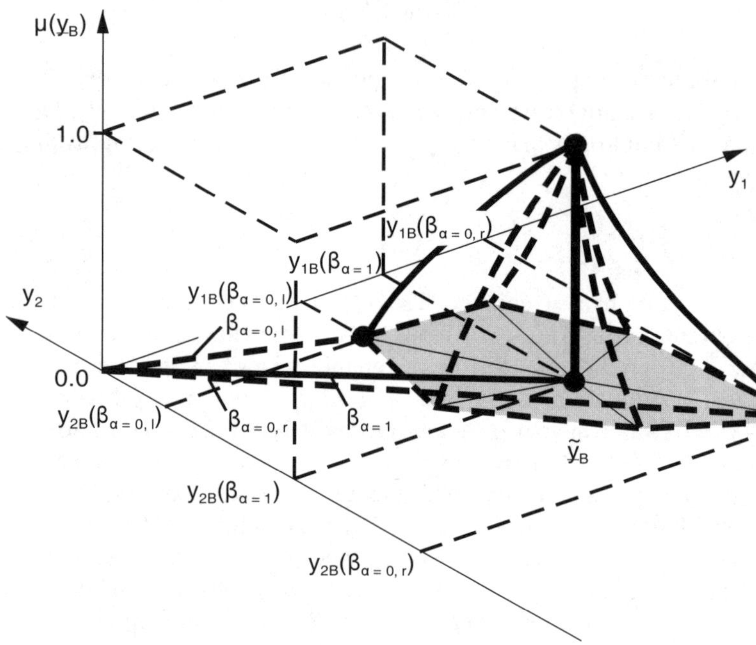

Fig. 6.10. Fuzzy design point $\tilde{y}_B$ and fuzzy reliability index $\tilde{\beta}$

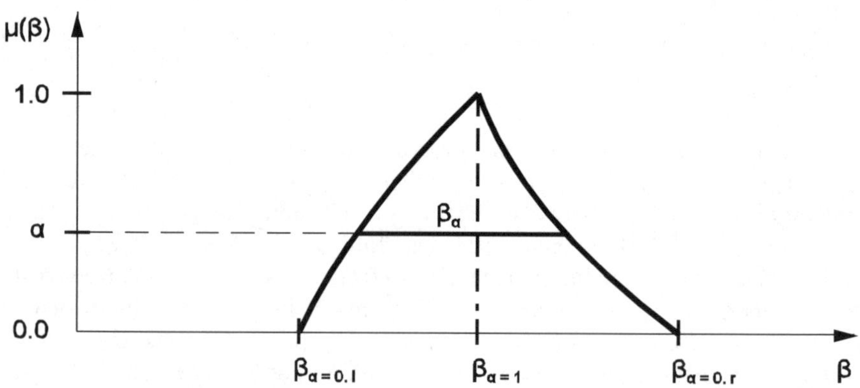

Fig. 6.11. Fuzzy reliability index $\tilde{\beta}$

The computed reliability index $\tilde{\beta}$ completely reflects the uncertainty with the characteristic fuzziness entering the investigation. The uncertainty of the safety level characterizes the sensitivity of the safety of the structure in relation to the introduced fuzziness. In order to assess the sensitivity of structural safety a measure for the uncertainty of the fuzzy reliability index is necessary. For this purpose, e.g., the modified Shannon entropy measure according to Sect. 2.2.4 may be applied, the uncertainty of $\tilde{\beta}$ is measured with

$$H_u(\tilde{\beta}) = -k \cdot \int_{\beta=\beta_{\mu=0,l}}^{\beta=\beta_{\mu=0,r}} \left[\mu(\beta) \cdot \ln(\mu(\beta)) + (1-\mu(\beta)) \cdot \ln(1-\mu(\beta))\right] d\beta \quad , \tag{6.44}$$

with a fixed k value.

By this means the sensitivity of the safety of the structure with regard to fuzziness becomes comparable for different problems. The modified entropy according to Eq. (6.44) enables a comparison to be made between the uncertainty determined in different investigations for $\tilde{\beta}$. This permits, for example, an assessment of the sensitivity of a structure with regard to different failure modes. Different structural designs may be directly compared in relation to failure criteria. Moreover, their robustness may be assessed in relation to the fuzzy values introduced.

6.5 Safety Verification

The safety verification has to ensure that the requirements concerning protection of persons and material assets are met. The measure values computed for the safety or reliability of a structure are compared with permissible or required values prescribed. For assessing the results from probabilistic investigations such comparative values exist in terms of the permissible failure probability zul_P_f and the required reliability index erf_β.

In fuzzy probabilistic safety assessment the existing safety level is described by the fuzzy values $\tilde{P}_f$ for the failure probability and $\tilde{\beta}$ for the reliability index. The assessment of these fuzzy values is carried out by comparison with the values req_β or $perm\_P_f$ to be complied with, as stipulated in currently applicable standards [217, 218]. The safety verification

$$\tilde{P}_f \leq perm\_P_f \quad , \tag{6.45}$$

or

$$\tilde{\beta} \geq req\_\beta \quad , \tag{6.46}$$

must be satisfied.

In Eq. (6.46) a real number is compared with a fuzzy value, i.e., the verification cannot be uniquely assessed as *fulfilled* or *not fulfilled* in all cases. Basically, a distinction must be made between three cases (Fig. 6.12):

1. The verification according to Eq. (6.46) is *fulfilled* when none of the elements of $\tilde{\beta}$ are smaller than *req_β*, i.e., when
$$\beta_{\alpha=0,l} \geq req\_\beta \tag{6.47}$$
holds.
2. The verification according to Eq. (6.46) is *not fulfilled* when all elements of $\tilde{\beta}$ are smaller than *req_β*, i.e., when
$$\beta_{\alpha=0,r} < req\_\beta \tag{6.48}$$
holds.
3. The verification according to Eq. (6.46) is *partially fulfilled* when the value for *req_β* is an element of the fuzzy set $\tilde{\beta}$, i.e., when
$$erf\_\beta \in \tilde{\beta} \ \wedge \ \beta_{\alpha=0,l} \neq req\_\beta \tag{6.49}$$
holds.

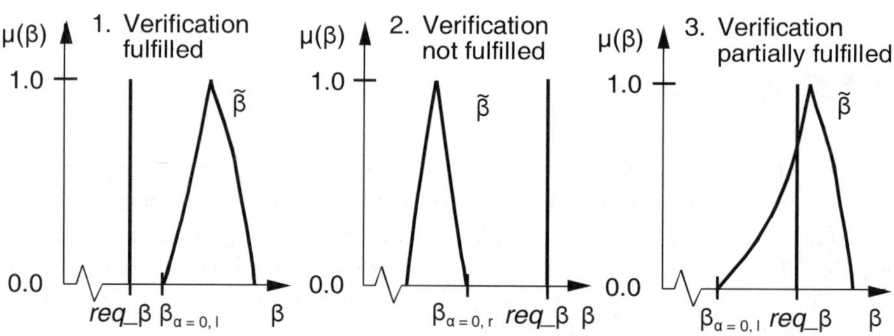

Fig. 6.12. Basic cases for the safety verification

If the safety verification is only *partially fulfilled* (Eq. (6.49)), a *subjective decision* must be made. The fuzzy set $\tilde{\beta}$ is separated into two subsets $\tilde{\beta}_1$ and $\tilde{\beta}_2$ at the point $\beta = req\_\beta$ (Fig. 6.13). The elements of the subset
$$\tilde{\beta}_1 = \{(\beta, \mu(\beta)) \mid \beta \in \tilde{\beta}; \beta \geq req\_\beta\} \tag{6.50}$$
to the right of *req_β* fulfill the verification according to Eq. (6.46), whereas the elements of the subset
$$\tilde{\beta}_2 = \{(\beta, \mu(\beta)) \mid \beta \in \tilde{\beta}; \beta < req\_\beta\} \tag{6.51}$$
to the left of *req_β* do not fulfill the verification. The subsets $\tilde{\beta}_1$ and $\tilde{\beta}_2$ are the

intersections of fuzzy set $\tilde{\beta}$ and the crisp set of the accepted (for $\tilde{\beta}_1$) and not accepted (for $\tilde{\beta}_2$) values for β according to the definition of the required safety level.

The subjective assessment of the computed fuzzy safety level $\tilde{\beta}$ may be backed up by the assessment of the subsets $\tilde{\beta}_1$ and $\tilde{\beta}_2$. It is known that the membership values μ(β) of the fuzzy sets $\tilde{\beta}_1$ and $\tilde{\beta}_2$ assess the degree to which the values of β belong to $\tilde{\beta}_1$ or to $\tilde{\beta}_2$. By applying the maximum operator it is possible to measure the fuzzy sets $\tilde{\beta}_1$ and $\tilde{\beta}_2$. The maximum membership values of $\tilde{\beta}_1$ and $\tilde{\beta}_2$ are introduced as measures for these subsets (Fig. 6.13)

$$\mu_1 = \sup_{\beta \in \tilde{\beta}_1} [\mu(\beta)] , \tag{6.52}$$

$$\mu_2 = \sup_{\beta \in \tilde{\beta}_2} [\mu(\beta)] . \tag{6.53}$$

The measure value μ_1 assesses the proposition "the safety verification is fulfilled" whereas the measure value μ_2 assesses the proposition "the safety verification is not fulfilled". The measure values μ_1 and μ_2 may be interpreted as the possibility with which the verification according to Eq. (6.46) is either fulfilled or not fulfilled, respectively (Sect. 2.2.3).

By subjectively prescribing a value for μ_1 that must be complied with or a tolerable value for μ_2 a decision may be made as to the acceptability of the partial fulfillment of the safety verification. If, for example, $\mu_1 = 0$ is demanded, the safety verification for the whole range of Eq. (6.49) is fulfilled. In the case of a conservative stipulation of the tolerable value $\mu_2 = 0$ the safety verification for the whole range of Eq. (6.49) is not fulfilled.

Moreover, starting from a partially fulfilled safety verification an uncertain structural design that completely fulfills the safety requirements may be derived as explained in Chap. 7.

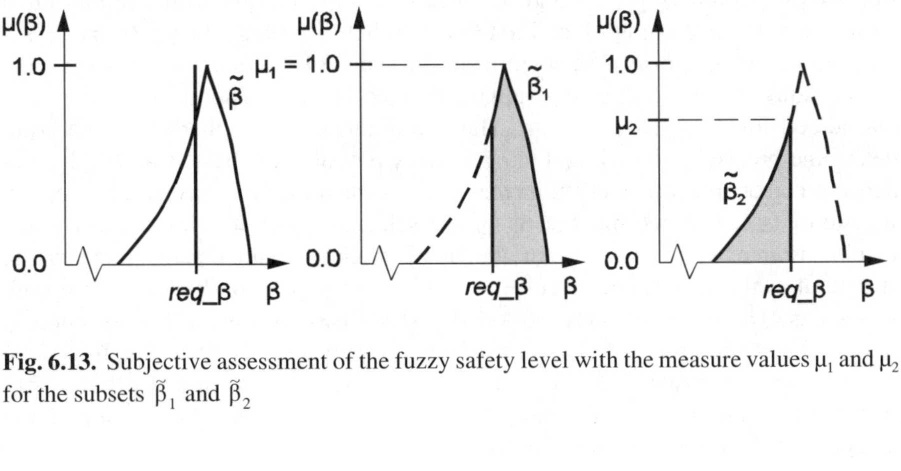

Fig. 6.13. Subjective assessment of the fuzzy safety level with the measure values μ_1 and μ_2 for the subsets $\tilde{\beta}_1$ and $\tilde{\beta}_2$

6.6 Numerical Realization

The Fuzzy First Order Reliability Method is described as a fuzzy analysis with the mapping of Eq. (6.43). The fuzzy parameters $\tilde{p}_t(\tilde{X}_i)$ of the fuzzy probability distributions of the fuzzy probabilistic basic variables and the fuzzy model parameters $\tilde{m}_j$ enter the fuzzy analysis as fuzzy input variables. The fuzzy result value is the fuzzy reliability index $\tilde{\beta}$. The algorithm that maps the fuzzy input values $\tilde{p}_t(\tilde{X}_i)$ and $\tilde{m}_j$ onto the fuzzy result $\tilde{\beta}$ is referred to as the mapping model. This represents the probabilistic fundamental solution; in the case of FFORM, the algorithms of the First Order Reliability Method.

Each of the fuzzy input variables $\tilde{p}_t(\tilde{X}_i)$ and $\tilde{m}_j$ is assigned to an axis of the multidimensional Cartesian coordinate system, that describes the space of the fuzzy input variables. Their Cartesian product forms the fuzzy input set for the mapping in Eq. (6.43). The mapping model assigns precisely one element of the fuzzy result $\tilde{\beta}$ to one point in each case from the space of the fuzzy input variables.

In order to determine the membership function of the fuzzy reliability index $\tilde{\beta}$ the α-level optimization (Sect. 5.2.2) is applied (Fig. 6.14). The fuzzy input variables $\tilde{p}_t(\tilde{X}_i)$ and $\tilde{m}_j$ are subdivided into α-level sets P_{t,i,α_k} and M_{j,α_k} by α-discretization (Sects. 2.1.3 and 2.1.9) and mapped onto the elements of the α-level sets β_{α_k} of the fuzzy result $\tilde{\beta}$. The α-level optimization solves this problem for all selected α-levels. The mapping in Eq. (6.43) possesses no special properties such as, e.g., monotonicity, which could otherwise be exploited. For the optimization problem, which has to be solved several times, the modified evolution strategy described in Sect. 5.2.3.4 is therefore applied. The general computation procedure is schematically shown in Fig. 6.15.

The numerical evaluation of the mapping in Eq. (6.43), i.e., the determination of the optima $\beta_{\alpha_k,l}$ and $\beta_{\alpha_k,r}$ for all selected α-levels $\alpha = \alpha_k$, requires the repeated computation of crisp values β for special combinations of (crisp) elements from the fuzzy input variables $\tilde{p}_t(\tilde{X}_i)$ and $\tilde{m}_j$. Each of these computations represents a system analysis according to the First Order Reliability Method (Fig. 6.15).

At first, a crisp point is selected from the space of the fuzzy input parameters $\tilde{p}_t(\tilde{X}_i)$ and $\tilde{m}_j$. In order to evaluate this point two disjoint subspaces are considered for the $\tilde{p}_t(\tilde{X}_i)$ and $\tilde{m}_j$. The coordinates $p_t(\tilde{X}_i)$ of the selected point determine precisely one original of each fuzzy probabilistic basic variable $\tilde{X}_i$ and hence precisely one element (trajectory) $f(\underline{x})$ of the joint fuzzy probability density function $\tilde{f}(\underline{x})$. The coordinates m_j of the selected point determine exactly one combination of crisp elements of the fuzzy model parameters for entering the structural analysis and hence precisely one element $g(\underline{x}) = 0$ of the fuzzy limit state surface $\tilde{g}(\underline{x}) = 0$. The element $g(\underline{x}) = 0$ is described with the aid of an approximation function, which takes account of several, purposefully computed limit state points in the x-space (Sect. 6.2.2). The approximation is thereby iteratively improved, in particular in the neighborhood of the assigned design point in the x-space, until a predefined termination limit is attained.

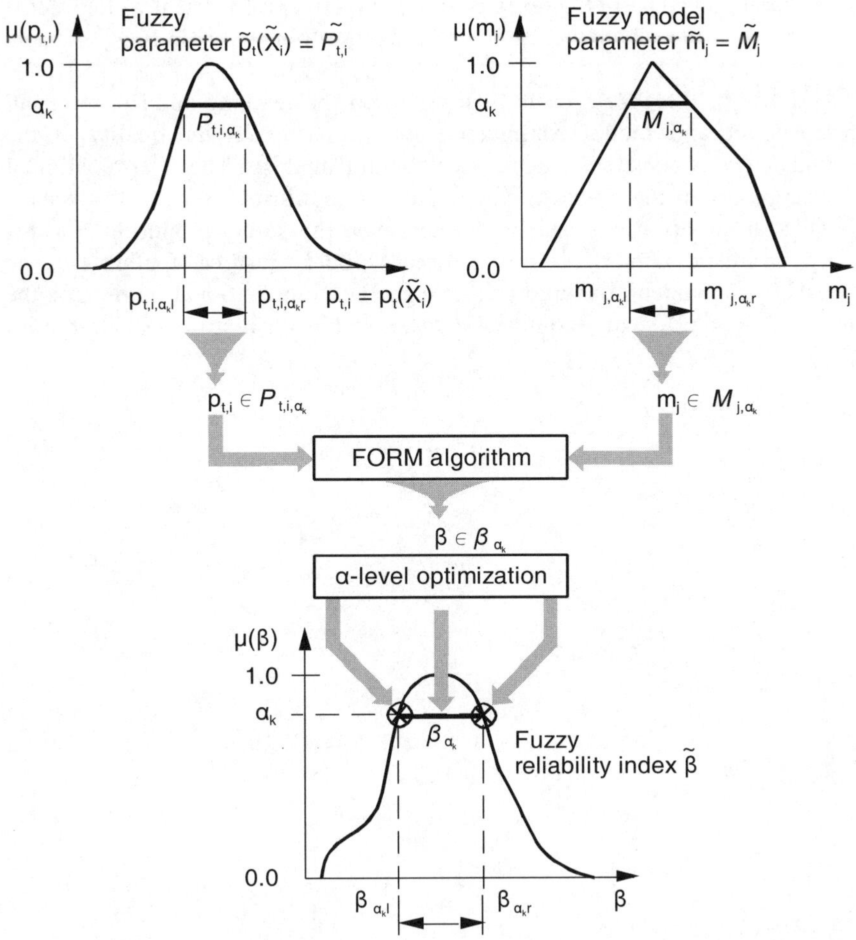

Fig. 6.14. Mapping of the fuzzy input parameters $\tilde{p}_t(\tilde{X}_i)$ and $\tilde{m}_j$ onto the fuzzy reliability index $\tilde{\beta}$ with α-level optimization

The problem described by the $f(\underline{x})$ and $g(\underline{x}) = 0$ is transformed into the standard normal space and solved. The result is a crisp value for the reliability index β. The computed value β is checked for optimality and taken to be a temporary result for $\beta_{\alpha_k,l}$ or $\beta_{\alpha_k,r}$ if indicated.

The computational procedure is stopped when special termination criteria of the α-level optimization are met. If all criteria are met, the fuzzy reliability index $\tilde{\beta}$ is determined. Otherwise, a new crisp point is directed selected from the space of the fuzzy input parameters $\tilde{p}_t(\tilde{X}_i)$ and $\tilde{m}_j$ and the computation of a crisp β is repeated (Fig. 6.15).

The general procedure of the fuzzy probabilistic safety assessment with FFORM is summarized in the schematically illustrated algorithm shown in Figs. 6.15 and 6.16.

The quality of the fuzzy result $\tilde{\beta}$ depends on the realistic description of all uncertain input and model parameters, and primarily on the quality of the deterministic fundamental solution for the structural analysis. The determination of limit state points in the x-space, which must be performed repeatedly in each computational sweep, has a decisive influence on the safety prognosis. For this reason, a realistic numerical computational model must be applied for the deterministic fundamental solution, like, e.g., the computational model for the geometrically and physically nonlinear analysis of plane bar structures after [116] and [133].

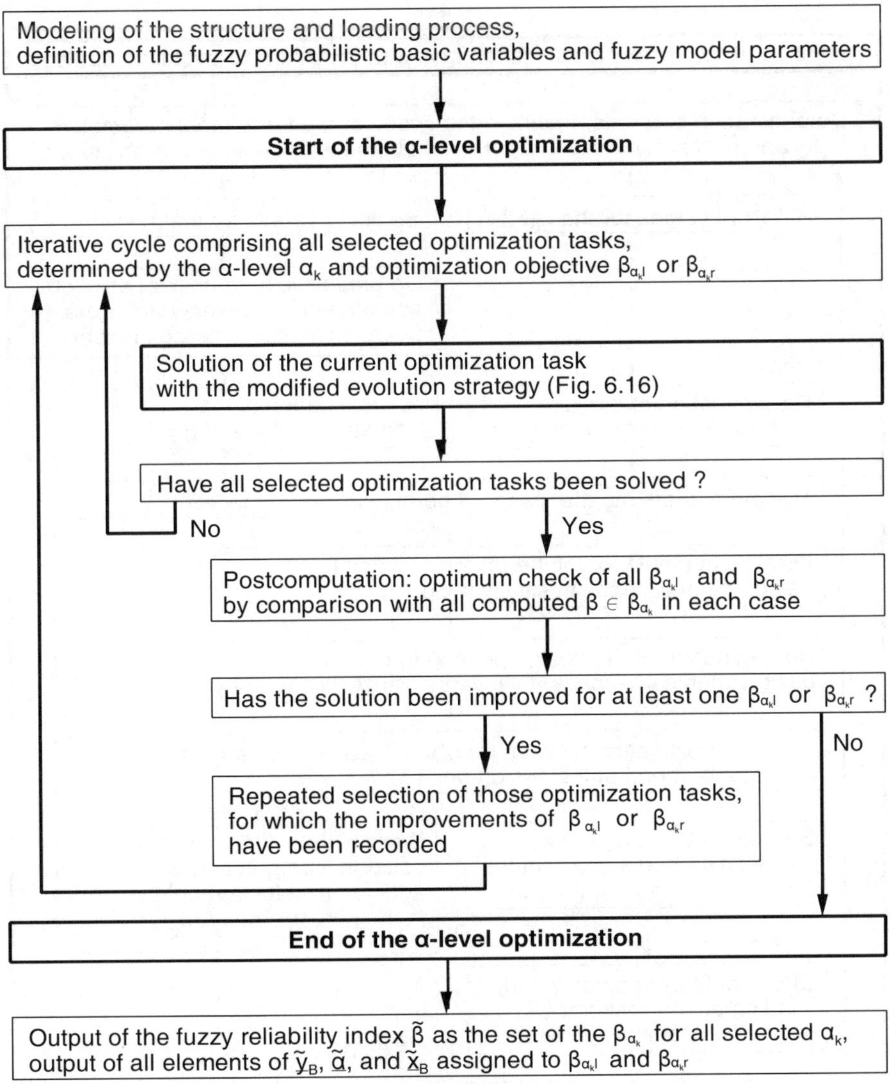

Fig. 6.15. Flowchart – FFORM algorithm with α-level optimization for the computation of the fuzzy reliability index $\tilde{\beta}$

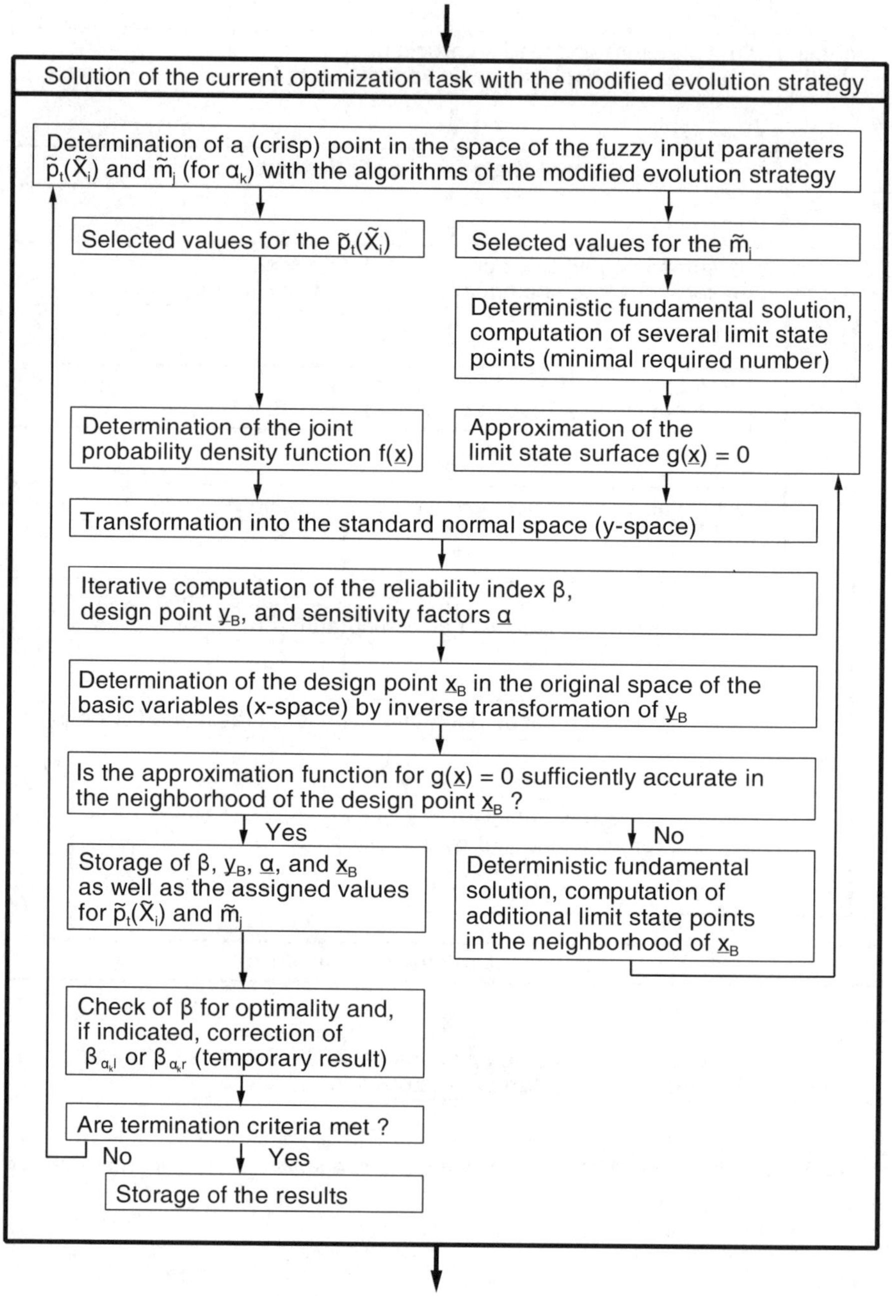

Fig. 6.16. Flowchart – solution of the optimization tasks according to the α-level optimization within the FFORM algorithm by means of the modified evolution strategy (simplified representation of the optimization algorithm)

6.7 Application of FFORM

6.7.1 Steel Girder, Physically Nonlinear Statical System Behavior

It is proposed to determine the structural reliability of the steel girder shown in Fig. 6.17 [119]. System failure is considered according to first-order plastic hinge theory, the ultimate limit state is described with the aid of the kinematic law of plasticity theory for irreversible loading. On attaining the system ultimate load, the cross section at point k is completely plasticized; Fig. 6.17 shows the corresponding failure mechanism.

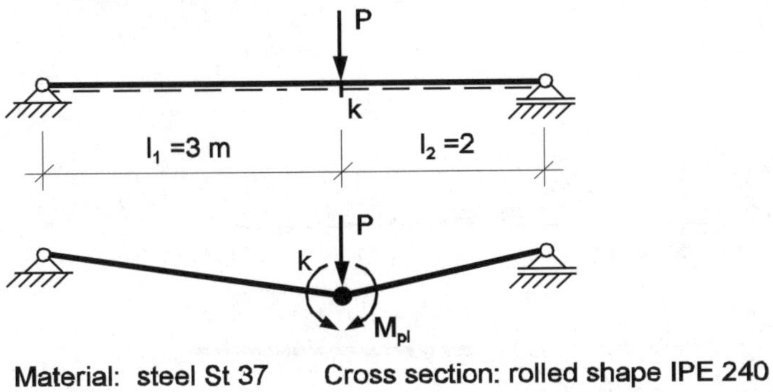

Fig. 6.17. Steel girder; statical system with loading; governing failure mechanism

The geometry of the structural model is determined by the deterministic values l_1 and l_2. The load that can be carried by the system is

$$P_{Tr} = M_{pl} \cdot \frac{l_1 + l_2}{l_1 \cdot l_2} . \tag{6.54}$$

Regarding the existent load P only informally uncertain statistical data are available, P is thus modeled as fuzzy random variable $\tilde{X}_1$.

The functional type of the fuzzy probability distribution of P is assumed to be an *extreme value distribution of Ex-Max Type I.* The expected value m_{X_1} and the standard deviation σ_{X_1} are obtained from the evaluation of the informally uncertain statistical data, they are determined as fuzzy triangular numbers $\tilde{m}_{X_1} = < 47, 50, 52 >$ kN and $\tilde{\sigma}_{X_1} = < 4.5, 5.0, 6.0 >$ kN (Fig. 6.18). The fuzzy probability density function $\tilde{f}_1(x_1)$ and the fuzzy probability distribution function $\tilde{F}_1(x_1)$ defined by these specifications are shown in Fig. 6.19.

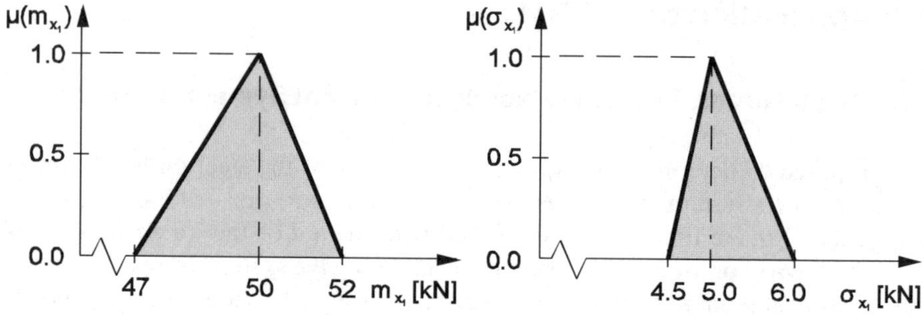

Fig. 6.18. Fuzzy expected value $\tilde{m}_{x_1}$ and fuzzy standard deviation $\tilde{\sigma}_{x_1}$ of the fuzzy probability distribution of the load P

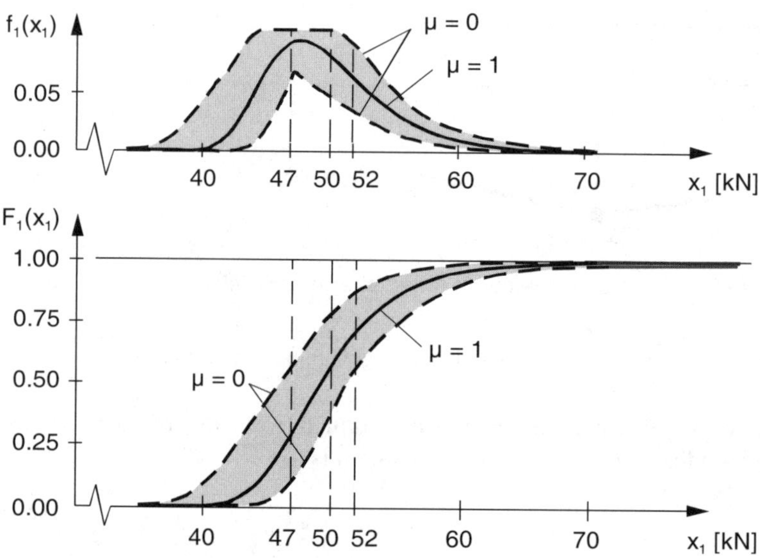

Fig. 6.19. Fuzzy probability density function $\tilde{f}_1(x_1)$ and fuzzy probability distribution function $\tilde{F}_1(x_1)$ of the load P

When assuming ideally plastic material behavior and a deterministic cross section, the structural resistance characterized by the fully plastic moment M_{pl} is exclusively determined by the yield stress f_y

$$M_{pl} = f_y \cdot W_{pl} \quad . \tag{6.55}$$

As for the yield stress f_y, statistical data are available, but these are associated with the manufacturing conditions (reproduction conditions for sample elements, influences on the production process) in different steel factories in the particular case. Place and time of manufacture of the girder cannot be identified without an element of doubt, the reproduction conditions and influences associated with the manufacture are therefore unknown. Owing to the uncertain information concerning reproduction conditions and influences the yield stress f_y is described as fuzzy random variable $\tilde{X}_2$.

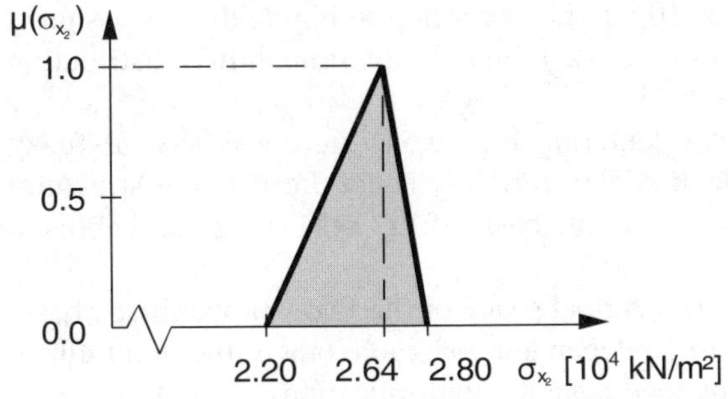

Fig. 6.20. Fuzzy standard deviation $\tilde{\sigma}_{X_2}$ of the fuzzy probability distribution of the yield stress f_y

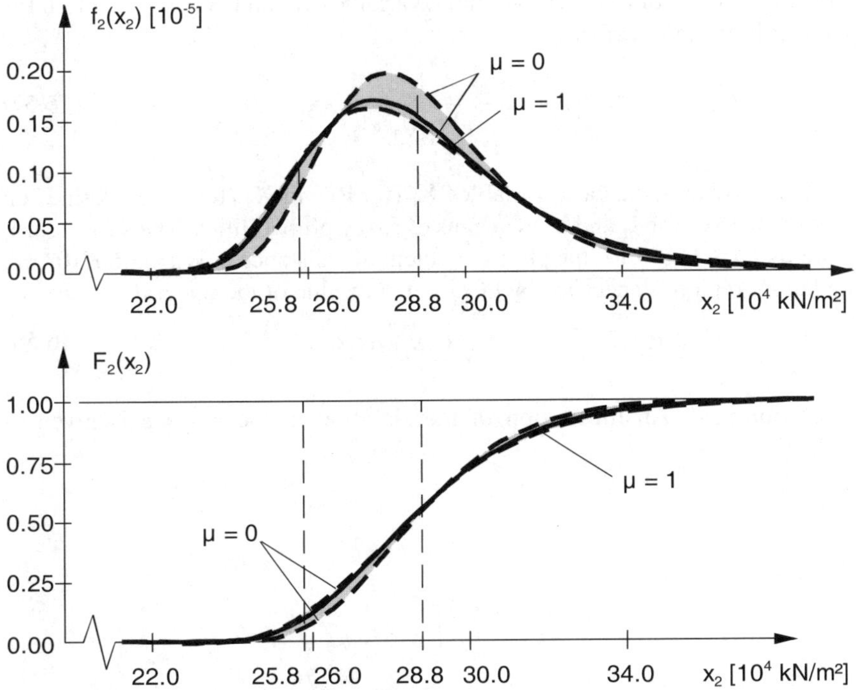

Fig. 6.21. Fuzzy probability density function $\tilde{f}_2(x_2)$ and fuzzy probability distribution function $\tilde{F}_2(x_2)$ of the yield stress f_y

Concerning the functional type of the fuzzy probability distribution of f_y, a *logarithmic normal distribution* is chosen. Additionally, the minimum value $x_{0,2} = 19.9 \cdot 10^4$ kN/m² and the expected value $m_{x_2} = 28.8 \cdot 10^4$ kN/m² are known. The standard deviation is computed by evaluating statistical data under consideration of nonconstant reproduction conditions. This leads to the fuzzy triangular number $\tilde{\sigma}_{x_2} = < 2.20, 2.64, 2.80 > \cdot 10^4$ kN/m² according to Fig. 6.20. The assigned fuzzy probability density function $\tilde{f}_2(x_2)$ and fuzzy probability distribution function $\tilde{F}_2(x_2)$ are shown in Fig. 6.21.

All nondeterministic parameters entering Eq. (6.54) are considered as fuzzy random variables in this example, model uncertainty is not present. The structural reliability of the system is calculated on the basis of FFORM (fuzzy probabilistic safety assessment).

The problem is formulated in the original space of the fuzzy probabilistic basic variables $\tilde{X}_1$ and $\tilde{X}_2$. These are independent of each other, the joint fuzzy probability density function is obtained from the multiplication

$$\tilde{f}(\underline{x}) = \tilde{f}(x_1, x_2) = \tilde{f}_1(x_1) \cdot \tilde{f}_2(x_2), \tag{6.56}$$

which is to be carried out original-by-original.

The load P and the yield stress f_y in Eq. (6.54) are replaced by the crisp coordinates x_1 und x_2 of the uncertain input variables $\tilde{X}_1$ and $\tilde{X}_2$. This leads to the crisp, linear limit state surface

$$g(\underline{x}) = g(x_1, x_2) = x_2 - \frac{l_1 \cdot l_2}{W_{pl}(l_1 + l_2)} \cdot x_1 = 0. \tag{6.57}$$

The fuzzy probabilistic basic variables $\tilde{X}_1$ (for P) and $\tilde{X}_2$ (for f_y) are defined on the coordinate axes for x_1 and x_2 by means of fuzzy probabilities. The values for l_1 and l_2 are listed in Fig. 6.17, the plastic moment of resistance is $W_{pl} = 3.66 \cdot 10^{-4}$ m³.

The failure region is specified by $g(\underline{x}) \leq 0$, the value of the integral

$$\tilde{P}_f = \int\limits_{x_1, x_2 \mid g(x_1, x_2) \leq 0} \int \tilde{f}(x_1, x_2)\, dx_2 dx_1 \tag{6.58}$$

is to be computed. An illustration of the situation in the x-space is given in Fig. 6.22.

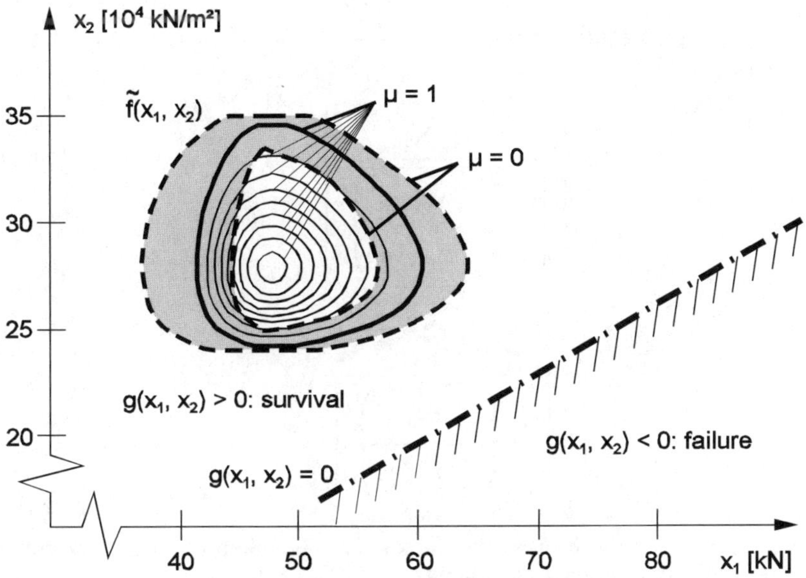

Fig. 6.22. Joint fuzzy probability density function $\tilde{f}(x_1, x_2)$ and limit state surface $g(x_1, x_2) = 0$ in the x-space

The fuzzy probabilistic basic variables $\tilde{X}_1$ and $\tilde{X}_2$ and the (crisp) limit state surface $g(x_1, x_2) = 0$ are transformed into standard normal space with

$$\tilde{y}_1 = \Phi^{NN^{-1}}\left[\exp\left(-\exp(-\tilde{a}\,(x_1 - \tilde{b})\right)\right] \quad , \tag{6.59}$$

and

$$\tilde{y}_2 = \frac{\ln\left(x_2 - x_{0,2}\right) - \tilde{m}_u}{\tilde{\sigma}_u} \ . \tag{6.60}$$

Results are the (crisp) joint probability density function $\varphi^{NN}(y_1, y_2)$ of the standard normal distribution and the nonlinear fuzzy limit state surface

$$\tilde{h}(\underline{y}) = x_{0,2} + e^{\tilde{\sigma}_u \cdot y_2 + \tilde{m}_u} + \frac{l_1 \cdot l_2}{W_{pl}\,(l_1 + l_2)} \cdot \left[\frac{1}{\tilde{a}} \cdot \ln\left[-\ln\left(\Phi^{NN}(y_1)\right)\right] - \tilde{b}\right] = 0 \ . \tag{6.61}$$

The function $\Phi^{NN}(y_1)$ thereby represents the probability distribution function of the standard normal distribution of y_1 and $\Phi^{NN^{-1}}$ is its inverse function. The fuzzy variables $\tilde{a}$, $\tilde{b}$, $\tilde{m}_u$, and $\tilde{\sigma}_u$ are the fuzzy parameters of the fuzzy probability distribution functions of the load and yield stress. The following holds

$$\tilde{a} = \frac{\pi}{\tilde{\sigma}_{x_1} \cdot \sqrt{6}} \ , \tag{6.62}$$

$$\tilde{b} = \tilde{m}_{x_1} - \frac{0.577216 \cdot \tilde{\sigma}_{x_1} \cdot \sqrt{6}}{\pi}\,, \tag{6.63}$$

$$\tilde{m}_u = \ln\left[\frac{m_{x_2} - x_{0,2}}{\sqrt{1 + \left(\dfrac{\tilde{\sigma}_{x_2}}{m_{x_2} - x_{0,2}}\right)^2}}\right], \tag{6.64}$$

and

$$\tilde{\sigma}_u = \sqrt{\ln\left[1 + \left(\frac{\tilde{\sigma}_{x_2}}{m_{x_2} - x_{0,2}}\right)^2\right]}\,. \tag{6.65}$$

The interaction between these fuzzy parameters is to be taken into consideration. The curve of $\tilde{h}(\underline{y}) = 0$ is shown in Fig. 6.23.

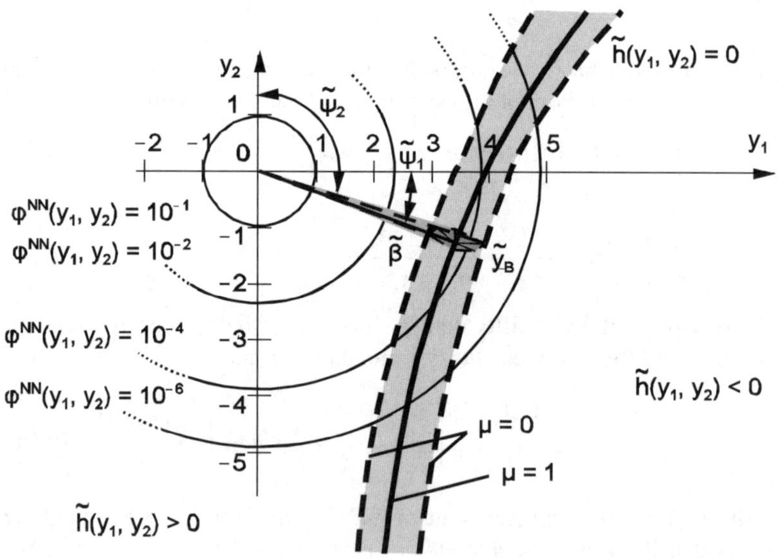

Fig. 6.23. Joint probability density function $\varphi^{NN}(y_1, y_2)$ of the standard normal distribution and fuzzy limit state surface $\tilde{h}(y_1; y_2) = 0$ in the y-space

The iterative evaluation of the function bunch according to Eq. (6.61) with the aid of the algorithm described in Sect. 6.6 leads to the fuzzy reliability index $\tilde{\beta}$ shown in Fig. 6.24 (see also Figs. 6.23 and 6.25). The computation was carried out for the α-levels $\alpha_1 = 0.0$, $\alpha_2 = 0.5$ and $\alpha_3 = 1.0$. The membership function $\mu(\beta)$ is weakly nonlinear, $\tilde{\beta}$ may thus be stated as fuzzy triangular number

$$\tilde{\beta} = <3.131,\ 3.614,\ 4.113> \tag{6.66}$$

as a good approximation. The assigned fuzzy failure probability is also determined approximately as fuzzy triangular number

$$\tilde{P}_f = <1.95 \cdot 10^{-5},\ 1.51 \cdot 10^{-4},\ 8.71 \cdot 10^{-4}>. \tag{6.67}$$

The elements of the parameters of the fuzzy probability distributions that belong to the value $\beta = 3.131$ are $m_{X_1} = 52$ kN and $\sigma_{X_1} = 6.0$ kN for the load $\tilde{X}_1$ and $\sigma_{X_2} = 2.80 \cdot 10^4$ kN/m² for the yield stress $\tilde{X}_2$. The value $\beta = 4{,}113$ is obtained with the parameter combination $m_{X_1} = 47$ kN, $\sigma_{X_1} = 4.5$ kN, and $\sigma_{X_2} = 2.20 \cdot 10^4$ kN/m².

Additionally, the fuzzy design point $\tilde{\underline{y}}_B$ was completely determined (Fig. 6.25) and entered in Fig. 6.23. The fuzzy sensitivity factors $\tilde{\alpha}_1$ and $\tilde{\alpha}_2$ may be recognized by means of the fuzzy angles ψ_1 and ψ_2 indicated in Fig. 6.23.

For interpreting the solution of the problem in the x-space (Fig. 6.22) the fuzzy design point $\tilde{\underline{y}}_B$ is inversely transformed into the original space of the fuzzy probabilistic basic variables under consideration of all interactive relationships. The result $\tilde{\underline{x}}_B$ is illustrated approximately in Fig. 6.26 as a fuzzy triangular number on the coordinate axis s running along $g(x_1, x_2) = 0$.

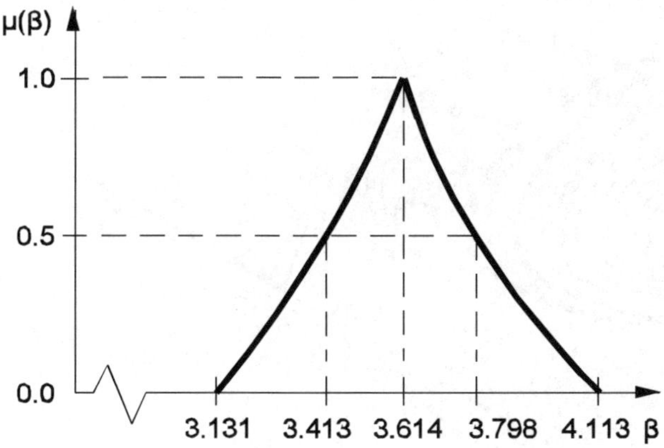

Fig. 6.24. Fuzzy reliability index $\tilde{\beta}$

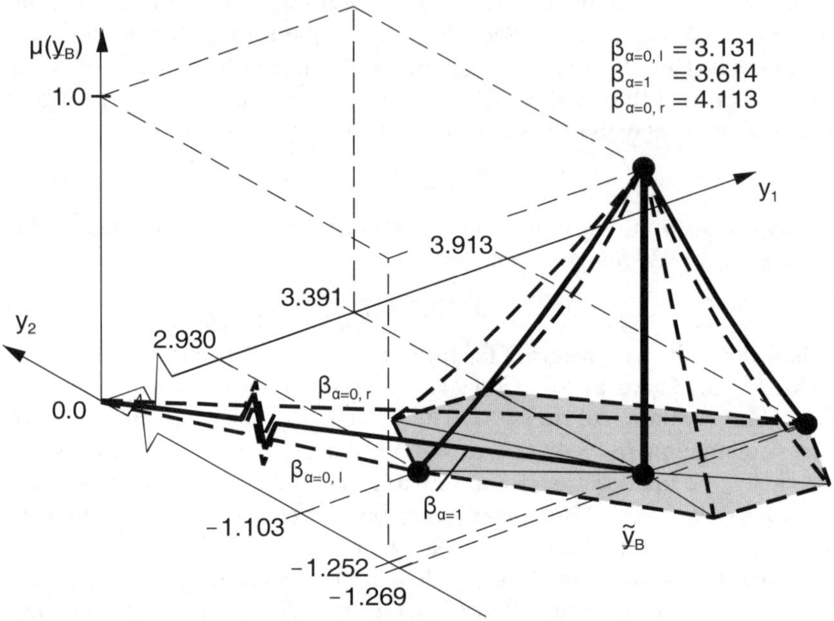

Fig. 6.25. Fuzzy design point $\tilde{\underline{y}}_B$ and fuzzy reliability index $\tilde{\beta}$

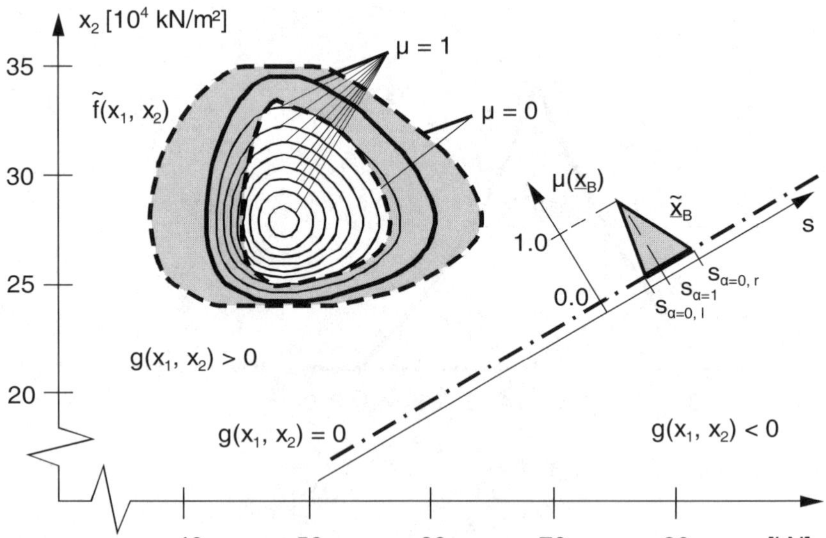

Fig. 6.26. Fuzzy design point $\tilde{\underline{x}}_B$ in the original space of the fuzzy probabilistic basic variables

The safety verification is carried out by comparing the fuzzy reliability index $\tilde{\beta}$ with required values *req_β*. The safety level to be complied with is specified as

$$req\_\beta = 3.8 \quad , \tag{6.68}$$

in compliance with [217] and [218].

The safety verification

$$\tilde{\beta} \geq req\_\beta \tag{6.69}$$

is only *partially fulfilled;* a *subjective assessment* of the comparison according to Eq. (6.69) is necessary. For this purpose, the fuzzy set $\tilde{\beta}$ is disjointed into the subsets (Fig. 6.27)

$$\tilde{\beta}_1 = \{(\beta, \mu(\beta)) \mid \beta \in \tilde{\beta}\,;\, \beta \geq req\_\beta\} \quad , \tag{6.70}$$

and

$$\tilde{\beta}_2 = \{(\beta, \mu(\beta)) \mid \beta \in \tilde{\beta}\,;\, \beta < req\_\beta\} \quad , \tag{6.71}$$

as described in Sect. 6.5. The fuzzy set $\tilde{\beta}_2$ comprising all elements $\beta \in \tilde{\beta}$ that do not satisfy the safety verification dominates in comparison with $\tilde{\beta}_1$. With the membership function $\mu(\beta)$ the measure values

$$\mu_1 = \sup_{\beta \in \tilde{\beta}_1} [\mu(\beta)] = 0.5 \quad , \tag{6.72}$$

and

$$\mu_2 = \sup_{\beta \in \tilde{\beta}_2} [\mu(\beta)] = 1.0 \tag{6.73}$$

are obtained. The safety verification is assessed as being fulfilled with $\mu_1 = 0.5$, but not fulfilled with $\mu_2 = 1.0$.

When subjectively assessing the partially fulfilled safety verification according to Eq. (6.69) a decision as to whether the safety verification may be considered to be fulfilled is made on the basis of expert knowledge with the aid of $\tilde{\beta}_1$, $\tilde{\beta}_2$, μ_1, and μ_2.

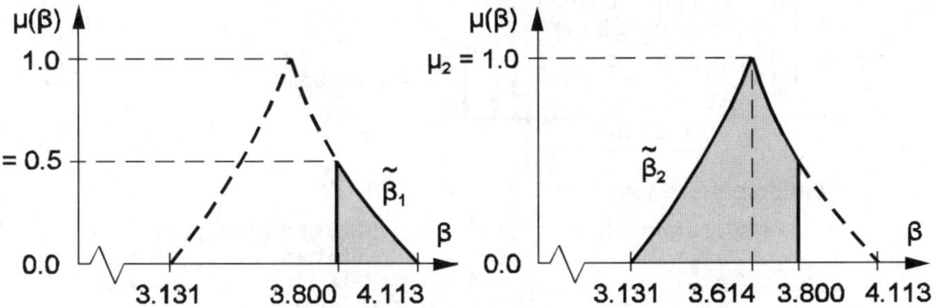

Fig. 6.27. Subjective assessment of the fuzzy safety level with the measure values μ_1 and μ_2 for the subsets $\tilde{\beta}_1$ and $\tilde{\beta}_2$

In *probabilistic safety assessment* particular subjective decisions are already made when determining the (crisp) probability distribution functions for the probabilistic basic variables, in fact without being able to evaluate the effects on the safety verification.

6.7.2 Reinforced-Concrete Frame, Nonlinear Statical System Behavior

The plane reinforced-concrete frame from Sect. 5.2.5.3 (Fig. 6.28) is investigated, the structural reliability in relation to global system failure is to be assessed [124]. For the deterministic fundamental solution the geometrically and physically nonlinear algorithm presented in [133] is chosen and the First Order Reliability Method matched to the latter according to [116] is applied as the probabilistic fundamental solution.

The effects of uncertainty with different characteristics are considered in the investigation and different modeling variants for uncertain parameters are compared. For assessing the structural reliability the fuzzy probabilistic safety concept FFORM presented in this chapter is chosen, the results are compared with those given by a common probabilistic method. The influence of the deterministic fundamental solution is demonstrated.

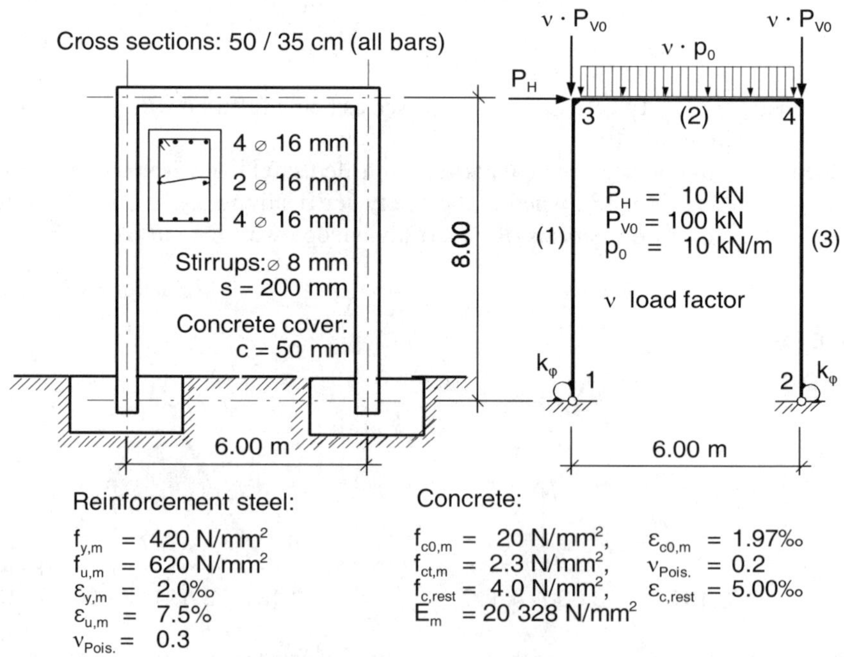

Fig. 6.28. Reinforced-concrete frame (plane); structure and statical system with loading

The system is modeled using three bars. Fifty integration increments are chosen for each bar and each cross section is subdivided into 60 layers. The geometrically and physically nonlinear analysis is carried out using the material laws for reinforcement steel and concrete after Oetes. Tension stiffening and the effects of stirrup reinforcement are accounted for in the concrete material law [133].

The simulated loading process is comprised of deadload, horizontal load P_H, vertical nodal loads $P_V = \nu \cdot P_{V0}$, and the line load $p = \nu \cdot p_0$. After applying dead load the horizontal load P_H is introduced; P_V and p are finally increased incrementally using the load factor ν until global system failure occurs.

Investigation I – Data Uncertainty. When modeling the frame, the rotational spring stiffness k_φ (for constraining the column bases in the foundation soil) must be defined. For this purpose an existing sample is considered, for which information concerning the recording of the sample and its numerical treatment is only available in an incomplete form.

The loading also possesses uncertainty; this is assigned to the load factor ν. For the loading and hence for ν only a sample may be considered the values of which cannot be assessed without an element of doubt regarding their correctness, owing to limited information.

The available samples for ν and k_φ are informally uncertain; both parameters are modeled as fuzzy random variables. Assuming that k_φ has the same value for both supports, the two fuzzy probabilistic basic variables $\tilde{X}_1$ for ν and $\tilde{X}_2$ for k_φ enter the safety assessment. Uncertainty with the characteristic fuzziness is not present and it is not necessary to take account of model uncertainty.

The functional types of the fuzzy probability distributions of ν and k_φ are taken to be specified. For the load factor ν an *extreme value distribution of the Ex-Max Type I* is adopted and the rotational spring stiffness k_φ is introduced into the investigation with a *logarithmic normal distribution.* The parameters of the fuzzy random variables are modeled as fuzzy triangular numbers. The load factor ν is specified by $\tilde{m}_{X_1} = < 5.7,\ 5.9,\ 6.0 >$ and $\tilde{\sigma}_{X_1} = < 0.08,\ 0.11,\ 0.12 >$; for the rotational spring stiffness k_φ the values $\tilde{m}_{X_2} = < 8.5,\ 9.0,\ 10.0 >$ MNm/rad and $\tilde{\sigma}_{X_2} = < 1.00,\ 1.35,\ 1.50 >$ MNm/rad are adopted. The minimum value of the logarithmic normal distribution is assumed to be a crisp value with $x_{0,2} = 0$ MNm/rad. The corresponding fuzzy probability density functions $\tilde{f}_1(x_1)$ for ν and $\tilde{f}_2(x_2)$ for k_φ are shown in Figs. 6.29 and 6.30, respectively.

In the original space of the fuzzy probabilistic basic variables $\tilde{X}_1$ and $\tilde{X}_2$ the $\tilde{f}_1(x_1)$ and $\tilde{f}_2(x_2)$ are combined to form the joint fuzzy probability density function $\tilde{f}(x_1, x_2)$. The (crisp) limit state surface $g(x_1, x_2) = 0$ is determined numerically (Fig. 6.31). The fuzzy design point $\tilde{\underline{x}}_B$, determined at a later stage, is also entered in the figure.

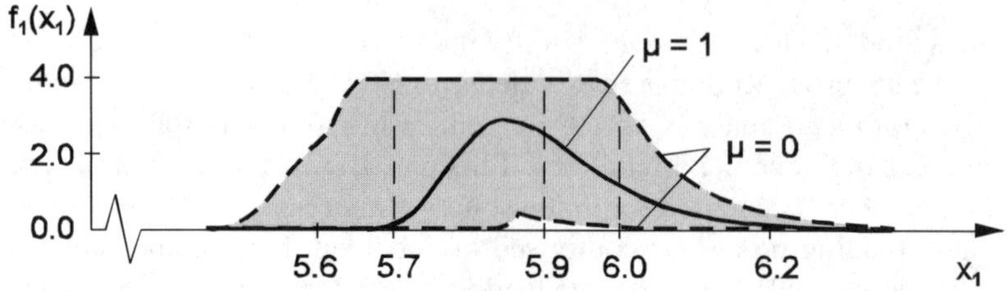

Fig. 6.29. Fuzzy probability density function $\tilde{f}_1(x_1)$ for the load factor ν

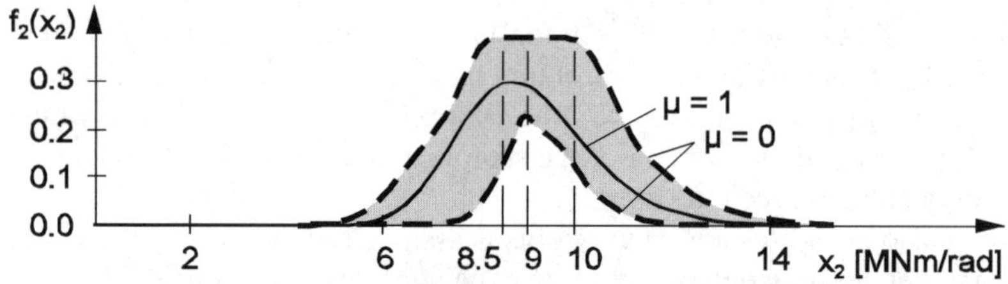

Fig. 6.30. Fuzzy probability density function $\tilde{f}_2(x_2)$ for the rotational spring stiffness k_φ

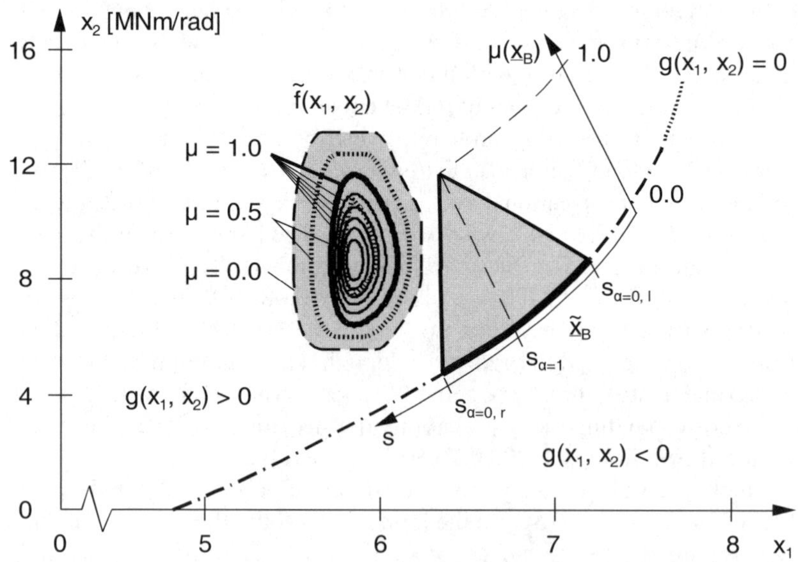

Fig. 6.31. Joint fuzzy probability density function $\tilde{f}(x_1, x_2)$, limit state surface $g(x_1, x_2) = 0$ and fuzzy design point $\tilde{\underline{x}}_B$

For the transformation of the problem into the standard normal space and the computation of the fuzzy reliability index the probabilistic and deterministic fundamental solution are applied in combination with the α-level optimization according to Sect. 6.6. The analysis is carried out for the five α-levels $\alpha_1 = 0.00$, $\alpha_2 = 0.25$, $\alpha_3 = 0.50$, $\alpha_4 = 0.75$, and $\alpha_5 = 1.00$. The fuzzy reliability index $\tilde{\beta}$ (Fig. 6.32) may be represented to a good approximation by the fuzzy triangular number

$$\tilde{\beta} = <3.851,\ 4.662,\ 6.684>. \tag{6.74}$$

The evaluation of further α-levels is not necessary. The corresponding fuzzy failure probability is computed as an approximation

$$\tilde{P}_f = <1.16\cdot 10^{-11},\ 1.57\cdot 10^{-6},\ 5.88\cdot 10^{-5}>. \tag{6.75}$$

The smallest element $\beta = 3.851$ of the fuzzy set $\tilde{\beta}$ is obtained with the parameter combination $m_{X_1} = 6.0$, $\sigma_{X_1} = 0.12$, $m_{X_2} = 8.5$ MNm/rad, and $\sigma_{X_2} = 1.5$ MNm/rad whereas the parameters $m_{X_1} = 5.7$, $\sigma_{X_1} = 0.08$, $m_{X_2} = 10.0$ MNm/rad, and $\sigma_{X_2} = 1.0$ MNm/rad belong to the largest element $\beta = 6.684$.

The fuzzy limit state surface $\tilde{h}(y_1, y_2) = 0$ and the fuzzy design point $\tilde{\underline{y}}_B$ are shown in Fig. 6.33. The fuzzy reliability index $\tilde{\beta}$ and the fuzzy angles ψ_1 and ψ_2 for determining the fuzzy sensitivity factors $\tilde{\alpha}_1$ and $\tilde{\alpha}_2$ are entered in the figure. The fuzzy design point $\tilde{\underline{x}}_B$ that was inversely transformed into the x-space is approximately represented in Fig. 6.31 by a fuzzy triangular number on the coordinate axis s running along $g(x_1, x_2) = 0$.

The safety verification is *fulfilled* with $req\_\beta = 3.8$

$$\tilde{\beta} = <3.851,\ 4.662,\ 6.684> \geq 3.8 = req\_\beta. \tag{6.76}$$

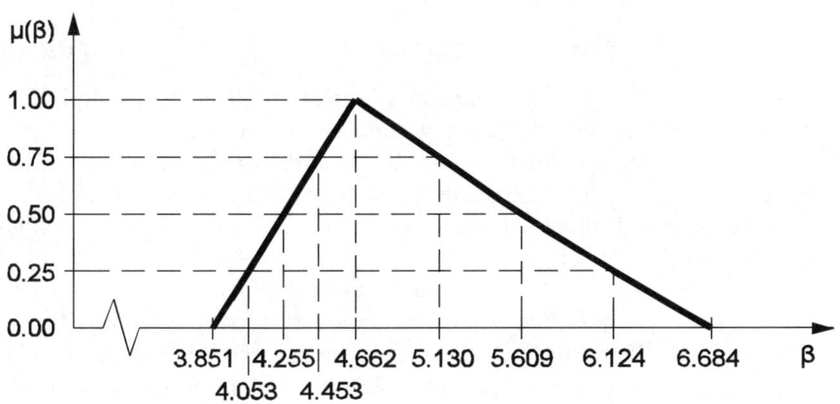

Fig. 6.32. Fuzzy reliability index $\tilde{\beta}$

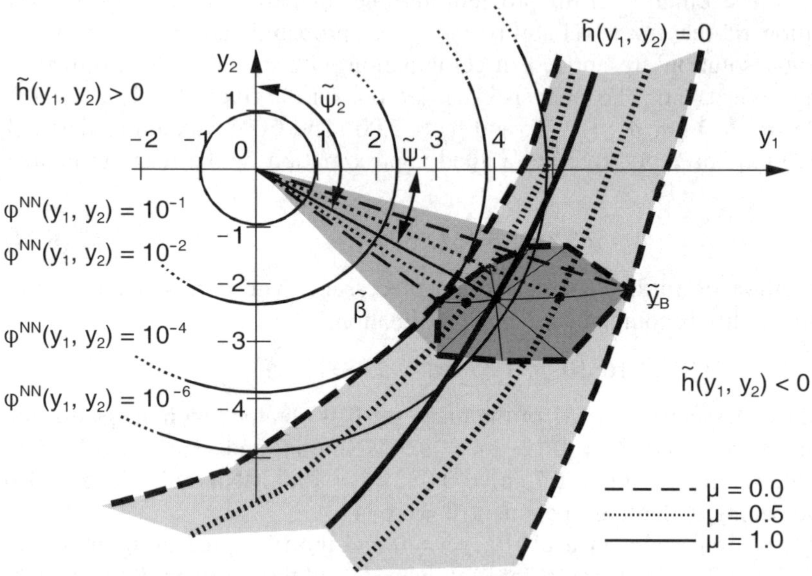

Fig. 6.33. Joint probability density function $\varphi^{NN}(y_1, y_2)$ of the standard normal distribution and fuzzy limit state surface $\tilde{h}(y_1, y_2) = 0$ in the standard normal space

Investigation II –Data and Model Uncertainty. The safety assessment of the frame is repeated with different modeling of the structural parameters. As an alternative, the rotational spring stiffness k_φ is described by the fuzzy triangular number

$$\tilde{k}_\varphi = <5,\ 9,\ 13> \ [\text{MNm/rad}]\,. \tag{6.77}$$

This modeling equals the problem description already used in the example for fuzzy structural analysis in Sect. 5.2.5.3, investigations II and III, Fig. 5.55.

In contrast to investigation I the rotational spring stiffness is now introduced as fuzzy model parameter $\tilde{k}_\varphi$ into the reliability analysis. Only the load factor ν is again treated as fuzzy probabilistic basic variable $\tilde{X}_1$; the modeling of ν remains unaltered.

The ultimate limit state is characterized by the load factor $\tilde{\nu}$ for the fuzzy failure load at global system failure. This is computed as a fuzzy result of a fuzzy structural analysis according to Sect. 5.2; the result obtained with the α-levels $\alpha_1 = 0.0$, $\alpha_2 = 0.2$, $\alpha_3 = 0.4$, $\alpha_4 = 0.6$, $\alpha_5 = 0.8$, and $\alpha_6 = 1.00$ is shown in Fig. 6.34 (see also Fig. 5.58).

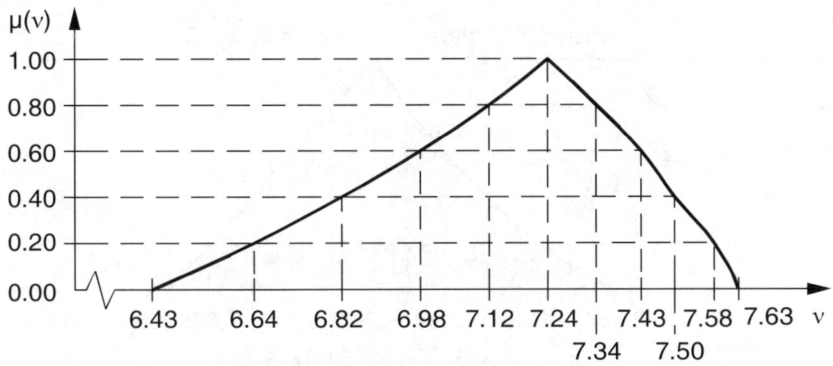

Fig. 6.34. Load factor $\tilde{v}$ for the fuzzy failure load at global system failure

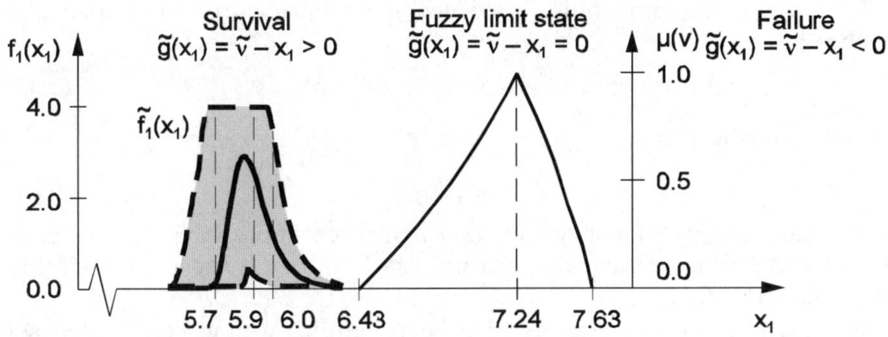

Fig. 6.35. Fuzzy probability density function $\tilde{f}_1(x_1)$ and fuzzy limit state surface

In the x-space a one-dimensional problem is formed (Fig. 6.35). The fuzzy failure region is bounded by $\tilde{v}$; the fuzzy limit state is described by

$$\tilde{g}(x_1) = \tilde{v} - x_1 = 0\,. \tag{6.78}$$

When integrating the fuzzy probability density function $\tilde{f}_1(x_1)$ the load factor of the fuzzy failure load $\tilde{v}$ is adopted as fuzzy integration limit

$$\tilde{P}_f = \int_{\tilde{x}_1 = \tilde{v}}^{\infty} \tilde{f}_1(x_1)\, dx_1\,. \tag{6.79}$$

The transformation into the standard normal space with the limit state

$$\tilde{h}(y_1) = \tilde{v}_y - y_1 = 0 \tag{6.80}$$

directly yields the fuzzy reliability index (Fig. 6.36)

$$\tilde{\beta} = \tilde{v}_y\,. \tag{6.81}$$

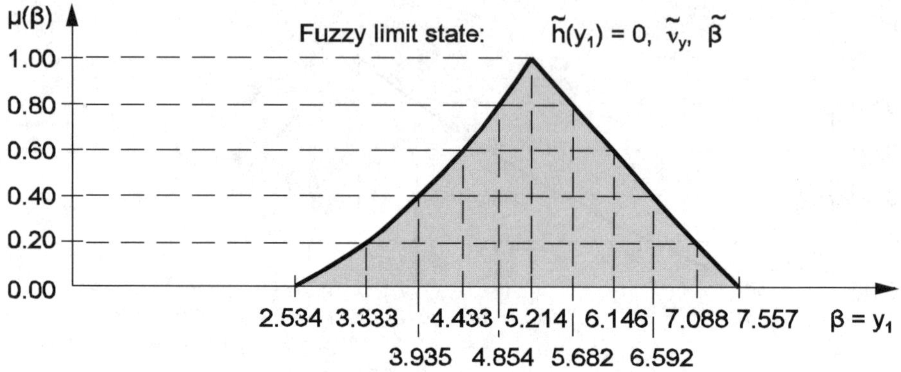

Fig. 6.36. Fuzzy reliability index $\tilde{\beta}$

The fuzzy failure probability, computed approximately as a fuzzy triangular number, is

$$\tilde{P}_f = <2.05 \cdot 10^{-14},\ 9.20 \cdot 10^{-8},\ 5.65 \cdot 10^{-3}> . \tag{6.82}$$

The safety verification

$$\tilde{\beta} \geq req\_\beta \tag{6.83}$$

is only *partially fulfilled*. For the *subjective assessment* of Eq. (6.83) the subsets $\tilde{\beta}_1$ and $\tilde{\beta}_2$ are constructed and the measure values μ_1 = 1.0 and μ_2 = 0.355 are determined (Fig. 6.37).

The *safety verification* is assessed as being fulfilled with μ_1 = 1.0 and as not fulfilled only with μ_2 = 0.355; in overall terms, however, it might be assessed as fulfilled.

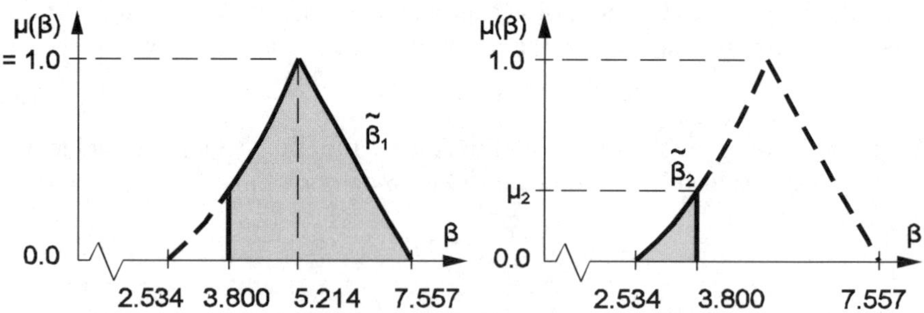

Fig. 6.37. Subjective assessment of the fuzzy safety level with the measure values μ_1 and μ_2 assigned to the subsets $\tilde{\beta}_1$ and $\tilde{\beta}_2$

A comparison of the fuzzy results $\tilde{\beta}_I$ from investigation I and $\tilde{\beta}_{II}$ from investigation II shows that the computed values of $\tilde{\beta}_I$ and $\tilde{\beta}_{II}$ are of comparable magnitude. The fuzzy reliability index $\tilde{\beta}_{II}$ from investigation II, however, possesses larger uncertainty (fuzziness) than $\tilde{\beta}_I$. In order to numerically compare the uncertainty of $\tilde{\beta}_I$ and $\tilde{\beta}_{II}$ the modified Shannon entropy measure according to Sect. 2.2.4 is applied. From Eq. (6.44) the results

$$H_u(\tilde{\beta}_I) = 1.410 \cdot k \ , \tag{6.84}$$

and

$$H_u(\tilde{\beta}_{II}) = 2.480 \cdot k \tag{6.85}$$

are obtained.

The reason for this effect is the different modeling of the rotational spring stiffness k_φ as a fuzzy random variable in investigation I and as a fuzzy variable in investigation II. The uncertainty of k_φ in investigation I is described with a component of randomness and a component of fuzziness. In contrast to this, only fuzziness is included in investigation II. Hence, more uncertainty with the characteristic fuzziness is introduced in investigation II. This fuzziness is reflected in the results $\tilde{\beta}_I$ and $\tilde{\beta}_{II}$.

Investigation III – Comparable Probabilistic Safety Assessment and Influence of the Deterministic Fundamental Solution. Investigation I is modified in such a way that the rotational spring stiffness k_φ as well as the load factor ν is modeled as real-valued random variables (without fuzziness). The *probabilistic safety assessment* may then be carried out using FORM. The parameters of the random variables, as assessed in investigation I with the membership value $\mu = 1$, are used. Hence, the load factor ν is characterized by $m_{X_1} = 5.9$ and $\sigma_{X_1} = 0.11$, and the rotational spring stiffness k_φ is determined with $m_{X_2} = 9.0$ MNm/rad, $\sigma_{X_2} = 1.35$ MNm/rad, and $x_{0,2} = 0$ MNm/rad. With an extreme value distribution of Ex-Max Type I for ν and a logarithmic normal distribution for k_φ the reliability index

$$\beta = 4.662 \tag{6.86}$$

is obtained (Fig. 6.32).

The influence of uncertainty in the determination of the parameters m_{X_1}, σ_{X_1}, m_{X_2}, and σ_{X_2} is neglected in the probabilistic method. The uncertainty when specifying the functional types of the probability distributions of the probabilistic basic variables is also not taken into consideration. This uncertainty also has a significant influence on the results of the safety assessment, however, as demonstrated by the following comparison.

For the load factor ν an *extreme-value distribution of the Ex-Max Type I* or a *normal distribution* is adopted according to choice, whereas a *normal distribution, a logarithmic normal distribution,* and a *beta distribution* are available as options for the rotational spring stiffness k_φ. For the beta distribution a maximum value of

$x_{max,2}$ = 12 MNm/rad is specified. The parameters m_{X_1}, σ_{X_1}, m_{X_2}, σ_{X_2}, and $x_{0,2}$ are retained with the (crisp) values stated above.

The FORM investigation leads to different safety prognoses for the different combinations of functional types of the probability distributions of ν and k_φ. The computed reliability index is shown in Table 6.1 for all possible combinations of functional types. The safety verification is *fulfilled* for the combinations one, two, four, and five whereas it is *not fulfilled* for the combinations three and six.

Table 6.1. Functional types of the probability distribution functions of ν and k_φ and assigned reliability index β

Functional type of the probability distribution for		Combi-nation	Reliability index β
the load factor ν	the rotational spring stiffness k_φ		
Extreme-value distribution Ex-Max Type I (Gumbel)	Normal distribution	1	4.172
	Logarithmic normal distribution	2	4.662
	Beta distribution	3	3.643
Normal distribution	Normal distribution	4	4.205
	Logarithmic normal distribution	5	5.778
	Beta distribution	6	3.663

An alternative fuzzy probabilistic safety assessment using FFORM may simultaneously account for all combinations given in Table 6.1. For this purpose fuzzy compound distributions for ν and k_φ are assumed. For the load factor ν the fuzzy probability density function

$$\tilde{f}_1(x_1) = \tilde{a}_1 \cdot f_{Ex\text{-}Max}(x_1) + \tilde{a}_2 \cdot f_{NV}(x_1) \tag{6.87}$$

with

$$\tilde{a}_1 = <0,\ 1,\ 1> \ , \tag{6.88}$$

and

$$\tilde{a}_2 = 1 - \tilde{a}_1 \tag{6.89}$$

is adopted. The rotational spring stiffness k_φ is described by

$$\tilde{f}_2(x_2) = \tilde{a}_3 \cdot f_{NV}(x_2) + \tilde{a}_4 \cdot f_{LNV}(x_2) + \tilde{a}_5 \cdot f_B(x_2) \ , \tag{6.90}$$

and

$$\tilde{a}_3 = <0,\ 0,\ 1> \ , \tag{6.91}$$

$$\tilde{a}_4 = <0,\ 1,\ 1> , \tag{6.92}$$

$$\tilde{a}_5 = <0,\ 0,\ 1>. \tag{6.93}$$

The interaction between $\tilde{a}_3$, $\tilde{a}_4$, and $\tilde{a}_5$ is specified by

$$\tilde{a}_3 + \tilde{a}_4 + \tilde{a}_5 = 1\,. \tag{6.94}$$

The evaluation of the α-levels $\alpha_1 = 0$ and $\alpha_2 = 1$ yields the fuzzy reliability index

$$\tilde{\beta} = <3.643,\ 4.662,\ 5.778>. \tag{6.95}$$

The uncertainty arising from the specification of the functional types of the probability distributions of ν and k_φ are included in this result.

The safety verification

$$\tilde{\beta} = <3.643,\ 4.662,\ 5.778> \ \geq\ 3.8 = req\_\beta \tag{6.96}$$

is only *partially fulfilled*.

Moreover, the *quality of the deterministic fundamental solution* has a significant influence on the safety assessment. The consideration of all essential geometrical and physical nonlinearities (in this case for reinforced concrete) is a necessary precondition in order to arrive at realistic results for the structural reliability.

Based on the example of the reinforced concrete frame, the influence of different nonlinearities on the results of the *FORM analysis* is investigated. For this purpose the reliability index β is determined with the aid of different mechanical models.

The load factor ν is again modeled using an *extreme-value distribution of Ex-Max Type I* with $m_{X_1} = 5.9$ and $\sigma_{X_1} = 0.11$. For the rotational spring stiffness k_φ a *beta distribution* with $m_{X_2} = 9.0$ MNm/rad, $\sigma_{X_2} = 1.35$ MNm/rad, a minimum value $x_{0,2} = 0$ MNm/rad, and a maximum value $x_{max,2} = 12$ MNm/rad is adopted.

Under consideration of all essential geometrical and physical nonlinearities the reliability index

$$\beta_{g+p} = 3.643 \tag{6.97}$$

is obtained (see combination three in Table 6.1); the safety verification is *not fulfilled*.

If, on the other hand, only the physical nonlinearities are taken into consideration, the safety verification given by

$$\beta_p = 6.845 \tag{6.98}$$

is *fulfilled*.

The consideration of geometrical nonlinearities alone in fact leads to

$$\beta_g > 10\,, \tag{6.99}$$

and hence also to a *fulfilled* safety verification.

The evaluated limit state surfaces $g(x_1, x_2) = 0$ are shown in Fig. 6.38 for the three investigated variants in the original space of the basic variables. The decisive influence of the deterministic computational model is reflected in the large differences between the corresponding limit state surfaces.

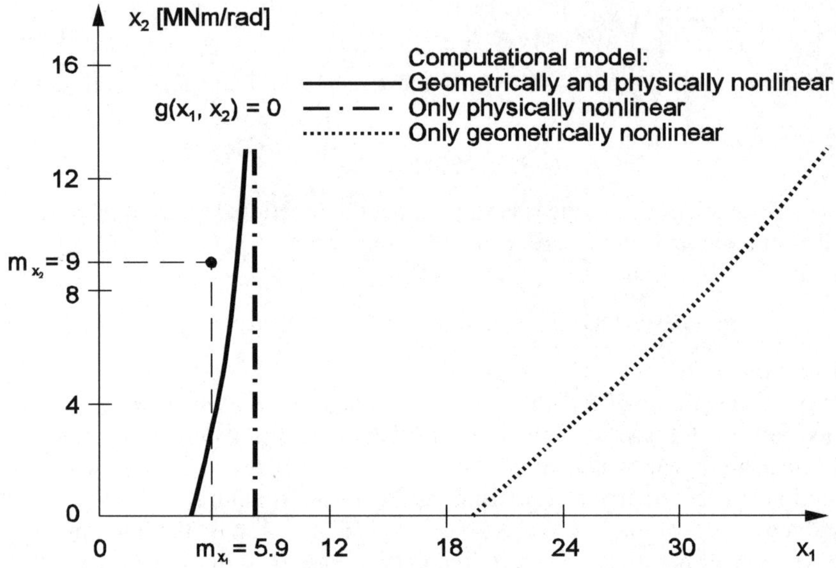

Fig. 6.38. Limit state surfaces $g(x_1, x_2) = 0$ in the x-space for different computational models

7 Structural Design Based on Clustering

7.1 Conceptual Idea

In fuzzy structural analysis according to Eq. (5.2) fuzzy input vectors $\underline{\tilde{x}}$ are mapped onto fuzzy result vectors $\underline{\tilde{z}}$, i.e.,

$$\underline{\tilde{x}} \rightarrow \underline{\tilde{z}} . \tag{7.1}$$

If α-level optimization is applied for solving Eq. (7.1), a set of input points $\underline{x}_i$ from the support subspace $\underline{X}_{\alpha=0}$ of the $\underline{\tilde{x}}$ and a set of result points $\underline{z}_i$ from the support subspace $\underline{Z}_{\alpha=0}$ of the $\underline{\tilde{z}}$ are obtained. The result points $\underline{z}_i$ are assigned to the input points $\underline{x}_i$ by means of the mapping model according to Eq. (5.6) on a point-by-point basis

$$\underline{z}_i = f(\underline{x}_i) . \tag{7.2}$$

As a result from α-level optimization, the inverse assignment is also available point-by-point

$$\underline{x}_i = f^{-1}(\underline{z}_i) . \tag{7.3}$$

The dependencies in Eqs. (7.2) and (7.3) virtually permit designing structures directly with the aid of cluster analysis methods. This initial situation is shown in Fig. 7.1 for points $\underline{x}_i$ and $\underline{z}_i$ in the two-dimensional support subspaces $\underline{X}_{\alpha=0}$ and $\underline{Z}_{\alpha=0}$.

The n-dimensional support subspace $\underline{X}_{\alpha=0}$ is formed by n one-dimensional supports $x_{k,\,\alpha=0}$, k = 1, ..., n, and the m-dimensional support subspace $\underline{Z}_{\alpha=0}$ is constituted by m one-dimensional supports $z_{j,\,\alpha=0}$, j = 1, ..., m. The elements of the supports $x_{i,1} \in x_{1,\alpha=0}, ..., x_{i,n} \in x_{n,\alpha=0}$ as well as $z_{i,1} \in z_{1,\alpha=0}, ..., z_{i,m} \in z_{m,\alpha=0}$ describe the points $\underline{x}_i = (x_{i,1}, ..., x_{i,n})$ and $\underline{z}_i = (z_{i,1}, ..., z_{i,m})$ in the support subspaces $\underline{X}_{\alpha=0}$ and $\underline{Z}_{\alpha=0}$. The points $\underline{x}_i$ are lumped together in the discrete point set M_x and the points $\underline{z}_i$ are combined to form the discrete point set M_z.

The elements $x_{i,1}$, ..., $x_{i,n}$ represent the *design parameters* $d_{i,1}$, ..., $d_{i,n}$ to be determined. These describe, e.g., geometrical properties or material parameters. The elements $z_{i,1}$, ..., $z_{i,m}$ are result variables, such as stresses, deformations, or a reliability index. These are referred to as *restricted parameters* $r_{i,1}$, ..., $r_{i,m}$ if they comply with the constraints CT_h, h = 1, ..., q according to the design requirements.

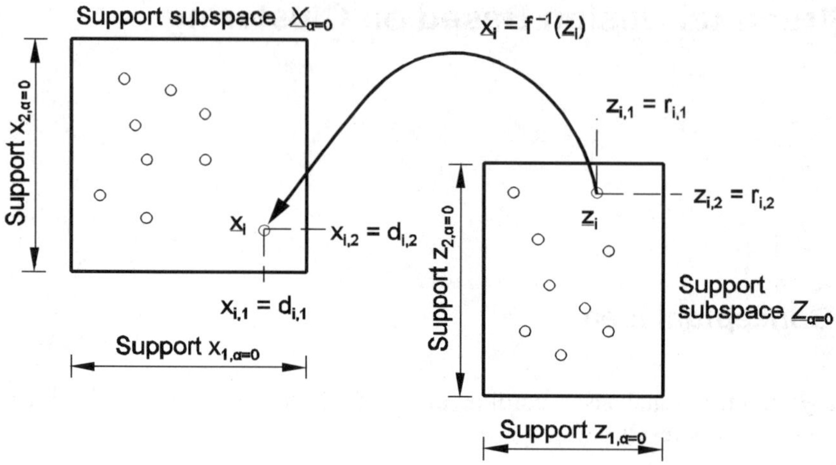

Fig. 7.1. Point sets M_x and M_z in the support subspaces $\underline{X}_{\alpha=0}$ and $\underline{Z}_{\alpha=0}$

As not all points $\underline{z}_i$ are restricted parameter vectors, the point set M_z is subdivided into two disjoint subsets comprising only permissible points $\underline{r}_i$ or only nonpermissible points $\overline{\underline{r}}_i$. As a consequence the point set M_x also becomes decomposed due to the dependencies in Eqs. (7.2) and (7.3). The resulting disjoint subsets of M_x contain only permissible points $\underline{d}_i$ or only nonpermissible points $\overline{\underline{d}}_i$.

A cluster analysis is separately applied to both the permissible points $\underline{d}_i$ and the nonpermissible points $\overline{\underline{d}}_i$. The obtained clusters comprised of permissible points represent alternative structural design variants [13, 126]. The clustering in a two-dimensional support subspace $\underline{X}_{\alpha=0}$ is shown in Fig. 7.2.

The conceptual idea presented may also be applied to fuzzy probabilistic safety assessment. In this case the reliability index or failure probability is the restricted parameter and structural model parameters as well as parameters of probability distributions represent design parameters. Furthermore, the concept of structural design based on clustering for the first time provides a tool for designing structures directly and independent of the computational model, i.e., every arbitrary nonlinear structural analysis algorithm may be taken as a basis.

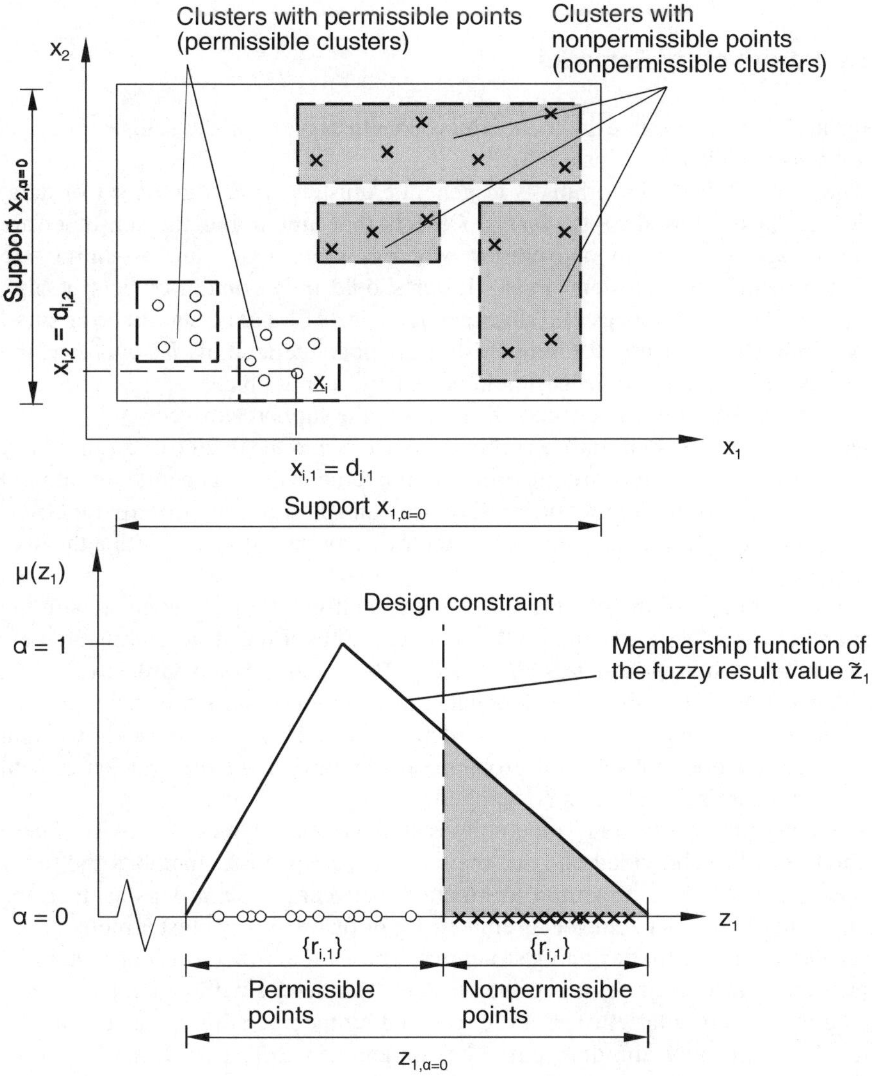

Fig. 7.2. Constraining result variables and clustering of the assigned permissible and nonpermissible input points

7.2 On the Application of Cluster Analysis Methods

7.2.1 Introductory Remarks

For generating permissible and nonpermissible clusters established cluster analysis algorithms are taken as a basis [76, 85].

The aim of these algorithms is to generate clusters from a given set of data, which are also referred to as *objects*. Objects that are similar to each other are lumped together to form one cluster, whereas objects that are dissimilar are assigned to different clusters. Each cluster should only contain objects of high *similarity*, i.e., the distances d(i,j) between its objects $\square_i$ and $\square_j$ should be as small as possible. Objects from different clusters are characterized by *dissimilarity*, the distances between these objects should be as large as possible.

In structural design with clusters the points in the support subspace $\underline{X}_{\alpha=0}$ are the objects. The degree of similarity within the clusters and the degree of dissimilarity between different clusters are maximized simultaneously to generate an optimal cluster configuration. For measuring similarity and dissimilarity different metrics, such as the Euclidean distance or an L_1-metric, may be adopted to define the distances d(i,j).

The distribution of the objects and thus the configuration of the clusters in the support subspace $\underline{X}_{\alpha=0}$ depend on the physical units of the structural parameters representing design parameters $d_{i,1}$, ..., $d_{i,n}$. By means of standardization, i.e., transformation of the unit-dependent data into a dimensionless representation, an attempt may be made to bypass this dependency. However, it is known that standardization does not solve the problem satisfactorily. For structural design with clusters standardization is thus not applied.

For determining clusters from point sets a variety of methods exist. These methods may be subdivided into two main groups, crisp cluster methods and fuzzy cluster methods. The essential distinction between these groups is that the assignment of objects to clusters is either crisp or fuzzy. Fuzzy cluster methods are particularly suitable for designing structures. These yield fuzzy clusters containing objects that uncertainly belong to the clusters. The degree with which the objects belong to the particular clusters are expressed by membership values $\mu \in [0, 1]$. One object $\square_i$ may simultaneously be assigned to different clusters C_v with different membership values. These membership values must comply with the requirement

$$\sum_{v=1}^{k} \mu_{iv} = 1 . \tag{7.4}$$

In addition to fuzzy cluster methods the application of crisp cluster methods is also shown subsequently. The latter is characterized by the fact that each object belongs precisely to one cluster. Two selected methods for determining crisp clusters as well as fuzzy clusters, which have proved to be suitable for structural design with clusters, are described in the following two sections. These may be replaced by other cluster methods if required.

Cluster analysis algorithms require the definition of a particular number of clusters in advance. The existing data set is usually clustered repeatedly for a different number of clusters. The results are then compared based on quality measures to determine the most suitable number of clusters. When assessing the quality of a cluster configuration, however, the absolute value of the quality measure should not be taken into consideration as the sole criterion. Additionally, the relative improvement in the quality measure with an altering, prescribed number of clusters should be accounted for.

7.2.2 k-medoid Cluster Method

The k-medoid cluster method [85] belongs to the partitioning methods, it is a crisp cluster method. Applying this method k clusters are constructed. The number k must be predefined in the particular case considered. In the result each object belongs to exactly one of the k clusters. As not all values of k lead to a useful clustering, it is advisable to run the algorithm several times with different values of k and to select the "best" k using numerical criteria.

After having defined the number k clusters, k representative objects are selected in a first stage (Fig. 7.3). The remaining objects are then assigned to the representative objects selected to form k clusters. This clustering is the basis for initializing the second stage.

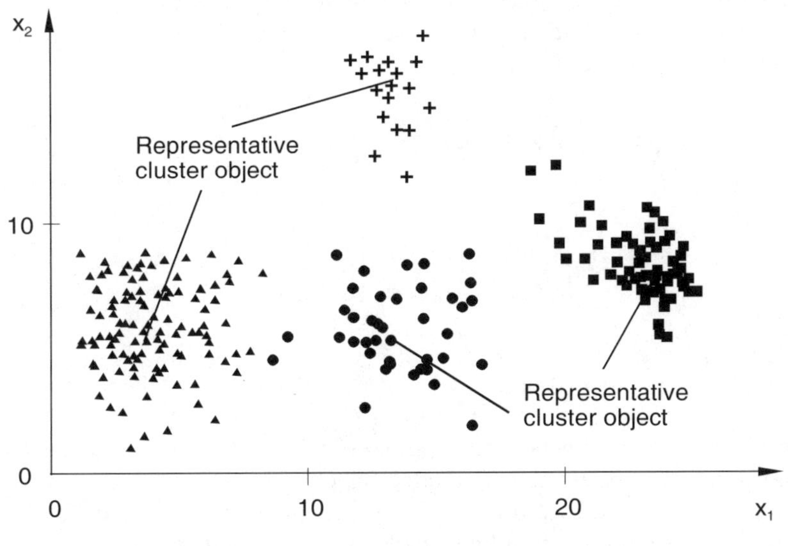

Fig. 7.3. Clustering based on the k-medoid method

In the second stage an attempt is made to improve the set of representative objects and hence to improve the clustering based on this set. This is done by considering all pairs of objects ($\square_i$, $\square_j$) for which object $\square_i$ has been selected as a representative object and object $\square_j$ has not. The effect of a swap on the quality of the clustering is then analyzed, i.e., object $\square_i$ is no longer selected as a representative object but object $\square_j$ is.

The result of a clustering based on the k-medoid method is shown in Fig. 7.3 for the two-dimensional case. The objects and the obtained clusters for $k = 4$ are illustrated graphically. For measuring dissimilarity the Euclidean distance was applied.

In the following, two numerical criteria are discussed, which permit assessment of the quality of clustering gained by applying the k-medoid cluster method.

Separation. According to [85] the separation of a cluster is defined by

$$SP(C_v) = \min d(\square_i, \square_j) \mid \square_i \in C_v; \square_j \notin C_v, \tag{7.5}$$

in which d ($\square_i$, $\square_j$) is the distance between objects $\square_i$ and $\square_j$ from different clusters.

The average separation of all k clusters may be computed with

$$ASP = \frac{1}{k} \sum_{v=1}^{k} SP(C_v). \tag{7.6}$$

This is a measure of the quality of the clustering, higher values of ASP characterize a better cluster configuration.

Silhouette Coefficient. The silhouette coefficient SC [85] assesses both the similarity of objects within the same cluster and the degree of separation between different clusters. It is defined by

$$SC = \frac{1}{k} \sum_{v=1}^{k} \frac{1}{m_v} \sum_{i=1}^{m_v} \frac{a_i - b_i}{\max [a_i, b_i]}, \tag{7.7}$$

with

$$a_i = \frac{1}{m_v} \sum_{j=1}^{m_v} d(\square_i, \square_j) \mid \square_i, \square_j \in C_v, \tag{7.8}$$

$$b_i = \min a_j ; \ i \neq j, \tag{7.9}$$

m_v representing the number of objects of cluster C_v, and k being the number of clusters.

The closer to the upper bound unity the value SC is obtained, the better the cluster configuration is assessed, i.e., a value SC close to unity indicates a clear cluster structure. Moreover, according to [85], values of $SC < 0.25$ signal that the k-medoid method does not represent a suitable cluster method with regard to the data set considered.

7.2.3 Fuzzy Cluster Method

For designing structures with the aid of cluster analysis the fuzzy cluster method presented in [85] is used as a basis. In this method the determination of the cluster configuration is formulated as a nonlinear optimization problem with equality and inequality constraints. Again this method aims at the determination of clusters for a predefined number k of clusters.

To obtain simultaneously a high similarity between objects within the same cluster and a high dissimilarity between objects from different clusters the objective function

$$CF = \sum_{v=1}^{k} \frac{\sum_{i=1}^{n} \sum_{j=1}^{n} \mu_{iv}^{r} \cdot \mu_{jv}^{r} \cdot d(\Box_i, \Box_j)}{2 \sum_{j=1}^{n} \mu_{jv}^{r}} \tag{7.10}$$

is minimized by iteratively improving the membership values μ_{iv} and μ_{jv}. These membership values denote the degree with which the objects $\Box_i$ and $\Box_j$ belong to the cluster C_v, k is the number of clusters, and n is the number of all existent objects. The exponent $r \in [1, \infty)$ controls the influence of the membership values sought. For large values r, i.e., $r \to \infty$, the result of the cluster analysis exhibits a strong fuzziness characterized by equal membership values $\mu_{iv} = 1/k$ for all objects. In contrast to this, a value r close to unity results in an almost crisp clustering. For applications the exponent $r = 2$ is frequently chosen.

The membership values μ_{iv} and μ_{jv} of the objects $\Box_i$ and $\Box_j$ in Eq. (7.10) are the results to be determined. An object $\Box_i$ may be assigned to different clusters with different membership values.

The constraints of the optimization problem are given by

$$\mu_{iv} \geq 0 \; ; \; i = 1, \ldots, n \; ; \; v = 1, \ldots, k \; , \tag{7.11}$$

and

$$\sum_{v=1}^{k} \mu_{iv} = 1 \; ; \; i = 1, \ldots, n \; . \tag{7.12}$$

The cluster configuration is iteratively determined. At first, the membership values μ_{iv} for $i = 1, \ldots, n$ and $v = 1, \ldots, k$ are initialized properly, e.g., with $\mu_{iv} = 1/k$ for all objects. Then, the value of the objective function according to Eq. (7.10) is minimized iteratively. As a result, each object gets its membership values and each cluster possesses a prototype in the sense of an (artificial) "main object". In the case that an object belongs to several clusters, the constraint according to Eq. (7.12) takes effect. The distribution of the membership values computed for the data set from Fig. 7.3 and k = 2 is illustrated graphically in Figs. 7.4 and 7.5.

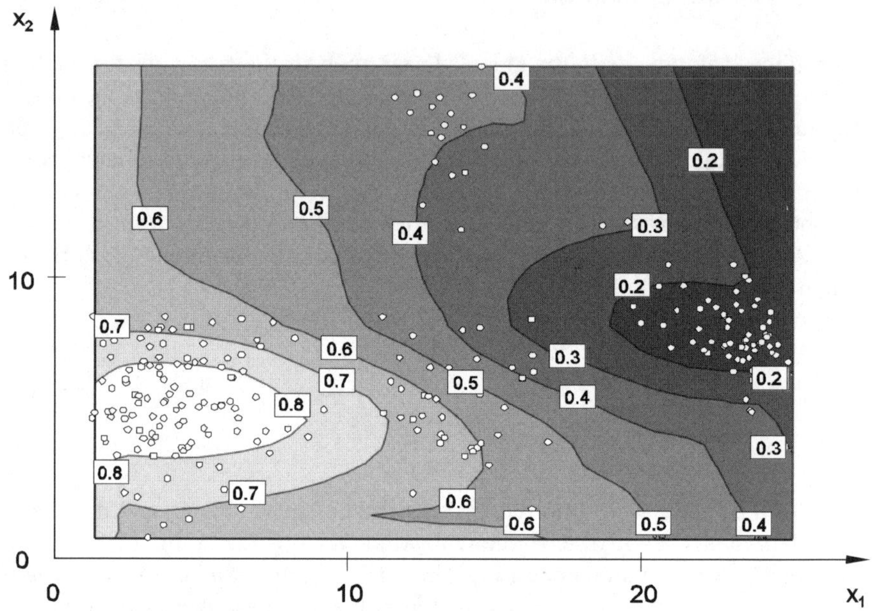

Fig. 7.4. Membership values for the assignment of the objects to the first cluster

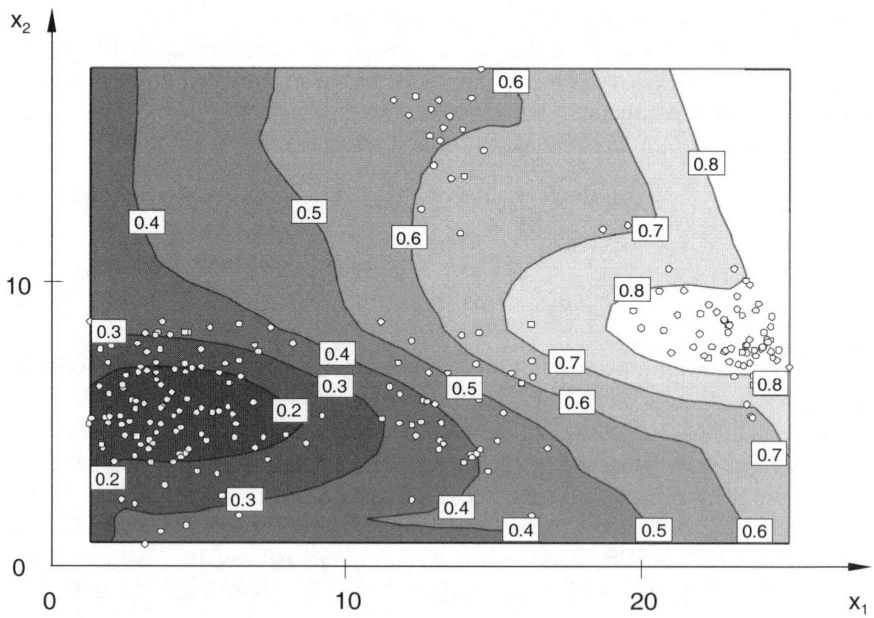

Fig. 7.5. Membership values for the assignment of the objects to the second cluster

The quality of a fuzzy clustering may be assessed by means of the following numerical criteria.

Partition Coefficient. This very simple criterion checks the membership values of the objects for values close to unity and zero. These membership values characterize a predominant assignment of the objects to only one cluster each. The more uncertain an assignment is obtained, the worse the cluster configuration is assessed to be.

The partition coefficient [76] is defined as

$$PC = \frac{1}{n} \sum_{v=1}^{k} \sum_{i=1}^{n} \mu_{iv}^{2}, \tag{7.13}$$

in which k is the number of clusters and n denotes the number of all existent objects.

The *normalized partition coefficient*, PCN, takes values from the interval [0, 1]

$$PCN = 1 - \frac{k}{k-1} \left(1 - PC\right) \quad | \quad k > 1\,. \tag{7.14}$$

An appropriate clustering is characterized by a normalized partition coefficient close to unity.

Separation Degree. With the aid of this measure the separation of the clusters is assessed. The distance between the clusters and the data assigned to the particular clusters, respectively, is related to the distance between different clusters. The separation degree [76] is computed with

$$SD = \frac{\sum_{i=1}^{n} \sum_{v=1}^{k} \mu_{iv}^{2} \cdot d\left(\square_i, \square_{p_v}\right)}{n \cdot \min\left[d^{2}\left(\square_{p_v}, \square_{p_w}\right) \;|\; \square_{p_v}, \square_{p_w} \in C_1, \ldots, C_k;\, \square_{p_v} \neq \square_{p_w}\right]}\,. \tag{7.15}$$

The objects $\square_{p_v}$ and $\square_{p_w}$ denote prototypes of clusters, k is the number of clusters, and n is the number of objects $\square_i$. The numerator in Eq. (7.15) evaluates the similarity of objects assigned to the same cluster in each case, this is desired to take a small value preferably. On the other hand, the denominator evaluates the dissimilarity between data from different clusters. For this purpose, use is made of the dissimilarity between the cluster prototypes $\square_{p_v}$ and $\square_{p_w}$, which is desired to be large. Combining these facts, a small value SD indicates a suitable cluster configuration.

7.3 Composition of Fuzzy Cluster Design

7.3.1 Mapping Spaces

Structural design with clusters is based on two spaces containing points known from fuzzy structural analysis or fuzzy probabilistic safety assessment based on α-level optimization; it may be applied in combination with arbitrary, nonlinear computational models.

Space of the Design Parameters. The n design parameters to be defined are determined in the space $\underline{D}^n$. The n-dimensional space $\underline{D}^n$ is formed by the supports $x_{k,\alpha=0}$, k = 1, ..., n of the fuzzy input variables $\tilde{x}_k$ of the α-level optimization (Sect. 5.2.3). For this purpose the assignment $D_1 = x_{1,\alpha=0}$, ..., $D_n = x_{n,\alpha=0}$ holds, i.e.,

$$\underline{D}^n = D_1 \times \ldots \times D_n = x_{1,\alpha=0} \times \ldots \times x_{n,\alpha=0} \,. \tag{7.16}$$

In the space $\underline{D}^n$ a distinction is made between permissible and nonpermissible design-parameter values. Permissible design-parameter values yield result values that comply with all design constraints CT_h. The elements of each one-dimensional set D_1, ..., D_n may represent both permissible design-parameter values d_1, ..., d_n and nonpermissible design-parameter values $\bar{d}_1$, ..., $\bar{d}_n$, i.e., the following holds

$$d_1 \in D_1, \ldots, d_n \in D_n \,, \tag{7.17}$$

$$\bar{d}_1 \in D_1, \ldots, \bar{d}_n \in D_n \,. \tag{7.18}$$

Moreover, discrete (known), real-valued permissible design-parameter values $d_{i,1}$, ..., $d_{i,n}$ and discrete (known), real-valued nonpermissible design-parameter values $\bar{d}_{i,1}$, ..., $\bar{d}_{i,n}$ are defined on the sets D_1, ..., D_n

$$d_{i,1} \in D_1, \ldots, d_{i,n} \in D_n \,, \tag{7.19}$$

$$\bar{d}_{i,1} \in D_1, \ldots, \bar{d}_{i,n} \in D_n \,. \tag{7.20}$$

Permissible design-parameter values and nonpermissible design-parameter values may be lumped together in n-tuples defining points in the space $\underline{D}^n$:

- n-tuple $\underline{d}$ = $(d_1, \ldots, d_n)$: arbitrary permissible point $\underline{d}$ in $\underline{D}^n$
- n-tuple $\bar{\underline{d}}$ = $(\bar{d}_1, \ldots, \bar{d}_n)$: arbitrary nonpermissible point $\bar{\underline{d}}$ in $\underline{D}^n$
- n-tuple $\underline{d}_i$ = $(d_{i,1}, \ldots, d_{i,n})$: discrete (known) permissible point $\underline{d}_i$ in $\underline{D}^n$
- n-tuple $\bar{\underline{d}}_i$ = $(\bar{d}_{i,1}, \ldots, \bar{d}_{i,n})$: discrete (known) nonpermissible point $\bar{\underline{d}}_i$ in $\underline{D}^n$

All discrete permissible points $\underline{d}_i$ are lumped together in the discrete set M_d, whereas all discrete nonpermissible points $\bar{\underline{d}}_i$ are lumped together in the discrete set $\bar{M}_d$:

$$M_d = \left\{\underline{d}_1, \ldots, \underline{d}_i, \ldots, \underline{d}_{n_d}\right\}, \tag{7.21}$$

$$\bar{M}_{\mathrm{d}} = \{\underline{\bar{\mathrm{d}}}_1, \ldots, \underline{\bar{\mathrm{d}}}_i, \ldots, \underline{\bar{\mathrm{d}}}_{\bar{n}_d}\}, \tag{7.22}$$

in which n_d is the number of permissible points $\underline{d}_i$ and $\bar{n}_d$ denotes the number of nonpermissible points $\underline{\bar{d}}_i$.

The elements of the sets M_d and $\bar{M}_d$ comprise all discrete (known) points in the space $\underline{D}^n$, which are lumped together in the point set

$$M_x = \{ M_d, \bar{M}_d \}. \tag{7.23}$$

This point set is known as an intermediate result from α-level optimization.

In structural design based on clustering the sets of known points M_d and $\bar{M}_d$ are separately introduced into cluster analysis, i.e., clusters exclusively formed by permissible points and clusters exclusively formed by nonpermissible points are obtained. Permissible design-parameter values $d_1, \ldots, d_n$ then lie within the clusters generated from the set M_d.

Space of the Restricted Parameters. The m-dimensional space $\underline{R}^m$ is constituted by the supports $z_{j,\,\alpha=0}$, $j = 1, \ldots, m$ of the fuzzy result values $\tilde{z}_j$ of the α-level optimization. With the assignment $R_1 = z_{1,\,\alpha=0}, \ldots, R_m = z_{m,\,\alpha=0}$ the space $\underline{R}^m$ is obtained from

$$\underline{R}^m = R_1 \times \ldots \times R_m = z_{1,\alpha=0} \times \ldots \times z_{m,\alpha=0}\,. \tag{7.24}$$

The elements of every one-dimensional set $R_1, \ldots, R_m$ may be restricted parameter values $r_1, \ldots, r_m$ as well as nonrestricted parameter values $\bar{r}_1, \ldots, \bar{r}_m$. Hence, the following holds

$$r_1 \in R_1, \ldots, r_m \in R_m, \tag{7.25}$$

$$\bar{r}_1 \in R_1, \ldots, \bar{r}_m \in R_m. \tag{7.26}$$

Furthermore, discrete (known), real-valued restricted parameter values $r_{i,1}, \ldots, r_{i,m}$ as well as discrete (known), real-valued nonrestricted parameter values $\bar{r}_{i,1}, \ldots, \bar{r}_{i,m}$ are defined on the sets $R_1, \ldots, R_m$

$$r_{i,1} \in R_1, \ldots, r_{i,m} \in R_m, \tag{7.27}$$

$$\bar{r}_{i,1} \in R_1, \ldots, \bar{r}_{i,m} \in R_m. \tag{7.28}$$

The restricted parameter values meet all requirements according to the design constraints CT_h.

The values of the restricted parameters and the nonrestricted parameters are lumped together in n-tuples characterizing permissible and nonpermissible points in the space $\underline{R}^m$:

- n-tuple $\underline{r} = (r_1, \ldots, r_m)$: arbitrary permissible point $\underline{r}$ in $\underline{R}^m$
- n-tuple $\underline{\bar{r}} = (\bar{r}_1, \ldots, \bar{r}_m)$: arbitrary nonpermissible point $\underline{\bar{r}}$ in $\underline{R}^m$
- n-tuple $\underline{r}_i = (r_{i,1}, \ldots, r_{i,m})$: discrete (known) permissible point $\underline{r}_i$ in $\underline{R}^m$
- n-tuple $\underline{\bar{r}}_i = (\bar{r}_{i,1}, \ldots, \bar{r}_{i,m})$: discrete (known) nonpermissible point $\underline{\bar{r}}_i$ in $\underline{R}^m$

All permissible points $\underline{r}_i$ are lumped together in the discrete set M_r, whereas all nonpermissible points $\overline{\underline{r}}_i$ are lumped together in the discrete set $\overline{M}_r$:

$$M_r = \left\{\underline{r}_1, \ldots, \underline{r}_i, \ldots, \underline{r}_{n_r}\right\}, \tag{7.29}$$

$$\overline{M}_r = \left\{\overline{\underline{r}}_1, \ldots, \overline{\underline{r}}_i, \ldots, \overline{\underline{r}}_{\overline{n}_r}\right\}, \tag{7.30}$$

in which n_r denotes the number of permissible points $\underline{r}_i$ and $\overline{n}_r$ is the number of nonpermissible points $\overline{\underline{r}}_i$.

The sets M_r and $\overline{M}_r$ contain all discrete points in the space $\underline{R}^m$ that are known as intermediate results from α-level optimization. At this stage of the structural-design process the separation of these points regarding permissibility and nonpermissibility has not yet been realized. This assignment is carried out directly with the aid of design constraints CT_h. After having separated all known points into $\underline{r}_i$ and $\overline{\underline{r}}_i$ with the aid of the design constraints, the overall discrete set

$$M_z = \{M_r, \overline{M}_r\} \tag{7.31}$$

is constituted.

The total number of discrete points known from α-level optimization is the same in both spaces $\underline{D}^n$ and $\underline{R}^m$, i.e., the following holds

$$n_d + \overline{n}_d = n_r + \overline{n}_r. \tag{7.32}$$

The following relationships exist between the discrete point sets M_x and M_z:

1. The elements $\underline{d}_i$ and $\underline{r}_i$ as well as the elements $\overline{\underline{d}}_i$ and $\overline{\underline{r}}_i$ are linked pairwise by the mapping model according to Eq. (5.6)

$$\underline{r}_i = f(\underline{d}_i), \tag{7.33}$$

$$\overline{\underline{r}}_i = f(\overline{\underline{d}}_i). \tag{7.34}$$

2. The inverse assignment

$$\underline{d}_i = f^{-1}(\underline{r}_i), \tag{7.35}$$

$$\overline{\underline{d}}_i = f^{-1}(\overline{\underline{r}}_i) \tag{7.36}$$

is present as a result from α-level optimization.

7.3.2 Algorithmic Procedure

The numerical procedure for fuzzy cluster design may be summarized in an algorithm comprising the following seven steps:

Step I. After the initialization of the spaces $\underline{D}^n$ and $\underline{R}^m$ the final results and intermediate results from α-level optimization are taken as the basis for fuzzy cluster design. The supports of the fuzzy input variables $\tilde{x}_k$ are assigned to the one-dimensional sets $D_1, \ldots, D_n$

$$\begin{aligned} D_1 &= x_{1,\alpha=0} \\ &\vdots \\ D_n &= x_{n,\alpha=0} \end{aligned} \quad . \tag{7.37}$$

The elements of the sets $D_1, \ldots, D_n$ taken into account by this means must represent values that are technically reasonable and practically realizable.

The supports of the fuzzy result variables $\tilde{z}_j$ are assigned to the one-dimensional sets $R_1, \ldots, R_m$

$$\begin{aligned} R_1 &= z_{1,\alpha=0} \\ &\vdots \\ R_m &= z_{m,\alpha=0} \end{aligned} \quad . \tag{7.38}$$

Step II. The points in $\underline{R}^m$ known from α-level optimization are evaluated by checking the design constraints CT_h. This permits the separation of these points into the subset M_r comprising all points $\underline{r}_i$ proved to be permissible and the subset $\overline{M}_r$ containing all nonpermissible points $\underline{\overline{r}}_i$. As a result the discrete set $M_z = \{M_r, \overline{M}_r\}$ is known.

Step III. The permissible points $\underline{d}_i$ and the nonpermissible points $\underline{\overline{d}}_i$, which belong to the points $\underline{r}_i$ and $\underline{\overline{r}}_i$, are determined in $\underline{D}^n$. For this purpose the inverse assignment according to Eqs. (7.35) and (7.36) is used, which is stored as a result from α-level optimization. Hence the discrete sets M_d and $\overline{M}_d$ according to Eqs. (7.21) and (7.22) are known.

Step IV. The set M_d is decomposed into k_1 permissible clusters and on the set $\overline{M}_d$ k_2 nonpermissible clusters are determined. To find an appropriate cluster configuration requires a procedure comprised of two stages. At first, clustering is carried out for a varying number of clusters $k_1, k_2 \in \mathbb{N}$. For this purpose, e.g., the k-medoid cluster method described in Sect. 7.2.2 or the fuzzy cluster method according to Sect. 7.2.3 may be applied. In the second stage the obtained cluster configurations, which differ from each other regarding the number of clusters, are assessed based on numerical quality measures to select the most suitable clustering. Cluster configurations obtained by applying the k-medoid method may be assessed with the aid of the criteria according to Eqs. (7.6) and (7.7). Cluster configurations resulting from fuzzy cluster analysis may be assessed by means of the criteria in Eqs. (7.14) and (7.15). In addition to the values of these quality measures, their relative variation when changing the number of clusters is taken into consideration.

Step V. The k_I permissible clusters C_v are taken as the basis for constructing modified one-dimensional sets $D_1^{[v]}$, ..., $D_n^{[v]}$ with $v = 1, ..., k_I$ of the *first generation*. These clusters are reduced by the intersections between the permissible clusters and the nonpermissible clusters. The remaining, reduced clusters $C_{v,red}$ comprise only permissible parameter combinations to a high probability. The boundaries of the k_I reduced permissible clusters are then used to form the sets $D_1^{[v]}$, ..., $D_n^{[v]}$. Each of the k_I reduced permissible clusters yields one sequence of sets $D_1^{[v]}$, ..., $D_n^{[v]}$, i.e., a total of k_I sequences are generated

- Reduced permissible cluster $C_{1,red} \rightarrow D_1^{[1]}, ..., D_n^{[1]}$
- Reduced permissible cluster $C_{2,red} \rightarrow D_1^{[2]}, ..., D_n^{[2]}$

$\vdots$

- Reduced permissible cluster $C_{k_I,red} \rightarrow D_1^{[k_I]}, ..., D_n^{[k_I]}$

Each sequence $D_1^{[v]}$, ..., $D_n^{[v]}$ represents an alternative structural design variant. These design variants may intersect each other.

Step VI. Clustering permissible points and generating reduced permissible clusters do not guaranty that all design variants include permissible design-parameter values exclusively. For the purpose of verification an α-level optimization is thus carried out for each design variant. This procedure starts with assigning the sets $D_1^{[v]}$, ..., $D_n^{[v]}$ with $v = 1, ..., k_I$ to supports of modified fuzzy input variables $\tilde{x}_1^{[v]}$, ..., $\tilde{x}_n^{[v]}$

$$\begin{aligned} x_{1,\alpha=0}^{[v]} &= D_1^{[v]} \\ &\vdots \\ x_{n,\alpha=0}^{[v]} &= D_n^{[v]} \end{aligned} \quad . \tag{7.39}$$

The supports $x_{1,\alpha=0}^{[v]}$, ..., $x_{n,\alpha=0}^{[v]}$ are taken as the basis for defining modified membership functions for the $\tilde{x}_k^{[v]}$. The results of structural design based on clustering do not primarily depend on these membership functions but on the supports. Arbitrary shapes of membership functions may thus be constructed in general. As simple shapes are usually preferred, e.g., fuzzy triangular numbers may be used. Their mean values may be determined, e.g., by considering the cluster centers, the representative objects, the prototypes, or other specifically selected points. Moreover, the curves of the membership functions may be oriented to, e.g., the membership values of the points resulting from fuzzy cluster analysis, the local density of points in the cluster, or other characteristics and properties of the cluster. Generally speaking, all concepts of fuzzification may be applied (Chap. 3).

Based on the modified fuzzy input variables $\tilde{x}_1^{[v]}$, ..., $\tilde{x}_n^{[v]}$ generated in this way modified fuzzy result variables $\tilde{z}_1^{[v]}$, ..., $\tilde{z}_m^{[v]}$ for $v = 1, ..., k_I$ are computed with the aid of α-level optimization.

Step VII. For each of the k_I structural design variants the support subspace $\underline{Z}_{\alpha=0}^{[v]}$ is determined and all points in $\underline{Z}_{\alpha=0}^{[v]}$ obtained from α-level optimization are evaluated by means of the design constraints CT_h. If not all points comply with the requirements according to the design constraints CT_h, the design variant is considered as not permissible. If all design variants are not permissible, the cluster configuration has to be improved. In this case, the procedure of structural design based on clustering is repeated starting with Step II. This yields modified design variants of the *second generation*.

If the design constraints CT_h are complied with by all points from the support subspaces $\underline{Z}_{\alpha=0}^{[v]}$ of several design variants, $k_d = k_1 - \bar{k}_d$ permissible design variants exist with $\bar{k}_d$ indicating the number of nonpermissible design variants. Each element of the sets $D_1^{[v]}, ..., D_n^{[v]}$ represents one permissible design parameter value.

7.3.3 Assessment of the Results from Fuzzy Cluster Design

Structural design based on clustering leads to k_d permissible design variants. Each of these design variants is characterized by the following variables (Sect. 7.3.2, Steps VI, VII):

- Fuzzy input variables $\tilde{x}_1^{[v]}, ..., \tilde{x}_n^{[v]}$ with the supports $x_{1,\alpha=0}^{[v]}, ..., x_{n,\alpha=0}^{[v]}$ for $v = 1, ..., k_d$
- Fuzzy result variables $\tilde{z}_1^{[v]}, ..., \tilde{z}_m^{[v]}$ with the supports $z_{1,\alpha=0}^{[v]}, ..., z_{m,\alpha=0}^{[v]}$ for $v = 1, ..., k_d$

All elements of the supports $z_{1,\alpha=0}^{[v]}, ..., z_{m,\alpha=0}^{[v]}$ meet the requirements according to the design constraints CT_h and all elements of the supports $x_{1,\alpha=0}^{[v]}, ..., x_{n,\alpha=0}^{[v]}$ are permissible design-parameter values.

For assessing the alternative, permissible structural design variants arbitrary criteria may be formulated and combined with each other. The following two criteria represent examples.

Criterion I – Constraint Distance. The fuzzy result variables $\tilde{z}_1^{[v]}, ..., \tilde{z}_m^{[v]}$ are defuzzified with the aid of suitably chosen methods (Sect. 2.1.10). This leads to the crisp values $z_{10}^{[v]}, ..., z_{m0}^{[v]}$ for the results. The distances between these crisp values and the design constraints CT_h represented by $perm\_z_j^{[v]}$ are used as a criterion

$$A_v = \sum_{j=1}^{m} w_j \cdot \left| perm\_z_j^{[v]} - z_{j0}^{[v]} \right| \tag{7.40}$$

for the distinction between the design variants. The weighting factors w_j are introduced for taking account of the importance of the assigned constraints.

A small weighted distance A_v characterizes a structural design that exploits the constraints to a high degree. On the other hand, a large weighted distance indicates a structural design providing some reserves, e.g., in load-bearing capacity.

Criterion II – Robustness. The robustness of a structural design is assessed on the basis of the sensitivity of the result variables (structural responses or safety measures) with regard to fluctuations of the input variables (structural parameters). With the aid of Shannon's entropy measure (Sect. 2.2.4) the criterion of robustness is defined as

$$B_v = \sum_{j=1}^{m} \sum_{k=1}^{n} \frac{H_u(\tilde{z}_j^{[v]})}{H_u(\tilde{x}_k^{[v]})}, \tag{7.41}$$

with $H_u(.)$ representing the modified Shannon's entropy according to Eq. (2.149). A small value B_v indicates a low sensitivity, i.e., a robust structural design.

7.4 Examples

7.4.1 Steel Girder, Structural Design for Time-dependent Load

Structural design based on clustering is demonstrated for the example of the steel girder shown in Fig. 7.6. The design is carried out in a two-dimensional space $\underline{D}^2$ of design parameters and a one-dimensional space $\underline{R}^1$ of restricted parameters. The procedure of fuzzy cluster design may therefore be illustrated graphically in an almost complete manner.

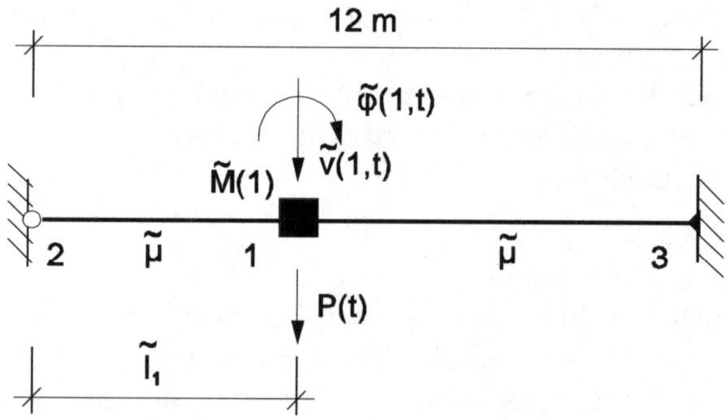

Dynamic load

$P(t) = P_1 \cdot \cos(\Omega_1 \cdot t) + P_2 \cdot \cos(\Omega_2 \cdot t)$
$P_1 = P_2 = 10$ kN
$\Omega_1 = 44\ s^{-1}$
$\Omega_2 = 66\ s^{-1}$

Steel girder

Young's modulus $E = 2.1 \cdot 10^8$ kN/m²
Moment of inertia $I = 1.5 \cdot 10^{-3}$ m⁴

Rotational masses are neglected

Fig. 7.6. Steel girder, statical system, and loading

The steel girder is excited to vibrate by a time-dependent load P(t). Fuzzy structural parameters are the nodal mass $\tilde{M}(1) = < 2, 10/3, 6 >$ t, the continuously distributed mass $\tilde{\mu} = < 1/3, 5/9, 1 >$ t/m, and the distance $\tilde{l}_1 = < 3, 4, 7 >$ m. Assuming full interaction between $\tilde{M}(1)$ and $\tilde{\mu}$ according to

$$M(1)\cdot 2 = \mu\cdot 12\mathrm{m}\,, \tag{7.42}$$

these masses may be lumped together to the global mass of the structure

$$\tilde{M} = \tilde{M}(1) + \tilde{\mu}\cdot 12\mathrm{m} = <6, 10, 18>\mathrm{t}\,. \tag{7.43}$$

Hence, only two fuzzy input variables enter the fuzzy structural analysis.

Fuzzy structural analysis with the fuzzy input variables $\tilde{M}$ and $\tilde{l}_1$ yields the time-dependent fuzzy results $\tilde{v}(1,t)$ and $\tilde{\varphi}(1,t)$ (Fig. 7.6). These are combined to compute the displacement norm

$$\tilde{v}(t) = \sqrt{\frac{\tilde{v}(1,t)^2}{l^2} + \frac{\tilde{\varphi}(1,t)^2}{r^2}}\,, \tag{7.44}$$

with l = 1 m and r = 1 rad. The maximum value $max\_\tilde{v}$ of $\tilde{v}(t)$ appears periodically in time intervals of 0.2856 s.

The global mass M and the distance l_1 are chosen as design parameters, for which permissible values are determined with the aid of fuzzy cluster design. A requirement is stated concerning the maximum displacement norm $max\_\tilde{v}$, this must comply with the design constraint

$$CT_1{:}\quad max\_\tilde{v} \le 5\cdot 10^{-3}\,. \tag{7.45}$$

The membership function of $max\_\tilde{v}$ and the constraint CT_1 are shown in Fig. 7.7.

The space of the restricted parameters is one-dimensional, i.e., $\underline{R}^1 = R_1$ holds. The set R_1 follows from the support $max\_v_{\alpha=0}$ according to

$$R_1 = max\_v_{\alpha=0}\,. \tag{7.46}$$

Permissible points $r_i \in \underline{R}^1$ and nonpermissible points $\overline{r}_i \in \underline{R}^1$ resulting from α-level optimization are illustrated symbolically in Fig. 7.7.

The space of the design parameters $\underline{D}^2$ is initialized by the supports of the fuzzy input variables $\tilde{M}$ and $\tilde{l}_1$ (Fig. 7.8)

$$\underline{D}^2 = D_1 \times D_2 \mid D_1 = M_{\alpha=0}\,;\; D_2 = l_{1,\alpha=0}\,. \tag{7.47}$$

From fuzzy structural analysis 385 permissible points $\underline{d}_i$ and 164 nonpermissible points $\overline{\underline{d}}_i$ are known in the space $\underline{D}^2$. The same number of permissible points r_i and nonpermissible points $\overline{r}_i$ simultaneously exist in $\underline{R}^1$.

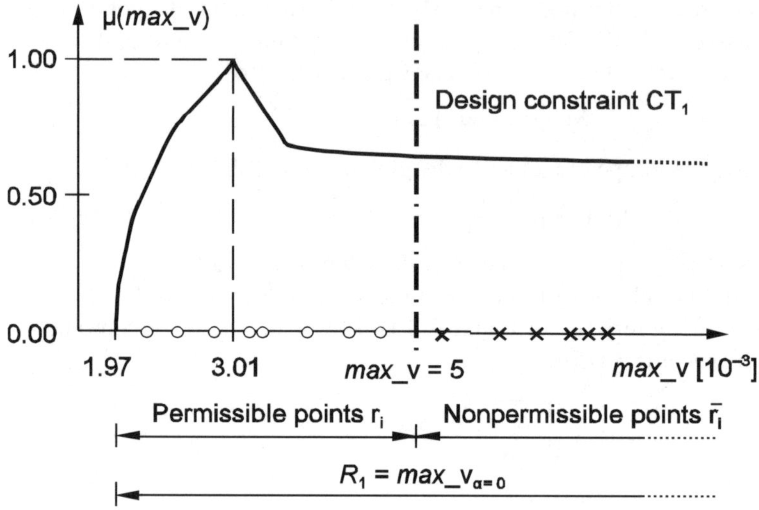

Fig. 7.7. Fuzzy result $max\_\tilde{v}$, design constraint CT_1, and restricted parameter values

Structural Design Using the k-medoid Cluster Method. With the aid of the k-medoid cluster method an appropriate cluster configuration is intended to be determined separately for the permissible points $\underline{d}_i$ and the nonpermissible points $\overline{\underline{d}}_i$. In order to find these configurations the prescribed number k of clusters is varied within a reasonable interval, e.g., between k = 2 and k = 20. The cluster configurations obtained are assessed with the numerical quality measures discussed in Sect. 7.2.2. Regarding the permissible points $\underline{d}_i$ the number of clusters is chosen to be k_1 = 12 (Fig. 7.9). For the nonpermissible points $\overline{\underline{d}}_i$ k_2 = 6 clusters are determined. The intersections between the permissible clusters and the nonpermissible clusters are taken into consideration when generating alternative design variants. The configurations of the permissible and nonpermissible clusters are plotted in Fig. 7.8, the intersections may be recognized clearly.

The bounds of the k_1 = 12 permissible clusters reduced by the intersections are taken as the basis for generating modified sets $D_1^{[v]}$ and $D_2^{[v]}$ with v = 1, ..., 12 and hence for defining alternative design variants.

However, not all permissible clusters are used to derive design alternatives, but clusters possessing large dimensions are preferably taken as a basis. Furthermore, clusters may also be merged artificially to obtain a supercluster with larger cluster dimensions. In the example clusters C_5, C_7, and C_{10} are considered suitable for further treatment. Cluster C_8 is not included in this consideration as partial nonpermissibility is expected to be discovered in a subsequent fuzzy structural analysis owing to its position close to a considerable number of nonpermissible points. Moreover, merging clusters C_4 and C_5 the larger supercluster $C_{4,5}$ is constituted (Figs. 7.8 and 7.10).

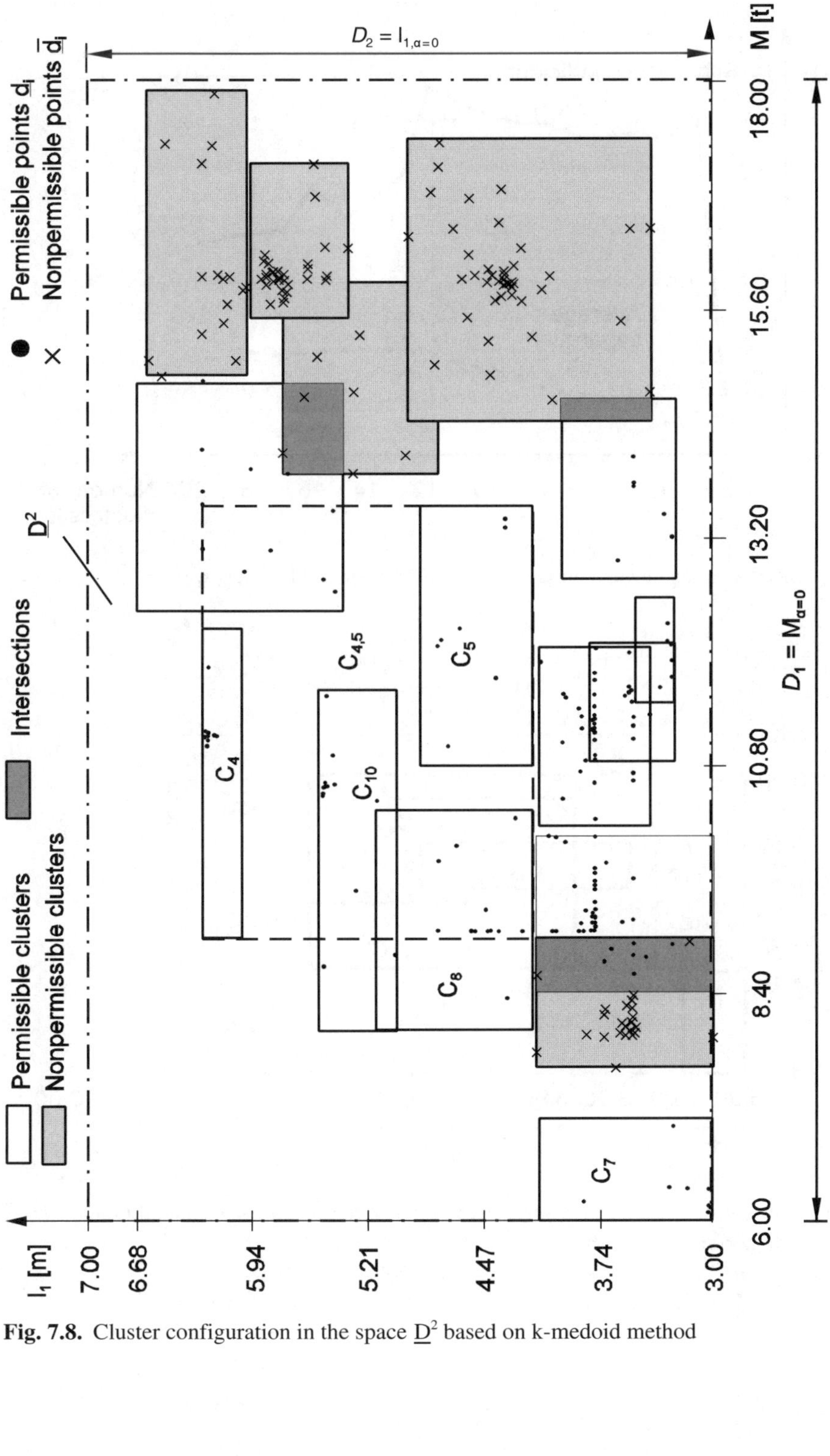

Fig. 7.8. Cluster configuration in the space $\underline{D}^2$ based on k-medoid method

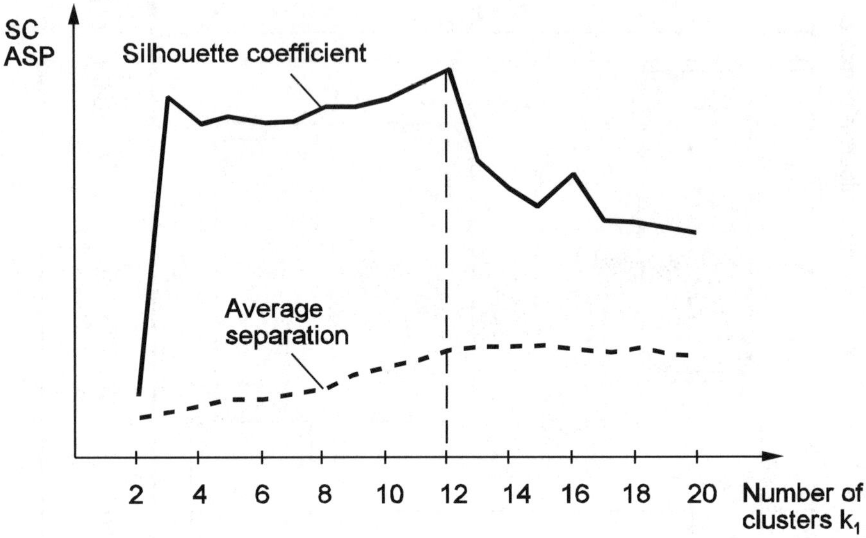

Fig. 7.9. Assessing cluster configurations for the permissible points $\underline{d}_i$

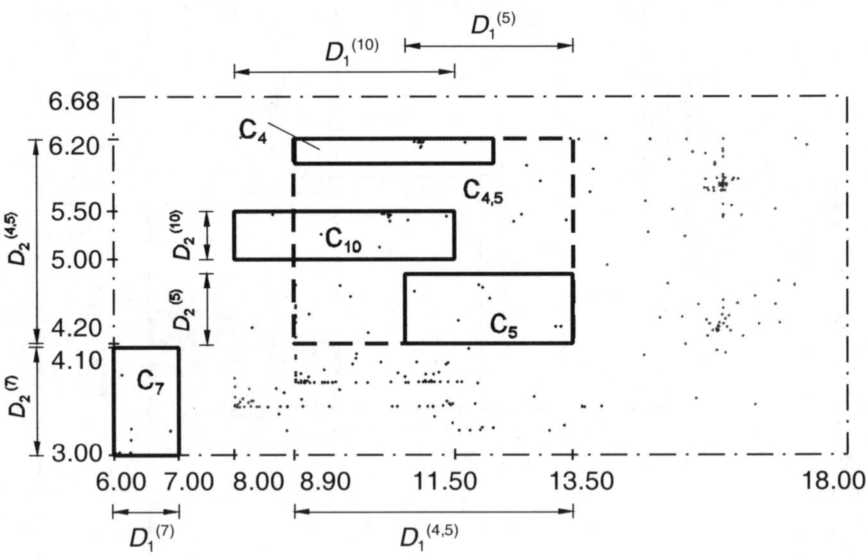

Fig. 7.10. Clusters selected for alternative design variants and assigned modified sets $D_1^{[v]}$, $D_2^{[v]}$, first generation

As a result, four sequences of modified sets $D_1^{[v]}$ and $D_2^{[v]}$ exist, which represent four alternative design variants of the first generation:

- Cluster C_5: $D_1^{[5]}$, $D_2^{[5]}$
- Cluster C_7: $D_1^{[7]}$, $D_2^{[7]}$
- Cluster C_{10}: $D_1^{[10]}$, $D_2^{[10]}$
- Cluster $C_{4,5}$: $D_1^{[4,5]}$, $D_2^{[4,5]}$

For verifying these structural design alternatives modified fuzzy result variables are computed with the aid of α-level optimization. For this purpose the modified fuzzy input variables

- $\tilde{M}^{[5]}$ and $\tilde{l}_1^{[5]}$
- $\tilde{M}^{[7]}$ and $\tilde{l}_1^{[7]}$
- $\tilde{M}^{[10]}$ and $\tilde{l}_1^{[10]}$
- $\tilde{M}^{[4,5]}$ and $\tilde{l}_1^{[4,5]}$

shown in Fig. 7.11 are constructed as fuzzy triangular numbers on the basis of the clusters selected (Sect. 7.3.2, Step VI). The mean values of the modified fuzzy input variables, i.e., the values assessed with the membership $\mu = 1$, are defined with regard to the center of gravity of the point sets. Applying α-level optimization yields the modified fuzzy result variables

- $max\_\tilde{v}^{[5]}$
- $max\_\tilde{v}^{[7]}$
- $max\_\tilde{v}^{[10]}$
- $max\_\tilde{v}^{[4,5]}$

shown in Fig. 7.12. All elements of the assigned supports

- $max\_v_{\alpha=0}^{[5]}$
- $max\_v_{\alpha=0}^{[7]}$
- $max\_v_{\alpha=0}^{[10]}$
- $max\_v_{\alpha=0}^{[4,5]}$

must comply with the design constraint CT_1, i.e., the maximum displacement norm of $max\_v = 5 \cdot 10^{-3}$ must not be exceeded. The elements of the support $max\_v_{\alpha=0}^{[7]}$ meet this requirement only partially.

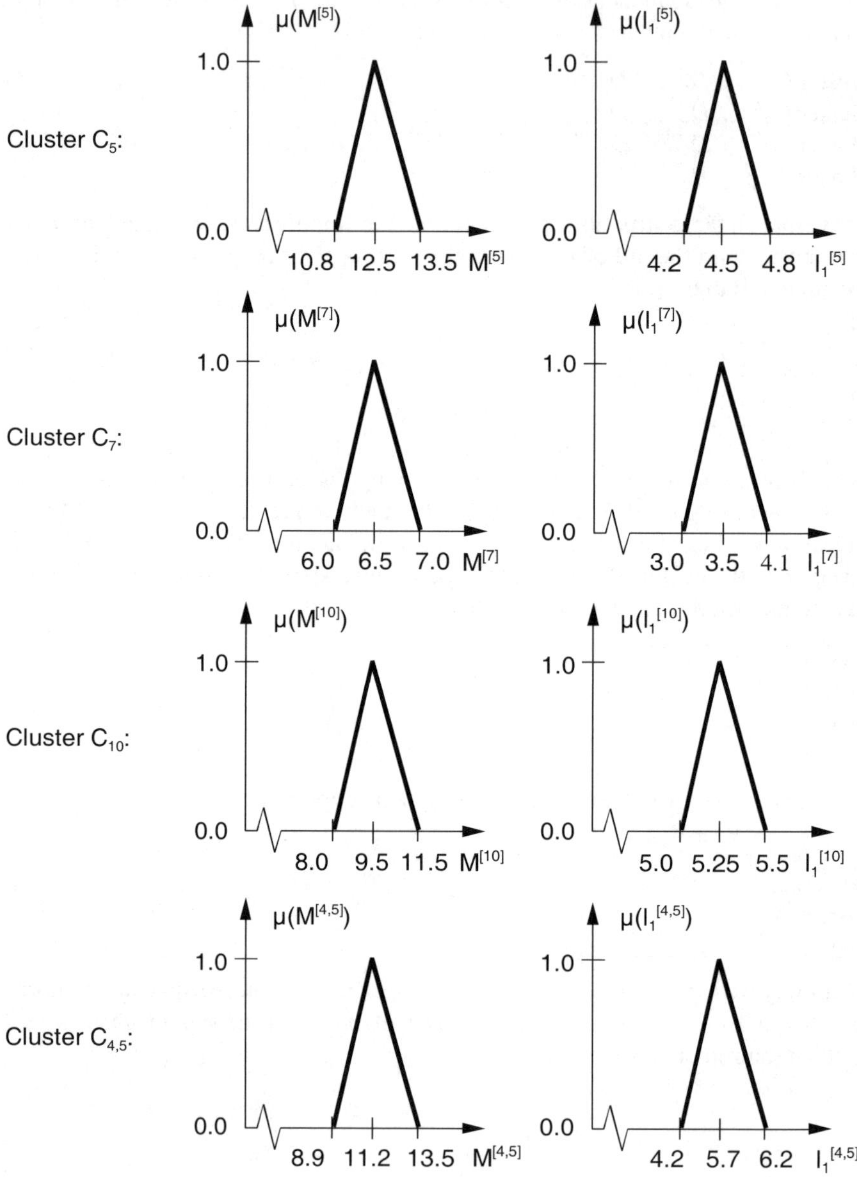

Fig. 7.11. Modified fuzzy input variables of the alternative design variants

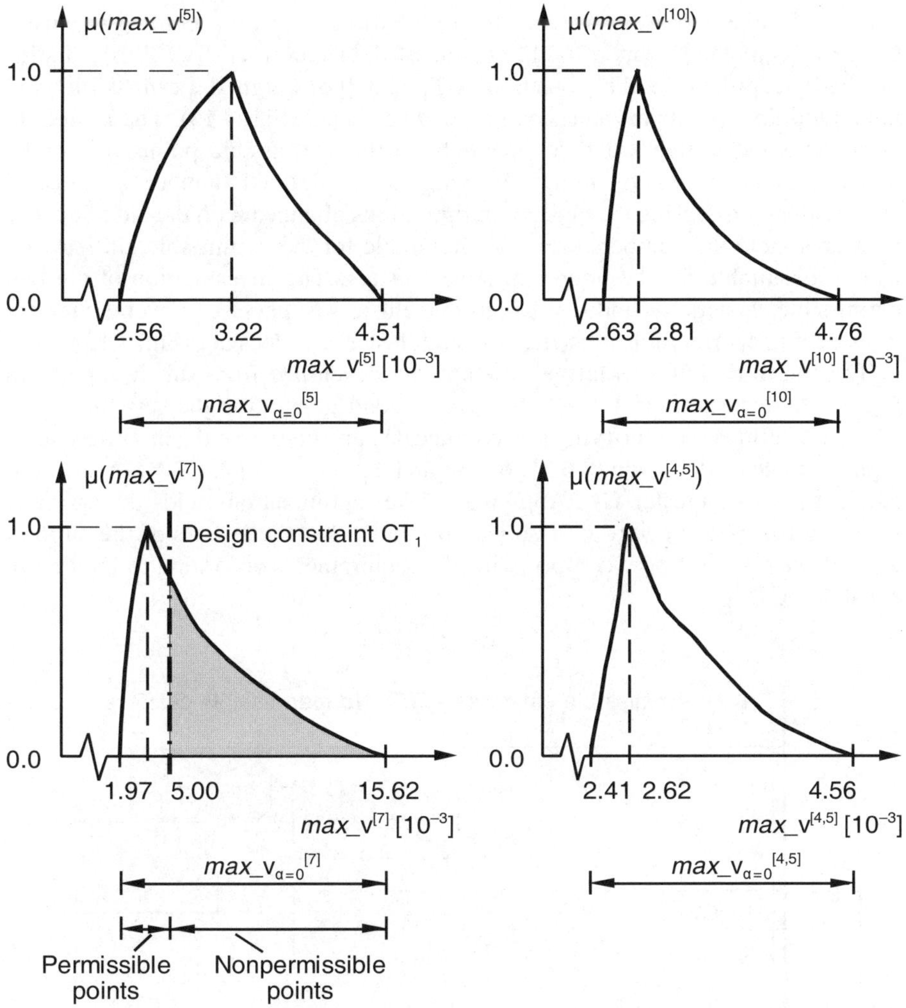

Fig. 7.12. Modified fuzzy result variables

Based on these results two ways may be followed. First, the further consideration may be limited to the design variants derived from clusters C_5, C_{10}, and $C_{4,5}$, which have proved permissible. Second, all points known in cluster C_7 are again introduced into a cluster analysis to obtain a new design variant (second generation) within the bounds of cluster C_7 and hence to keep four design variants for the further design process.

Following the second way, cluster C_7 is introduced into a clustering of second generation to obtain permissible subclusters, which satisfy the design constraint CT_1 completely. This subclustering starts exclusively with the points known in cluster C_7 from both the initial fuzzy structural analysis and the fuzzy structural analysis for the purpose of verification of the design variant according to cluster C_7. Permissible and nonpermissible points of the support $max\_v_{\alpha=0}^{[7]}$ are distinguished according to the constraint CT_1 and the assigned permissible and nonpermissible points in the space $\underline{D}^2$ are determined (Fig. 7.13). The k-medoid cluster method is then applied separately to the permissible points $\underline{d}_i$ and the nonpermissible points $\overline{\underline{d}}_i$ in $\underline{D}^2$. Varying the predefined number of clusters systematically assessing the cluster configurations obtained with the aid of quality measures yield the numbers $k_1 = 2$ as preferable for the permissible clusters and $k_2 = 5$ as suitable for the nonpermissible clusters. The investigation of the two permissible clusters C_{7_1} and C_{7_2} reveals that cluster C_{7_2} possesses several intersections with nonpermissible clusters, whereas cluster C_{7_1} does not (Fig. 7.13).

The fourth design alternative may now be developed from subcluster C_{7_1} in Fig. 7.13. According to this subcluster of second generation the sets $D_1^{[7_1]}$ and $D_2^{[7_1]}$ are defined. For verifying the assigned design alternative the modified fuzzy input variables $\tilde{M}^{[7_1]} = < 6.0, 6.35, 6.7 >$ t and $\tilde{l}_1^{[7_1]} = < 3.0, 3.3, 3.6 >$ m are constructed from subcluster C_{7_1}. Applying α-level optimization yields the assigned modified fuzzy result $max\_\tilde{v}^{[7_1]}$ shown in Fig. 7.14. All elements of the support $R_1^{[7_1]} = max\_v_{\alpha=0}^{[7_1]}$ now comply with the requirements according to the design constraint CT_1.

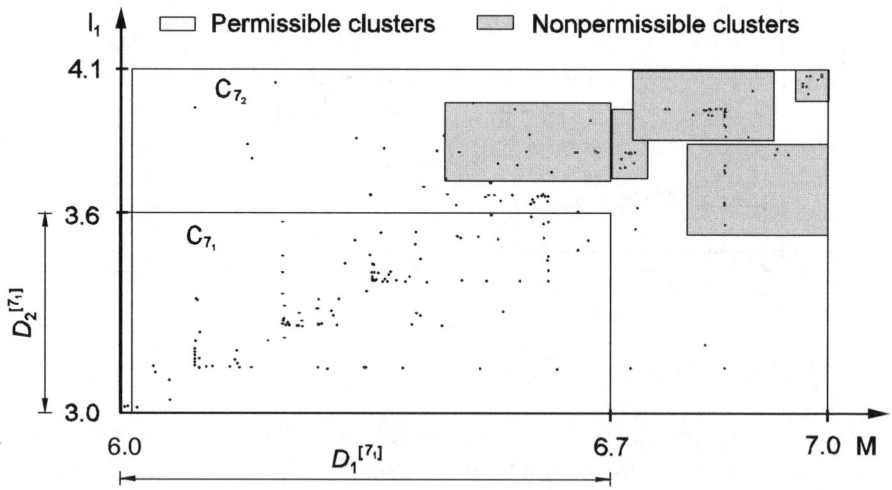

Fig. 7.13. Clustering of the points in cluster C_7, new design variant of second generation with $D_1^{[7_1]}$ and $D_2^{[7_1]}$

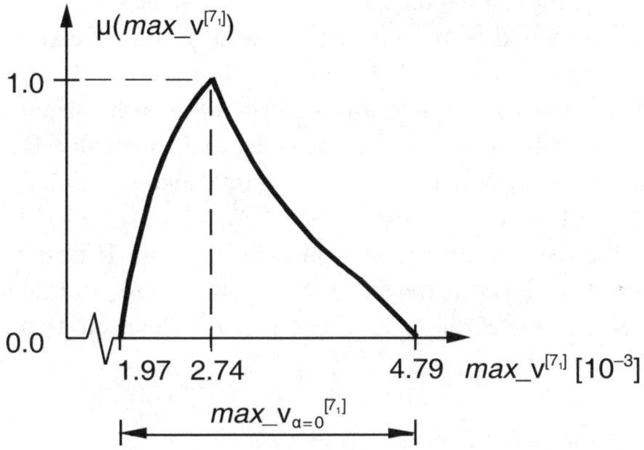

Fig. 7.14. Modified fuzzy result variable $max\_\tilde{v}^{[7_1]}$

The four design variants gained from cluster analysis are characterized by the following design parameters:

Design Variant I – Obtained from Cluster C_5.

- $M \in [10.8, 13.5]$ t $\rightarrow d_1 \in \tilde{M}^{[5]} = < 10.8, 12.5, 13.5 >$ t
- $l_1 \in [4.2, 4.8]$ m $\rightarrow d_2 \in \tilde{l}_1^{[5]} = < 4.2, 4.5, 4.8 >$ m

Design Variant II – Obtained from Cluster C_{7_1}.

- $M \in [6.0, 6.7]$ t $\rightarrow d_1 \in \tilde{M}^{[7_1]} = < 6.0, 6.35, 6.7 >$ t
- $l_1 \in [3.0, 3.6]$ m $\rightarrow d_2 \in \tilde{l}_1^{[7_1]} = < 3.0, 3.3, 3.6 >$ m

Design Variant III – Obtained from Cluster C_{10}.

- $M \in [8.0, 11.5]$ t $\rightarrow d_1 \in \tilde{M}^{[10]} = < 8.0, 9.5, 11.5 >$ t
- $l_1 \in [5.0, 5.5]$ m $\rightarrow d_2 \in \tilde{l}_1^{[10]} = < 5.0, 5.25, 5.5 >$ m

Design Variant IV – Obtained from Cluster $C_{4,5}$.

- $M \in [8.9, 13.5]$ t $\rightarrow d_1 \in \tilde{M}^{[4,5]} = < 8.9, 11.2, 13.5 >$ t
- $l_1 \in [4.2, 6.2]$ m $\rightarrow d_2 \in \tilde{l}_1^{[4,5]} = < 4.2, 5.7, 6.2 >$ m

These design variants are assessed with the aid of the criteria discussed in Sect. 7.3.3.

According to criterion I the distances between the defuzzified values of the fuzzy input variables $max\_\tilde{v}^{[5]}$, $max\_\tilde{v}^{[7_1]}$, $max\_\tilde{v}^{[10]}$, and $max\_\tilde{v}^{[4,5]}$ and the value $max\_v = 5 \cdot 10^{-3}$ to be complied with according to the design constraint CT_1 are compared. These distances are listed in Table 7.1 for different defuzzification methods. The smallest and the largest constraint distances obtained in each case are marked. These indicate the design variants that may be assessed as being the most conservative one and the less conservative one. In overall terms, design variant IV obtained from cluster $C_{4,5}$ represents the most conservative design alternative, whereas design variant III obtained from cluster C_{10} yields defuzzification results closest to the design constraint CT_1.

The robustness assessed with criterion II is also computed for all design variants and listed in Table 7.2. The largest robustness, i.e., the smallest measure value B_v, is determined for design variant IV assigned to cluster $C_{4,5}$ and design variant I belonging to cluster C_5 proves to be the most sensitive one.

Combining the results from the assessments based on criteria I and II design variant IV derived from cluster $C_{4,5}$ is considered as being the most suitable alternative, which may be selected to be taken as the basis for designing the structure.

Table 7.1. Assessment of the design variants, criterion I – constraint distance

	Defuzzification results			
	$max\_v_0^{[5]}$	$max\_v_0^{[7_1]}$	$max\_v_0^{[10]}$	$max\_v_0^{[4,5]}$
Centroid method	3.37	3.17	3.39 (max)	3.11 (min)
Chen method	4.00 (min)	4.10	4.60 (max)	4.30
Level rank method	3.30 (max)	3.00	3.10	2.97 (min)

Table 7.2. Assessment of the design variants, criterion II – robustness

	Design variant (cluster)			
	I (C_5)	II (C_{7_1})	III (C_{10})	IV ($C_{4,5}$)
Robustness measure B	0.039 33 (max)	0.005 35	0.004 58	0.001 37 (min)

Structural Design Using Fuzzy Cluster Method. The two-dimensional space $\underline{D}^2$ of the design parameters $\tilde{M}$ and $\tilde{l}_1$ is again formed by the supports $D_1 = M_{\alpha=0} = [6.0,\ 18.0]$ t and $D_2 = l_{1,\alpha=0} = [3.0,\ 7.0]$ m of the initial fuzzy input variables (Figs. 7.8 and 7.15). For fuzzy cluster analysis the same set of permissible points $\underline{d}_i$ and nonpermissible points $\underline{\bar{d}}_i$, which has already been introduced into k-medoid clustering, is taken as the basis.

Alternative cluster configurations are determined for a prescribed number of 2, 3, ..., 20 clusters by applying the fuzzy cluster method according to Sect. 7.2.3. Based on the numerical quality measures discussed in Sect. 7.2.3 the number $k_1 = 7$

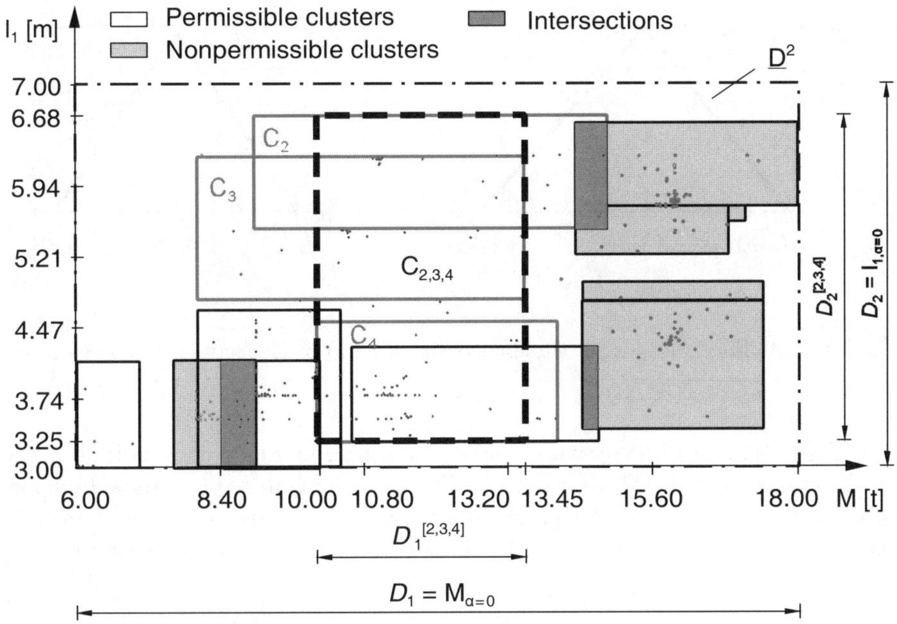

Fig. 7.15. Cluster configuration in the space $\underline{D}^2$ based on fuzzy cluster method

is assessed as being suitable for the clusters of the permissible points $\underline{d}_i$ and $k_2 = 6$ is selected as an appropriate number of clusters comprised of nonpermissible points $\overline{\underline{d}}_i$.

Considering a minimum value of $min\_\mu_{iv} = 0.25$ for the membership μ_{iv} assessing the degree with which the points $\underline{d}_i$ or $\overline{\underline{d}}_i$ belong to the particular clusters C_v the crisp cluster configuration shown in Fig. 7.15 is obtained.

Clusters C_2, C_3, and C_4 are closely adjacent and even partially intersecting; merging these clusters to form the supercluster $C_{2,3,4}$ is thus considered advantageous. The supercluster $C_{2,3,4}$ with the assigned sets $D_1^{[2,3,4]}$ and $D_2^{[2,3,4]}$ represents the only design variant that is investigated in this example.

For verifying this design variant modified fuzzy result variables are computed by means of α-level optimization. The modified fuzzy input variables $\tilde{M}^{[2,3,4]}$ and $\tilde{l}_1^{[2,3,4]}$ entering the fuzzy structural analysis for this purpose are illustrated graphically in Fig. 7.16.

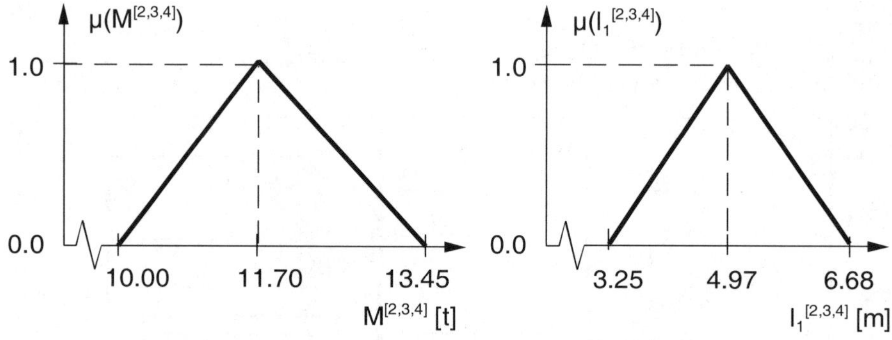

Fig. 7.16. Modified fuzzy input variables of the design variant according to supercluster $C_{2,3,4}$

Fuzzy structural analysis with α-level optimization yields the modified fuzzy result variable $max\_\tilde{v}^{[2,3,4]}$ shown in Fig. 7.17. All elements of the assigned support $max\_v_{\alpha=0}^{[2,3,4]}$ satisfy the design constraints CT_1, i.e., possess values less than $max\_v = 5 \cdot 10^{-3}$. Hence the design has proved permissible and the design parameters are

- $M \in [10.00, 13.45]$ t $\rightarrow$ $d_1 \in \tilde{M}^{[2,3,4]} = < 10.00, 11.70, 13.45 >$ t
- $l_1 \in [3.25, 6.68]$ m $\rightarrow$ $d_2 \in \tilde{l}_1^{[2,3,4]} = < 3.25, 4.97, 6.68 >$ m

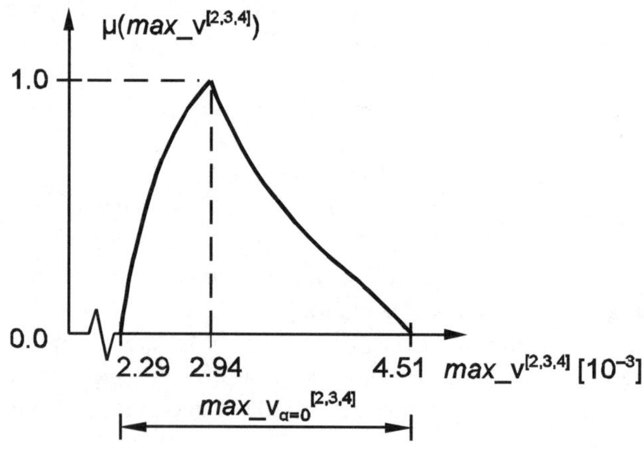

Fig. 7.17. Modified fuzzy result variable $max\_\tilde{v}^{[2,3,4]}$

7.4.2 Reinforced-concrete Frame, Structural Design with FFORM

The reinforced-concrete frame shown in Fig. 7.18 is designed on the basis of a safety analysis according to the Fuzzy First Order Reliability Method (FFORM) presented in Chap. 6. Two FFORM analyses (Sect. 6.7.2) as well as several fuzzy structural analyses (Sect. 5.2.5.3) were already carried out for this structure.

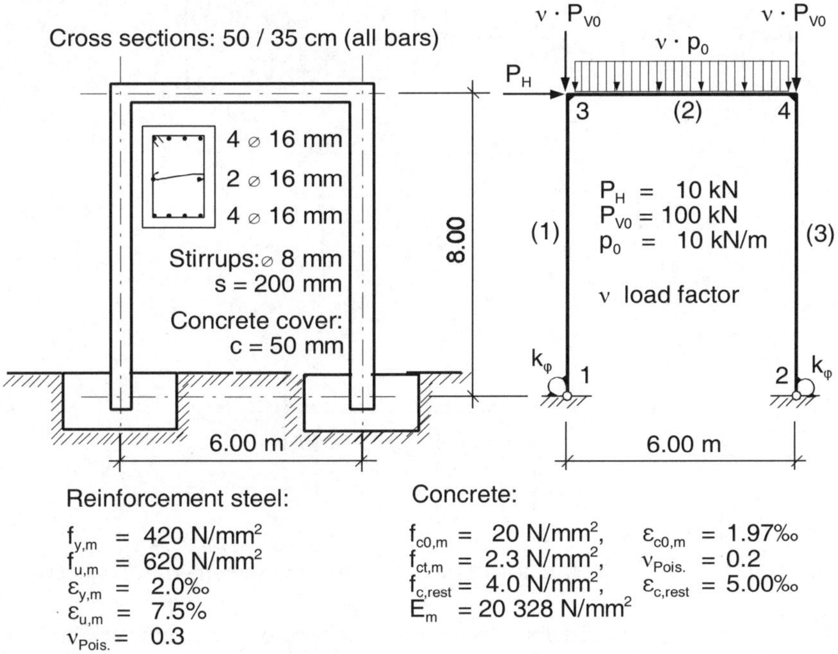

Fig. 7.18. Reinforced-concrete frame (plane); structure, cross sections, materials, statical system with loading

The structure is initially stressed by deadload and the horizontal load P_H. The line load $p = \nu \cdot p_0$ and the vertical nodal loads $P_V = \nu \cdot P_{V0}$ are simultaneously increased with the load factor ν up to global system failure. Geometrical nonlinearities as well as all essential physical nonlinearities related to reinforced concrete (e.g., crack formation, concrete compression cracking, hysteretic material behavior of concrete and reinforcement steel) are taken into consideration.

Based on the data available the load factor ν and the rotational spring stiffness k_φ are modeled as fuzzy random variables. Hence two fuzzy probabilistic basic variables enter the investigation

$$\nu \rightarrow \tilde{X}_1 \,, \quad k_\varphi \rightarrow \tilde{X}_2 \,. \tag{7.48}$$

For determining a permissible structural design the fuzzy reliability index $\tilde{\beta}$ must completely comply with the requirements stipulated in currently applicable standards [217, 218], i.e., the value *req_*β = 3.8 must not be exceeded.

The fuzzy random variable $\tilde{X}_1$ is assumed to follow a fuzzy probability distribution of Ex-Max Type I (Gumbel) according to Eqs.(4.3), (4.4), and (4.5) (Fig. 7.19). For describing the fuzzy random variable $\tilde{X}_2$ a logarithmic normal distribution

$$\tilde{F}(x_2) = \frac{1}{\sqrt{2\pi}} \int_{-\infty}^{x} \frac{1}{\tilde{\sigma}_U (t - x_{0,2})} \exp\left(- \frac{(\ln(t - x_{0,2}) - \tilde{m}_U)^2}{2\tilde{\sigma}_U^2} \right) dt , \tag{7.49}$$

with

$$\tilde{\sigma}_U = \sqrt{\ln\left(1 + \left(\frac{\tilde{\sigma}_{X_2}}{\tilde{m}_{X_2} - x_{0,2}}\right)^2\right)} , \tag{7.50}$$

and

$$\tilde{m}_U = \ln\left(\frac{\tilde{m}_{X_2} - x_{0,2}}{\sqrt{1 + \left(\frac{\tilde{\sigma}_{X_2}}{\tilde{m}_{X_2} - x_0}\right)^2}} \right) \tag{7.51}$$

is presumed (Fig. 7.20). Interaction of the functional parameters must be accounted for.

The fuzzy probability distribution functions depend on the fuzzy variables $\tilde{m}_{X_1}$, $\tilde{\sigma}_{X_1}$, $\tilde{m}_{X_2}$, and $\tilde{\sigma}_{X_2}$, which are specified by the following fuzzy triangular numbers:

$$\begin{aligned} \tilde{m}_{X_1} &= <5.70,\ 5.90,\ 6.30> , \\ \tilde{\sigma}_{X_1} &= <0.08,\ 0.11,\ 0.15> , \\ \tilde{m}_{X_2} &= <8.50,\ 9.00,\ 10.00>\ \text{MNm/rad} , \\ \tilde{\sigma}_{X_2} &= <1.00,\ 1.35,\ 1.50>\ \text{MNm/rad} . \end{aligned} \tag{7.52}$$

The minimum value of the logarithmic normal distribution of $\tilde{X}_2$ is assumed to be crisp with $x_{0,2} = 0$ MNm/rad.

In this design problem the parameters m_{X_1}, σ_{X_1}, m_{X_2}, and σ_{X_2} are chosen as design parameters.

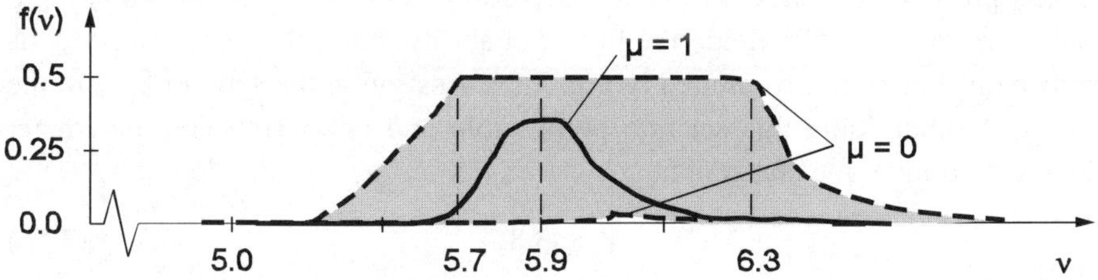

Fig. 7.19. Fuzzy probability density function of the fuzzy probabilistic basic variable $\tilde{X}_1$ representing the load factor ν

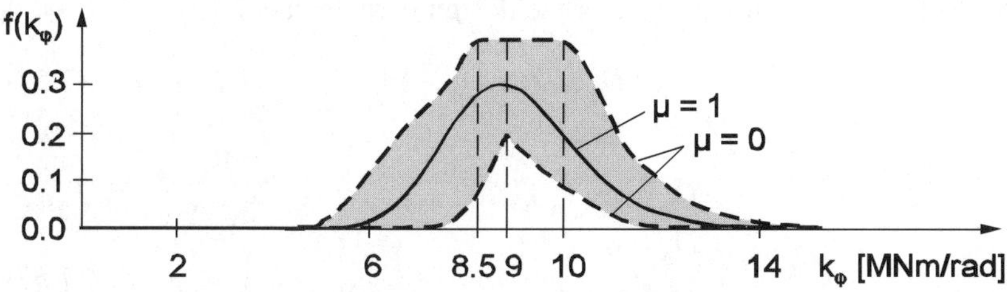

Fig. 7.20. Fuzzy probability density function of the fuzzy probabilistic basic variable $\tilde{X}_2$ representing the rotational spring stiffness k_φ

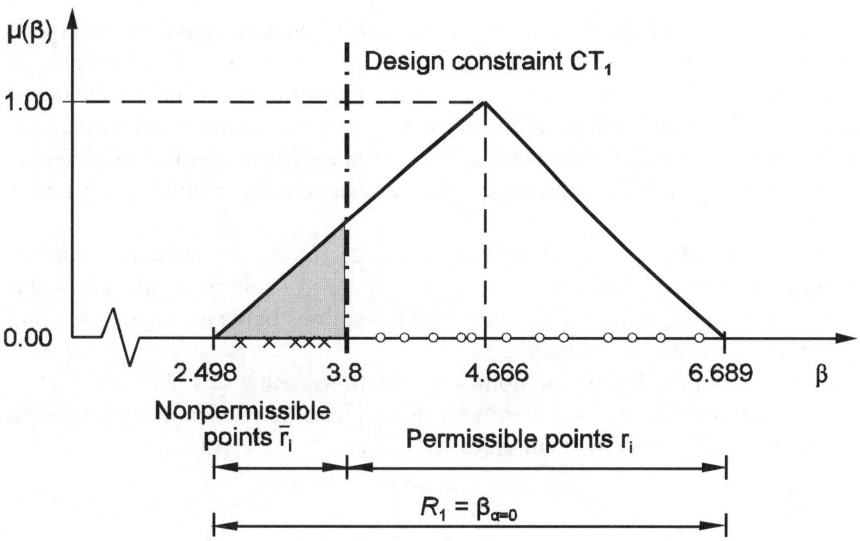

Fig. 7.21. Fuzzy result $\tilde{\beta}$, design constraint CT_1, and restricted parameter values

Fuzzy probabilistic safety assessment according to FFORM with the fuzzy input variables from Eq. (7.52) yields the fuzzy reliability index $\tilde{\beta}$ characterized by its membership function $\mu(\beta)$ plotted in Fig. 7.21. Assigning the support $\beta_{\alpha=0}$ to the set R_1 and subdividing this set into permissible and nonpermissible parameter values according to the constraint

$$\mathrm{CT}_1: \quad \tilde{\beta} \geq 3.8 \tag{7.53}$$

initialize the one-dimensional space of the restricted parameters $\underline{\mathrm{R}}^1 = R_1 = \beta_{\alpha=0}$. On this basis the set of crisp result values $\beta_i \in \tilde{\beta}$ obtained from α-level optimization is subdivided into permissible points $\underline{\mathrm{r}}_i$ and nonpermissible points $\overline{\underline{\mathrm{r}}}_i$.

The supports of the four fuzzy input variables from Eq. (7.52) are consequently comprised of both permissible and nonpermissible design-parameter values. These constitute the four-dimensional space of the design parameters

$$\underline{\mathrm{D}}^4 = D_1 \times D_2 \times D_3 \times D_4, \tag{7.54}$$

with

$$\begin{aligned} D_1 &= \mathrm{m}_{X_1,\alpha=0}, \\ D_2 &= \sigma_{X_1,\alpha=0}, \\ D_3 &= \mathrm{m}_{X_2,\alpha=0}, \\ D_4 &= \sigma_{X_2,\alpha=0}. \end{aligned} \tag{7.55}$$

From α-level optimization 414 permissible points $\underline{\mathrm{d}}_i$ and 195 nonpermissible points $\overline{\underline{\mathrm{d}}}_i$ are known in the space $\underline{\mathrm{D}}^4$.

The fuzzy cluster method according to Sect. 7.2.3 is now applied in the space $\underline{\mathrm{D}}^4$. The number of clusters is varied within the interval $\mathrm{k} \in [1, 10]$ and the cluster configurations obtained are assessed on the basis of the numerical quality measures stated in Sect. 7.2.3. According to the measures *separation degree* and *normalized partition coefficient* a clustering with $\mathrm{k}_1 = 6$ permissible clusters is considered as being suitable (Fig. 7.22). Furthermore, the number of nonpermissible clusters is chosen to be $\mathrm{k}_2 = 7$.

Prescribing a minimum value of $min\_\mu_{iv} = 0.25$ for the membership value μ_{iv} describing the degree with which the points $\underline{\mathrm{d}}_i$ or $\overline{\underline{\mathrm{d}}}_i$ belong to the particular clusters C_v a cluster configuration without intersections between permissible and nonpermissible clusters is obtained.

From the cluster configuration modified, one-dimensional sets $D_1^{[v]}$, $D_2^{[v]}$, $D_3^{[v]}$, and $D_4^{[v]}$ are generated, which contain only design parameters that meet the design requirement, i.e., that lead to result values $\beta_i \geq 3.8$.

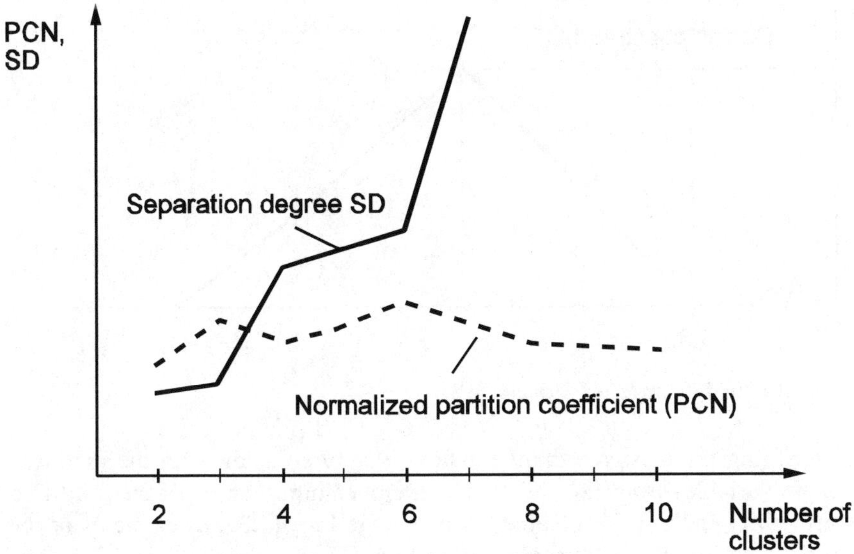

Fig. 7.22. Assessing cluster configurations for the permissible points $\underline{d}_i$

The one-dimensional sets $D_1^{[v]}$, $D_2^{[v]}$, $D_3^{[v]}$, and $D_4^{[v]}$ are thus known. As the one-dimensional sets assigned to clusters C_1, C_2, C_5, and C_6 are very small, i.e., their smallest and largest elements are very close to each other, these clusters are not taken as a basis for generating alternative design variants. Moreover, clusters C_3 and C_4 are approximately congruent, i.e., $D_1^{[3]} \approx D_1^{[4]}$, $D_2^{[3]} \approx D_2^{[4]}$, $D_3^{[3]} \approx D_3^{[4]}$, and $D_4^{[3]} \approx D_4^{[4]}$, these clusters are merged to constitute the supercluster $C_{3,4}$. This yields the sets

$$\begin{aligned} D_1^{[3,4]} &= [5.85, 6.00]\,, \\ D_2^{[3,4]} &= [0.08, 0.11]\,, \\ D_3^{[3,4]} &= [8.50, 10.00]\ \text{MNm/rad}\,, \\ D_4^{[3,4]} &= [1.30, 1.50]\ \text{MNm/rad}\,. \end{aligned} \tag{7.56}$$

Owing to the assignment in Eq. (7.55) modified supports $m_{X_1,\alpha=0}^{[v]}$, $\sigma_{X_1,\alpha=0}^{[v]}$, $m_{X_2,\alpha=0}^{[v]}$, and $\sigma_{X_2,\alpha=0}^{[v]}$ and hence modified fuzzy input variables $\tilde{m}_{X_1}^{[v]}$, $\tilde{\sigma}_{X_1}^{[v]}$, $\tilde{m}_{X_2}^{[v]}$, and $\tilde{\sigma}_{X_2}^{[v]}$ are also known with the modified sets $D_1^{[v]}$, $D_2^{[v]}$, $D_3^{[v]}$, and $D_4^{[v]}$:

$$\begin{aligned} \tilde{m}_{X_1}^{[3,4]} &= \langle 5.85,\ 5.90,\ 6.00 \rangle\,, \\ \tilde{\sigma}_{X_1}^{[3,4]} &= \langle 0.08,\ 0.10,\ 0.11 \rangle\,, \\ \tilde{m}_{X_2}^{[3,4]} &= \langle 8.50,\ 9.50,\ 10.00 \rangle\ \text{MNm/rad}\,, \\ \tilde{\sigma}_{X_2}^{[3,4]} &= \langle 1.30,\ 1.40,\ 1.50 \rangle\ \text{MNm/rad}\,. \end{aligned} \tag{7.57}$$

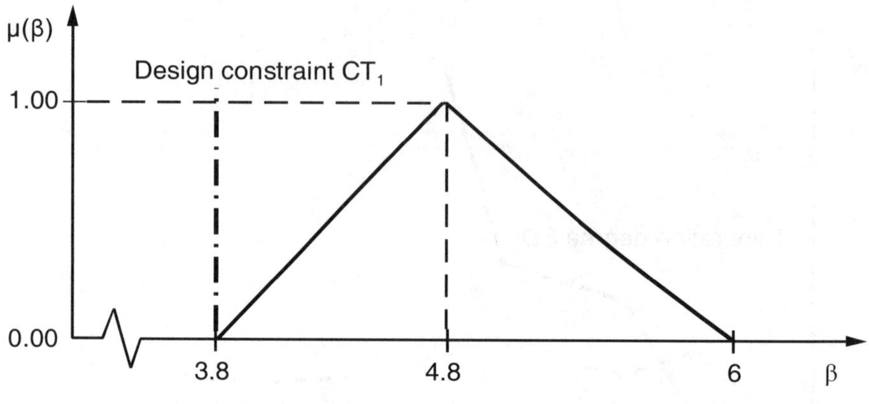

Fig. 7.23. Modified fuzzy result variable $\tilde{\beta}^{[3,4]}$

For verifying this design variant FFORM is applied again with the modified fuzzy input variables from Eq. (7.57). The membership function of the modified fuzzy reliability index $\tilde{\beta}^{[3,4]}$ obtained is plotted in Fig. 7.23. All elements of the support $\beta_{\alpha=0}^{[3,4]}$ comply with the design constraint CT_1.

The modified fuzzy input variables define modified fuzzy probability distribution functions for the modified fuzzy random variables $\tilde{X}_1$ and $\tilde{X}_2$ (Figs. 7.24 and 7.25). To ensure a permissible structural design, the final design parameters must be elements of the modified fuzzy input variables. This may be realized by acquiring more information about $\tilde{X}_1$ and $\tilde{X}_2$. For example, additional prior information or additional sample elements drawn subsequently may be accounted for with the aid of the Bayes theorem or a larger sample may be taken as the basis for an interval estimation of the parameters. On inverting or iteratively solving this problem it may be inferred to what extent particular additional information must be obtained to meet the design requirements for the parameters. This necessary information extent usually concerns a minimum sample size.

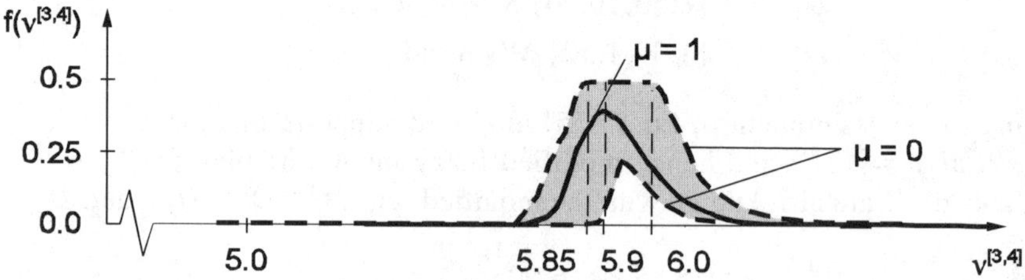

Fig. 7.24. Fuzzy probability density function of the fuzzy probabilistic basic variable $\tilde{X}_1^{[3,4]}$ representing the load factor $\nu^{[3,4]}$

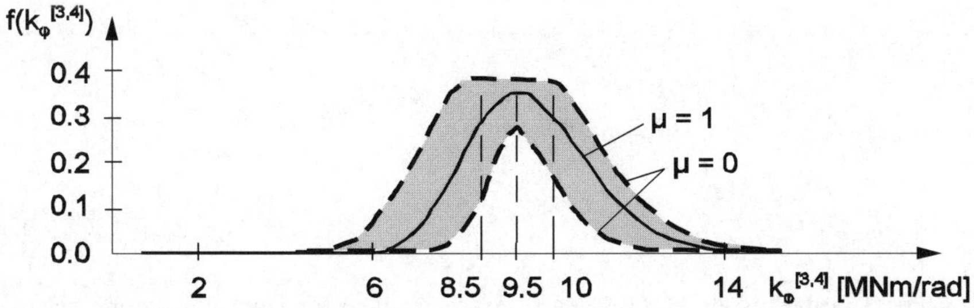

Fig. 7.25. Fuzzy probability density function of the modified fuzzy probabilistic basic variable $\tilde{X}_2^{[3,4]}$ representing the modified rotational spring stiffness $k_{\varphi}^{[3,4]}$

References

[1] Abdel-Tawab K, Noor AK (1998) A fuzzy-set analysis for a dynamic thermo-elasto-viscoplastic damage response. Computers and Structures 70(1) : 91-107

[2] Alefeld G, Herzberger J (1983) Introduction to Interval Computations. Academic Press, New York

[3] Argyris J, Mlejnek H-P (1988) Die Methode der Finiten Elemente. Einführung in die Dynamik III. Verlag F. Vieweg & Sohn, Braunschweig Wiesbaden

[4] Arnold BF (1998) Testing fuzzy hypotheses with crisp data. Fuzzy Sets and Systems 94 : 323–333

[5] Au SK, Beck JL (2001) First excursion probabilities for linear systems by very efficient importance sampling. Probabilistic Engineering Mechanics 16 : 193-207

[6] Ayyub BM (ed) (1998) Uncertainty Modeling and Analysis in Civil Engineering. CRC Press, Boston London New York

[7] Bandemer H (1992) Modelling Uncertain Data. Akademie-Verlag, Berlin

[8] Bandemer H (1997) Ratschläge zum mathematischen Umgang mit Ungewißheit. B.G. Teubner Verlagsgesellschaft, Stuttgart Leipzig

[9] Bandemer H, Gebhardt A (2000) Bayesian fuzzy kriging. Fuzzy Sets and Systems 112 : 405–418

[10] Bandemer H, Gottwald S (1989) Einführung in FUZZY-Methoden. Wissenschaftliche Taschenbücher Mathematik/Physik, Band 305. Akademie-Verlag, Berlin

[11] Bandemer H, Näther W (1992) Fuzzy Data Analysis. Kluwer Academic Publishers, Dordrecht

[12] Bardossy A, Bogardi I (1989) Fuzzy fatigue life prediction. Structural Safety 6 : 25–38

[13] Beer M, Möller B, Liebscher M (2003) Processing uncertainty in structural analysis, design and safety assessment. In: 4th Int. Symposium on Uncertainty Modeling and Analysis (ISUMA). University of Maryland, College Park, USA, pp. 34-39

[14] Ben-Haim Y (1996) Robust Reliability in the Mechanical Sciences. Springer, Berlin Heidelberg New York

[15] Benjamin JR, Cornell CA (1970) Probability, Statistics and Decision for Civil Engineers. McGraw-Hill, New York

[16] Bernardo JM, Smith AFM (1994) Bayesian Theory. Wiley, Chichester New York Brisbane Toronto Singapore

[17] Bharathidevi B, Sarma VV (1985) Estimation of fuzzy membership from histograms. Information Sciences 35 : 43–59

[18] Bodjanova S (2000) A generalized histogram. Fuzzy Sets and Systems 116 : 155–166

[19] Bonarini A, Bontempi G (1994) A Qualitative Simulation Approach for Fuzzy Dynamical Models. ACM Transactions on Modeling and Computer Simulation (TOMACS) 4 : 285–313

[20] Bontempi G (1995) Modeling with uncertainty in continuous dynamical systems: the probability and possibility approach. In: Berichte. IRIDIA, Université libre de Bruxelles

[21] Bosch K (ed) (1998) Statistik-Taschenbuch. Oldenbourg, München Wien

[22] Bothe HH (1993) Fuzzy Logic. Springer, Berlin Heidelberg New York

[23] Brenner CE, Schuëller GI (1998) Stochastic Finite Elements – Under the Perspectives of the Assumptions on Material Properties. In: Murakami H, Luco JE (eds) Proc. 12th Engineering Mechanics Conf. – Engineering Mechanics: A Force for the 21th Century. La Jolla, CA, USA. ASCE, Reston, VA, pp. 1441–1444

[24] Brown CB, Yao JTP (1983) Fuzzy sets and structural engineering. ASCE, J. Structural Engineering 109 : 1211–1225

[25] Brown CB, Johnson JL, Loftus JJ (1984) Subjective seismic safety assessments. ASCE, J. Structural Engineering 110 : 2212–2233

[26] Bucher CG (1999) Application of probability based concepts in computational mechanics. In: Wunderlich W, Stein E (eds) 1st Europ. Conf. on Computational Mechanics. TU München, pp. 1–11

[27] Buckley JJ, Hayashi Y (1994) Fuzzy Neural Network: A survey. Fuzzy Sets and Systems 66 : 1–13

[28] Butar FB, Lahiri P (2002) Empirical Bayes estimation of several population means and variances under random sampling variances model. J. Statistical Planning and Inference 102(1) : 59–69

[29] Butar FB, Lahiri P (2003) On measures of uncertainty of empirical Bayes small-area estimators. J. Statistical Planning and Inference 112 (1) : 63–76

[30] Caers J, van Dyck J (1998) Nonparametric tail estimation using a double bootstrap method. Computational Statistics and Data Analysis 29(2) : 191–211

[31] Cagno E, Caron F, Mancini M, Ruggeri F (2000) Using AHP in determining the prior distributions on gas pipeline failures in a robust Bayesian approach. Reliability Engineering and System Safety 67 : 275–284

[32] Capelle B, Kerre EE (1994) Possibilistic and Necessistic Reliability Functions: Fundamental Concepts and Theorems to Represent Non-Probabilistic Uncertainty in Reliability Theory. In: Ayyub BM, Gupta MM (eds) Uncertainty Modelling and Analysis: Theory and Applications. North-Holland-Elsevier Scientific Publishers, pp. 131–144

[33] Chao-Ming Hwang, Jing-Shing Yao (1996) Independent fuzzy random variables and their application. Fuzzy Sets and Systems 82 : 335–350

[34] Cheung-Bin Lee, Ju-Won Park (1997) Reliability Analysis Based on Fuzzy-Bayesian Approach. In: 7th Int. Conf. on Computing in Civil and Building Engineering. Seoul, pp. 589–593

[35] Ching-Chang Wong, Nine-Shen Lin (1997) Rule extraction for fuzzy modeling. Fuzzy Sets and Systems 88 : 23–30

[36] Ching-Hsue Cheng, Don-Lin Mon (1993) Fuzzy system reliability analysis by interval of confidence. Fuzzy Sets and Systems 56 : 29–35

[37] Civanlar MR, Trussell HJ (1986) Constructing membership functions using statistical data. Fuzzy Sets and Systems 18 : 1–13

[38] Cloete I, Zurada JM (eds) (1999) Knowledge-Based Neurocomputing. MIT Press, Cambridge, Massachusetts

[39] Cremona C, Gao Y (1997) The possibilistic reliability theory: theoretical aspects and applications. Structural Safety 19(2) : 173–201

[40] Dallmann H, Elster K-H (1991, 1992) Einführung in die höhere Mathematik. vol. I, 1991, vol. II, 1991, vol. III, 1992. Gustav Fischer, Jena

[41] DeFinetti B (1990) Theory of Probability: A Critical Introductory Treatment. Wiley, Chichester

[42] Der Kiureghian A, Liu P-L (1986) Structural reliability under incomplete probability information. ASCE, J. Engineering Mechanics 112(1) : 85–104

[43] Dessombz O, Thouverez F, Lainé J-P, Jézéquel L (2001) Analysis of Mechanical Systems using Interval Computations applied to Finite Elements Methods. J. Sound and Vibration, 239(5) : 949–968

[44] Dmitrienko A, Govindarajulu Z (2002) Sequential determination of the number of bootstrap samples. J. Statistical Planning and Inference 100 : 349–363

[45] Dong W, Shah HC (1987) Vertex method for computing fuzzy variables. Fuzzy Sets and Systems 24 : 65–78

[46] Dubois D, Prade H (1980) Fuzzy Sets and Systems Theory and Applications. Academic Press, New York London

[47] Dubois D, Prade H (1986) Possibility Theory. Plenum Press, New York London

[48] Efron B, Tibshirani RJ (1993) An Introduction to the Bootstrap. Chapman & Hall, New York

[49] Elishakoff I (1990) Convex versus probabilistic models of uncertainty in structural dynamics. Opening Keynote Lecture. In: Petyt M, Wolfe HF, Mei C (eds) Structural Dynamics: Recent Advances. Elsevier Applied Science, London, pp. 3–21

[50] Elishakoff I (1995) Essay on uncertainties in elastic and viscoelastic structures: from A.M. Freudenthal's criticisms to modern convex modeling. Computers and Structures 56 : 871–895

[51] Elishakoff I (1998) Editorial: Three facets of uncertainty. Computers and Structures 67 : 1–2

[52] Elishakoff I (ed) (1999) Whys and Hows in Uncertainty Modelling Probability, Fuzziness and Anti-Optimization. Springer, Wien New York

[53] Emam H, Pradlwarter HJ, Schuëller GI (2001) Reliability Analysis of General MDOF-Systems by Direct and Weight Controlled Monte Carlo Simulation. In: Schuëller GI, Spanos PD (eds) Proc. Int. Conf. on Monte Carlo Simulation MCS_2000, Monte Carlo, Monaco, 2000. Swets & Zeitlinger BV, Lisse, The Netherlands, pp. 415–422

[54] Fetz Th, Oberguggenberger M, Jager J, Koll D, Krenn G, Lessmann H, Stark RF (1999) Fuzzy Models in Geotechnical Engineering and Construction Management. Computer-Aided Civil and Infrastructure Engineering 14 : 93–106

[55] Fishman GS (1996) Monte Carlo: Concepts, Algorithms, and Applications. Springer, New York Berlin Heidelberg

[56] Fragola JR (1996) Reliability and risk analysis data base development: an historical perspective. Reliability Engineering and System Safety 51 : 125–136

[57] Frangopol DM, Maute K (2003) Life-cycle reliability-based optimization of civil and aerospace structures. Computers and Structures 81 : 397–410

[58] Franz J (1994) On estimation problems in random censored repair models. Economic Quality Control 9/3 : 125–142

[59] Gellert M (1979) A direct integration method for analysis of a certain class of non-linear dynamic problems. Ingenieur-Archiv 48(6) : 403–415

[60] Gellert W, Kästner H (eds) (1985) Lexikon der Mathematik. Bibliograph. Inst. Leipzig

[61] Geyskens P, Der Kiureghian A, Monteiro P (1998) Bayesian Prediction of Elastic Modulus of Concrete. ASCE, J. Structural Engineering 124(1) : 89–95

[62] Ghanem RG, Spanos PD (1991) Stochastic Finite Elements: A Spectral Approach. Springer, New York Berlin Heidelberg

[63] Gioffrè M, Gusellav V, Grigoriu M (2000) Simulation of non-Gaussian field applied to wind pressure fluctuations. Probabilistic Engineering Mechanics 15 : 339–345

[64] Gray HL, Schucany WR (1972) The Generalized Jackknife Statistic. Marcel Dekker, New York

[65] Grigoriu M (1995) Applied Non-Gaussian Processes: Examples, Theory, Simulation, Linear Random Vibration and MATLAB Solutions. Prentice Hall, Englewood Cliffs

[66] Grzegorzewski P (2000) Testing statistical hypotheses with vague data. Fuzzy Sets and Systems 112 : 501–510

[67] Grzegorzewski P (2001) Fuzzy sets – defuzzification and randomization. Fuzzy Sets and Systems 118 : 437–446

[68] Gupta MM, Sanchez E (eds) (1982) Approximate Reasoning in Decision Analysis. North-Holland, New York

[69] Ha Nguyen S, Möller B, Graf W (2002) Modeling of time-dependent loading as a fuzzy-process. In: Grundmann H, Schuëller GI (eds) 5th Europ. Conf. on Structural Dynamics, Munich. A.A. Balkema, EURODYN 2002, vol. 1, pp. 679– 684

[70] Haldar A, Reddy RK (1992) A random-fuzzy analysis of existing structures. Fuzzy Sets and Systems 48 : 201–210

[71] Hammersley JM, Handscomb DC (1964) Monte Carlo Methods. Methuen & Co LTD, London; Wiley, New York

[72] Hanss M (2002) The transformation method for the simulation and analysis of systems with uncertain parameters. Fuzzy Sets and Systems 130 : 277–289

[73] Hanss M, Willner K (2000) A fuzzy arithmetical approach to the solution of finite element problems with uncertain parameters. Mechanics Research Communications 27(3) : 257–272

[74] Hawkins DM, Olive DJ (1999) Improved feasible solution algorithms for high breakdown estimation. Computational Statistics and Data Analysis 30(1) : 1–11

[75] Hopfield JJ (1982) Neural Networks and Physical Systems with Emergent Collective Computational Abilities. Proc. National Academy of Sciences of the USA 79 : 2554–2558

[76] Höppner F, Klawonn F, Kruse R, Runkler T (1999) Fuzzy Cluster Analysis: Methods for Classification, Data Analysis and Image Recognition. Wiley, Chichester Weinheim

[77] Jesse F, Ortlepp R, Curbach M (2002) Tensile Stress-Strain Behaviour of Textile Reinforced Concrete. In: Proc. IABSE Symposium. Towards a better built environment – innovation, sustainability, information technology. Melbourne, IABSE ETH Zurich, www.iabse.ethz.ch, CD-ROM

[78] Jiazhong You (1999) A Monte Carlo comparison of several high breakdown and efficient estimators. Computational Statistics and Data Analysis 30(2) : 205–219

[79] Johnson NL, Kotz S (1970) Distributions in statistics: Continuous Univariate Distributions. Wiley, New York Chichester Brisbane Toronto Singapore

[80] Johnson NL, Kotz S (1972) Distributions in Statistics: Continuous Multivariate Distributions. Wiley, New York Chichester Brisbane Toronto Singapore

[81] Kadane JB, Moreno E, Perez ME, Pericchi LR (2002) Applying non-parametric robust Bayesian analysis to non-optionated judicial neutrality. J. Statistical Planning and Inference 102 : 425–439

[82] Kai-Yuan Cai (1996) System failure engineering and fuzzy methodology – An introductory overview. Fuzzy Sets and Systems 83(2) : 113–133

[83] Karen C Chou, Jie Yuan (1993) Fuzzy-Bayesian Approach to Reliability of Existing Structures. ASCE, J. Structural Engineering 119(11) : 3276–3290

[84] Kapitaniak T (2000) Chaos for Engineers: Theory, Applications, and Control. Springer, Berlin Heidelberg New York

[85] Kaufman L, Rousseeuw PJ (1990) Finding Groups in Data: An Introduction to Cluster Analysis. Wiley, New York

[86] Kaufmann A, Gupta MM (1991) Fuzzy Mathematical Models in Engineering and Management Science. Elsevier Science Publishers, Amsterdam

[87] Körner R (1997) On the variance of fuzzy random variables. Fuzzy Sets and Systems 92 : 83–93

[88] Körner R (1997) Linear Models with Random Fuzzy Variables. Diss., Bergakademie Freiberg, Fakultät für Mathematik und Informatik

[89] Kozine IO, Filimonov YV (2000) Imprecise reliabilities: experiences and advances. Reliability Engineering and System Safety 67 : 75–83

[90] Krätschmer V (2001) A unified approach to fuzzy random variables. Fuzzy Sets and Systems 123 : 1–9

[91] Krätschmer V (2002) Limit theorems for fuzzy-random variables. Fuzzy Sets and Systems 126 : 253–263

[92] Krug W, Schönfeld S (1981) Rechnergestützte Optimierung für Ingenieure. Verlag Technik, Berlin

[93] Kruse R, Meyer KD (1987) Statistics with Vague Data. Reidel, Dordrecht

[94] Kulczycki P, Wisniewski R (2002) Fuzzy controller for a system with uncertain load. Fuzzy Sets and Systems 131 : 185–195

[95] Kupfer H (1973) Das Verhalten des Betons unter mehrachsiger Kurzzeitbelastung unter besonderer Berücksichtigung der zweiachsigen Beanspruchung. Deutscher Ausschuß für Stahlbeton, H. 229

[96] Kwakernaak H (1978) Fuzzy random variables – I. Definitions and Theorems. Information Sciences 15 : 1–29

[97] Kwakernaak H (1979) Fuzzy random variables - II. Algorithms and Examples for the Discrete Case. Information Sciences 17 : 253–278

[98] Lee-Chae Jang, Joong-Sung Kwon (1998) A uniform strong law of large numbers for partial sum processes of fuzzy random variables indexed by sets. Fuzzy Sets and Systems 99 : 97–103

[99] Li Bing, Zhu Meilin, Xu Kai (2000) A practical engineering method for fuzzy reliability analysis of mechanical structures. Reliability Engineering and System Safety 67 : 311–315

[100] Lin YK, Elishakoff L (eds) (1991) Stochastic Structural Dynamics, vol 1, 2. Springer, Berlin Heidelberg New York Tokyo

[101] Lin YK, Schuëller GI (eds) (1988) Lecture Notes in Engineering 31: Stochastic Structural Mechanic. Springer, Berlin Heidelberg New York

[102] Link J (1976) Eine Formulierung des zweiaxialen Verformungs- und Bruchverhaltens von Beton und deren Anwendung auf die wirklichkeitsnahe Berechnung von Stahlbetonplatten. Deutscher Ausschuß für Stahlbeton, H. 270

[103] Lippmann RP (1987) An Introduction to Computing with Neural Nets. J. IEEE ASSP Magazine: 4–22

[104] Liu Yubin, Qiao Zhong, Wang Guangyuan (1997) Fuzzy random reliability of structures based on fuzzy random variables. Fuzzy Sets and Systems 86 : 345–355

[105] Lombardi M (1998) Optimization of uncertain structures using non-probabilistic models. Computers and Structures 67 : 99–103

[106] Ma SM, Bertero VV, Popov EP (1976) Experimental and analytical studies on hysteretic behavior of reinforced concrete rectangular and T-beams. EERC 76(2), University of California, Berkeley

[107] Madsen HO, Krenk S, Lind NC (1986) Methods of Structural Safety. Prentice-Hall, Englewood Cliffs, New Jersey

[108] Mahadevan S, Raghothamachar P (2000) Adaptive simulation for system reliability analysis of large structures. Computers and Structures 77(6) : 725-734

[109] Masson E, Wang Y-J (1990) Introduction to Computation and Learning in Artificial Neural Networks. Europ. J. Operational Research, 47 : 1–28

[110] McWilliam St (2001) Anti-optimisation of uncertain structures using interval analysis. Computers and Structures 79(4) : 421-430

[111] Meeden G (1999) Interval estimators for the population mean for skewed distributions with a small sample size. J. Applied Statistics 26(1) : 81-96

[112] Melnik-Melnikov PG, Dekhtyaruk ES (2000) Rare events probabilities estimation by "Russian Roulette and Splitting" simulation technique. Probabilistic Engineering Mechanics 15 : 125–129

[113] Meskouris K, Krätzig WB, Elenas A, Heiny L, Meyer IF (1988) Mikrocomputer-unterstützte Erdbebenuntersuchung von Tragwerken. SFB 151 - Tragwerksdynamik, Ruhr-Univ. Bochum, Berichte Nr. 8

[114] Möller B, Beer M (1997) Application of Fuzzy Modeling in Civil Engineering. Proc. 2nd Int. ICSC Symposium on Fuzzy Logic and Applications ISFL 97. ETH Zürich, pp. 345–351

[115] Möller B, Beer M (1998) Safety Assessment using Fuzzy Theory. Proc. Int. Computing Congress in Civil Engineering. ASCE, Annual Convention, Boston, pp. 756 - 759

[116] Möller B, Graf W, Schneider R (1999) Determination of the failure probability with implicite limit state function. In: Wunderlich W, Stein E (eds) 1st Europ. Conf. on Computational Mechanics. TU München, pp. 1–20

[117] Möller B, Graf W, Beer M (2000) Fuzzy-Tragwerksanalyse - Tragwerksanalyse mit unscharfen Parametern. Bauingenieur 75(11) : 697–708

[118] Möller B, Graf W, Beer M (2000) Fuzzy structural analysis using α-level optimization. Computational Mechanics 26(6) : 547–565

[119] Möller B, Graf W, Beer M, Sickert J-U (2001) Fuzzy Probabilistic Method and its Application for the Safety Assessment of Structures. In: Waszczyszyn Z, Pamin J (eds) 2nd Europ. Conf. on Computational Mechanics. Cracow, 2001, CD-ROM, Doc 94, pp. 1–20

[120] Möller B, Beer M, Graf W, Sickert J-U (2001) Fuzzy Finite Element Method and its Application. In: Wall WA, Bletzinger K-U, Schweizerhof K (eds) Trends in Computational Structural Mechanics. CIMNE, Barcelona, pp. 529–538

[121] Möller B, Graf W, Ha Nguyen S (2001) Uncertainty in damage assessment of structures and its numerical simulation. Computer-Aided Civil and Infrastructure Engineering 16(6) : 375–383

[122] Möller B, Graf W, Ha Nguyen S (2002) Modeling the life-cycle of a structure using fuzzy processes. In: Shang-Hsien Hsieh, Liang-Jeng Leu, Chuin-Shan Chen, Jenn-Feng Li (eds) 9th Int. Conf. in Civil and Building Engineering. Taipei, vol. 1, pp. 97–102

[123] Möller B, Graf W, Beer M, Sickert J-U (2002) Fuzzy Randomness – Towards a New Modeling of Uncertainty. In: Mang AH, Rammerstorfer FG, Eberhardsteiner J (eds) 5th World Congress on Computational Mechanics, WCCM V. Vienna, Doc 80170, pp. 1–10

[124] Möller B, Graf W, Beer M (2003) Safety assessment of structures in view of fuzzy randomness. Computers and Structures 81(15) : 1567–1582

[125] Möller B, Graf W, Beer M, Sickert J-U (2003) Fuzzy Stochastic Finite Element Method. In: Bathe KJ (ed) 2nd M.I.T. Conf. on Computational Fluid and Solid Mechanics, Cambridge, USA, CD-ROM and Hardbound ed., vol. 2. Elsevier Ltd, Oxford, pp. 2074–2077

[126] Möller B, Beer M, Liebscher M (2003) Fuzzy cluster design – a new way for structural design. In: Proc. 9th Int. Conf. on Civil and Structural Engineering Computing. Egmond aan Zee, The Netherlands, CD-ROM, pp. 1–6

[127] Möller B, Graf W, Beer M (2004) Discussion on "Structural reliability analysis through fuzzy number approach, with application to stability". Computers and Structures 82(2-3) : 325–327

[128] Mood AM, Graybill FA, Boes DC (1974) Introduction to the Theory of Statistics. McGraw-Hill, New York

[129] Moore RE (1979) Methods and applications of interval analysis. SIAM Society for Industrial and Applied Mathematics, Philadelphia

[130] Muc A, Kedziora P (2001) A fuzzy set analysis for a fracture and fatigue damage response of composite materials. Composite Structures 54 : 283–287

[131] Muc A, Kedziora P (2001) Fuzzy Set Approach to the Limit Load Carrying Capacity Analysis of Composite Shell and Plated Structures. In: Waszczyszyn Z, Pamin J (eds) 2nd Europ. Conf. on Computational Mechanics. Cracow, CD-ROM, Doc 545, pp. 1–11

[132] Muhanna RL, Mullen RL (1999) Formulation of Fuzzy Finite Element Methods for Solid Mechanics Problems. Computer-Aided Civil and Infrastructure Engineering 14 : 107–117

[133] Müller H, Graf W, Schneider R, Stanoev E (1996) Zum nichtlinearen mechanischen Verhalten von seilvorgespannten Stahlbeton-Stabtragwerken. 6. Fachtagung Baustatik-Baupraxis BB6, Weimar, pp. 1–20

[134] Näther W, Körner R (2002) On the Variance of Random Fuzzy Variables. In: Bertoluzza C, Gil MA, Ralescu DA (eds) Statistical Modelling, Analysis and Management of Fuzzy Data. Physica-Verlag, Heidelberg, pp. 25–42

[135] Noor AK (1999) Uncertainty Analysis of Composite Structures. In: Wunderlich W, Stein E (eds) 1st Europ. Conf. on Computational Mechanics. TU München, Doc 117, pp. 1–29

[136] Noor AK, Starnes Jr JH, Peters JM (2001) Uncertainty analysis of stiffened composite panels. Composite Structures 51 : 139–158

[137] Oden JT, Belytschko T, Babuska I, Hughes TJR (2003) Research directions in computational mechanics. Computer Methods in Applied Mechanics and Engineering 192 : 913–922

[138] Oh M-S (1999) Estimation of posterior density functions from a posterior sample. Computational Statistics and Data Analysis 29(4) : 411–427

[139] Pal NR (1999) On quantification of different facets of uncertainty. Fuzzy Sets and Systems 107 : 81–91

[140] Pan W, Louis TA (1999) Two semi-parametric empirical Bayes estimators. Computational Statistics and Data Analysis 30(2) : 185–196

[141] Pandey MD, van Gelder PHAJM, Vrijling JK (2001) The estimation of extreme quantiles of wind velocity using L-moments in the peaks-over-threshold approach. Structural Safety 23 : 179–192

[142] Papadimitriou C, Beck JL, Katafygiotis LS (2001) Updating robust reliability structural test data. Probabilistic Engineering Mechanics 16 : 103–113

[143] Pawlak Z (1991) Rough sets. Theoretical aspects of reasoning about data. Kluwer Academic Publishers, Dordrecht

[144] Pradlwarter HJ (2001) On the Significance of Deterministic FE-Procedures in Stochastic Structural Analysis. In: Waszczyszyn Z, Pamin J (eds) 2nd Europ. Conf. on Computational Mechanics. Cracow, CD-ROM, Doc 579, pp. 1–18

[145] Pradlwarter HJ (2002) Deterministic integration algorithms for stochastic response computations of FE-systems. Computers and Structures 80 : 1489–1502

[146] Proppe C (2001) Distributed Computing with Weight Controlled Monte Carlo Simulation Algorithms. In: Schuëller GI, Spanos PD (eds) Proc. Int. Conf. on Monte Carlo Simulation MCS_2000, Monte Carlo, Monaco, 2000. Swets & Zeitlinger BV, Lisse, The Netherlands, pp. 49–54

[147] Proppe C, Schuëller GI (2001) Effects of Uncertainties on Lifetime Prediction of Aircraft Components. In: Bathe KJ (ed) Proc. 1st M.I.T. Conf. on Computational Fluid and Solid Mechanics. M.I.T., Cambridge, MA, USA. Elsevier, pp. 425–428

[148] Pulkkinen U, Simola K (2000) Bayesian models and ageing indicators for analysing random changes in failure occurence. Reliability Engineering and System Safety 68 : 255–268

[149] Puri ML, Ralescu D (1982) A possibility measure is not a fuzzy measure. Fuzzy Sets and Systems 7 : 311–313

[150] Puri ML, Ralescu D (1986) Fuzzy random variables. J Math Anal Appl 114: 409–422

[151] Qiang Song, Robert P Leland, Brad S Chissom (1997) Fuzzy stochastic fuzzy time series and its models. Fuzzy Sets and Systems 88 : 333–341

[152] Rackwitz R (2000) Reliability Analysis – Past, Present and Future. In: Kareem A, Haldar A, Spencer BF, Johnson EA (eds) 8th ASCE Specialty Conf. on Probabilistic Mechanics and Structural Reliability (PMC2000). University of Notre Dame, CD-ROM

[153] Rao SS, Sawyer JP (1995) Fuzzy Finite Element Approach for the Analysis of Imprecisely Defined Systems. AIAA J. 33(12) : 2364 –2370

[154] Rechenberg I (1973) Evolutionsstrategie: Optimierung technischer Systeme nach Prinzipien der biologischen Evolution, Problemata. Friedrich Frommann Verlag, Stuttgart

[155] Rommelfanger H (1988) Fuzzy Decision Support-Systeme. Springer, Berlin Heidelberg

[156] Rosqvist T (2000) Bayesian aggregation of experts' judgements on failure intensity. Reliability Engineering and System Safety 70 : 283–289

[157] Rubinstein RY (1981) Simulation and the Monte Carlo Method. Wiley, New York

[158] Ruge P (2001) Restricted Padè-scheme in computational structural dynamics. Computers and Structures 79 : 1913–1921

[159] Samarasooriya VNS, Varshney PK (2000) A fuzzy modeling approach to decision fusion under uncertainty. Fuzzy Sets and Systems 114 : 59–69

[160] Sang Yeol Joo, Yun Kyong Kim (2001) Kolmogorov's strong law of large numbers for fuzzy random variables. Fuzzy Sets and Systems 120 : 499–503

[161] Savchuk VP (1995) Estimation of structures reliability for nonprecise limit state models and vague data. Reliability Engineering and System Safety 47 : 47–58

[162] Savoia M (2002) Reliability analysis of structures against buckling according to fuzzy number theory. Computers and Structures 80 : 1087–1102

[163] Schenk CA, Bergman LA (2001) Vibration of Elastic Continuum with Stochastically Varying Surface Roughness Carrying an Accelerating Oscillator. In: Proc. Int. Mechanical Engineering Congress & Exposition. New York, NY, USA. ASME, CD-ROM, vol. 2

[164] Schnellenbach-Held M, Albert A (2000) Representation of Crisp and Fuzzy Knowledge for Preliminary Structural Design. Proc. 8th Int. Conf. on Computing in Civil and Building Engineering, Stanford, USA

[165] Schorling Y, Bucher C, Purkert G (2000) Stability analysis for randomly imperfect structures. In: Melchers RE, Stewart MG (eds) Applications of Statistics and Probability. Balkema Rotterdam, Brookfield, pp. 1027–1032

[166] Schuëller GI (1991) Einführung in die Sicherheit und Zuverlässigkeit von Tragwerken. W. Ernst & Sohn, Berlin

[167] Schuëller GI (ed) (1997) A state-of-the-art report on computational stochastic mechanics. Probabilistic Engineering Mechanics 12(4) : 197 –321

[168] Schuëller GI (1999) Computational Stochastic Mechanics – Recent Advances. In: Wunderlich W, Stein E (eds) 1st Europ. Conf. on Computational Mechanics. München, CD-ROM, Doc 170, pp. 1–19

[169] Schuëller GI (2000) On The Rationale For Treating Uncertainties In Structural Analysis. In: Papadrakakis M, Samartin A, Onate E (eds) IASS-IACM, Proc. 4th Int. Colloquium on Computation of Shell & Spatial Structures. Chania-Crete, Greece, CD-ROM, Doc 311, pp. 1–21

[170] Schuëller GI (2001) On Computational Procedures for Processing Uncertainties in Structural Mechanics. In: Waszczyszyn Z, Pamin J (eds) 2nd Europ. Conf. on Computational Mechanics. Cracow, CD-ROM, Doc 608, pp. 1–24

[171] Schuëller GI (2001) Computational stochastic mechanics – recent advances. Computers and Structures 79(22-25) : 2225–2234

[172] Schuëller GI, Shinozuka M (eds) (1987) Stochastic Methods in Structural Dynamics. Martinus Nijhoff Publication, Dordrecht, The Netherlands

[173] Schuëller GI, Spanos PD (eds) (2001) Proc. Int. Conf. on Monte Carlo Simulation MCS_2000, Monte Carlo, Monaco, 2000. Swets & Zeitlinger BV, Lisse, The Netherlands

[174] Schuëller GI, Spanos PD (eds) (2001) Spec. Issue on Monte Carlo Simulation. Probabilistic Engineering Mechanics 16(4) : 261–397

[175] Schuëller GI, Bucher CG, Bourgund U, Ouypornprasert W (1989) On efficient computational schemes to calculate failure probabilities. Probabilistic Engineering Mechanics 4(1) : 10–18

[176] Schwefel H (1977) Numerische Optimierung von Computermodellen mittels der Evolutionsstrategie. Birkhauser, Berlin

[177] Seising R (ed) (1999) Fuzzy Theorie und Stochastik: Modelle und Anwendungen in der Diskussion. Vieweg, Braunschweig Wiesbaden

[178] Shafer G (1976) A Mathematical Theory of Evidence. Princeton University Press, Princeton

[179] Shao Jun, Tu Dongsheng (1995) The Jackknife and Bootstrap. Springer, New York, NY

[180] Sickert J-U, Beer M, Graf W, Möller B (2003) Fuzzy Probabilistic Structural Analysis Considering Fuzzy Random Functions. In: Der Kiureghian A, Madanat S, Pestana JM (eds) Proc. 9th Int. Conf. on Applications of Statistics and Probability in Civil Engineering, San Francisco. Millpress, Rotterdam, vol. 1, pp. 379–386

[181] Simões LMC (2001) Fuzzy optimization of structures by the two-phase method. Computers and Structures 79(26–28) : 2481–2490

[182] Spaethe G (1987, 1992) Die Sicherheit tragender Baukonstruktionen. Verlag Bauwesen, Berlin, 1987; Springer, Berlin Heidelberg New York, 1992

[183] Spanos PD, Rao VRS (2001) Random field representation in a biorthogonal wavelets basis. ASCE, J Engineering Mechanics 127 : 194–205

[184] Spanos PD, Tsavachidis S (2001) Monte Carlo Simulations of a Nonlinear System with Biot Hysteretic Damping. In: Schuëller GI, Spanos PD (eds) Proc. Int. Conf. on Monte Carlo Simulation MCS_2000, Monte Carlo, Monaco, 2000. Swets & Zeitlinger BV, Lisse, The Netherlands, pp. 383–389

[185] Spanos PD, Zeldin BA (1998) On Random Field Discretization in Stochastic Finite Elements. J. Applied Mechanics 65 : 320–327

[186] Sugeno M (1974) Theory of Fuzzy Integral and Its Applications. Ph. D. Thesis, Tokyo Institute of Technology, Tokyo

[187] Szekely GS, Schuëller GI (2001) Computational procedure for a fast calculation of eigenvectors and eigenvalues of structures with random properties. Computer Methods in Applied Mechanics and Engineering 191(8–10) : 799–816

[188] Taheri SM, Behboodian J (2001) A Bayesian approach to fuzzy hypotheses testing. Fuzzy Sets and Systems 123 : 39–48

[189] Thoft-Christensen P, Baker MJ (1982) Structural Reliability Theory and Its Applications. Springer, Berlin Heidelberg New York Tokyo

[190] Thoft-Christensen P, Murotsu Y (1985) Application of Structural System Reliability Theory. Springer, Berlin Heidelberg New York Tokyo

[191] Tonon F, Bernardini A (1998) How many Different Faces of Uncertainty in Civil Engineering? Proc. Int. Computing Congress in Civil Engineering. ASCE, Annual Convention, Boston, pp. 95–97

[192] Tonon F, Bernardini A (1998) A random set approach to the optimization of uncertain structures. Computers and Structures 68(6) : 583–600

[193] Viertl R (1988) Statistical Methods in Accelerated Life Testing. Vandenhoeck & Ruprecht, Göttingen

[194] Viertl R (1996) Statistical Methods for Non-Precise Data. CRC Press, Boca Raton New York London Tokyo

[195] Viertl R (1997) Einführung in die Stochastik mit Elementen der Bayes-Statistik und Ansätzen für die Analyse unscharfer Daten. Springer, Wien

[196] Vlachos PK, Gelfand AE (2003) On the calibration of Bayesian model choice criteria. J.Statistical Planning and Inference 111(1) : 223–234

[197] Wang Guangyuan, Qiao Zhong (1994) Convergence of sequences of fuzzy random variables and its application. Fuzzy Sets and Systems 63 : 187–199

[198] Wang Guangyuan, Zhang Yue (1992) The theory of fuzzy stochastic processes. Fuzzy Sets and Systems 51 : 161–178

[199] Wen-Liang Hung (2001) Bootstrap method for some estimators based on fuzzy data. Fuzzy Sets and Systems 119 : 337–341

[200] Wood KL, Antonsson EK (1989) Computations with imprecise parameters in engineering design: background and theory. J. Mechanisms, Transmissions and Automation in Design 111 : 616–625

[201] Wood KL, Otto KN, Antonsson EK (1992) Engineering design calculations with fuzzy parameters. Fuzzy Sets and Systems 52 : 1–20

[202] Wright G (ed) (1994) Subjective probability. Wiley, Chichester

[203] Yager RR (1984) A representation of the probability of fuzzy subsets. Fuzzy Sets and Systems, 13 : 273–283

[204] Yuhu Feng, Liangjian Hu, Huisheng Shu (2001) The variance and covariance of fuzzy random variables and their applications. Fuzzy Sets and Systems 120(3) : 487–497

[205] Zadeh LA (1965) Fuzzy sets. Information and Control 8 : 338–353

[206] Zadeh LA (1968) Probability measures of fuzzy events. J. Mathematical Analysis and Applications. 23 : 421–427

[207] Zadeh LA (1975) The concept of a linguistic variable and its application to approximate reasoning. Information Sciences 8 : 199–249, 301–357; Information Sciences 9 : 43–80

[208] Zadeh LA (1975) Fuzzy logic and approximate reasoning. Synthese 30 : 407–428

[209] Zhang R, Mahadevan S (2000) Model uncertainty and Bayesian updating in reliability-based inspection. Structural Safety 22 : 145–160

[210] Zhang Yue, Wang Guangyuan, Su Fen (1996) The general theory for response analysis of fuzzy stochastic dynamical systems. Fuzzy Sets and Systems 83 : 369–405

[211] Zhao Y-X, Gao Q, Wang J-N (2000) An approach for determining an appropriate assumed distribution of fatigue life under limited data. Reliability Engineering and System Safety 67 : 1–7

[212] Zimmermann H-J (1992) Fuzzy set theory and its applications. Kluwer Academic Publishers, Boston London

[213] Zong Z, Lam KY (1998) Estimation of complicated distributions using B-spline functions. Structural Safety 20 : 341–355

[214] Zong Z, Lam KY (2000) Bayesian estimation of complicated distributions. Structural Safety 22 : 81–95

[215] Zong Z, Lam KY (2001) Bayesian estimation of 2-dimensional complicated distributions. Structural Safety 23 : 105–121

[216] Zurada JM (1992) Introduction to Artificial Neural Systems. West Publishing Company, St. Paul, Minnesota

[217] Eurocode 1. Grundlagen von Entwurf, Berechnung und Bemessung sowie Einwirkung auf Tragwerke. Deutsche Fassung DIN ENV 1991-1

[218] DIN 1055-100, Ausgabe 10/2002. Beuth Verlag, Berlin

Index

Printing: Saladruck, Berlin
Binding: Stein+Lehmann, Berlin